D1797339

OXFORD PURE MATHEMATICS
2/3 for Cambridge International AS & A Level

Series editor: David Rayner

**Jean Linsky, James Nicholson,
Brian Western**

OXFORD
UNIVERSITY PRESS

OXFORD
UNIVERSITY PRESS

Great Clarendon Street, Oxford, OX2 6DP, United Kingdom

Oxford University Press is a department of the University of Oxford.
It furthers the University's objective of excellence in research,
scholarship, and education by publishing worldwide. Oxford is a registered
trade mark of Oxford University Press in the UK and in certain other countries

British Library Cataloguing in Publication Data
Data available

978-0-19-830690-0

10 9 8 7 6 5 4 3 2 1

Paper used in the production of this book is a natural, recyclable product
made from wood grown in sustainable forests.
The manufacturing process conforms to the environmental regulations
of the country of origin.

Printed in Great Britain by Bell and Bain Ltd, Glasgow

Acknowledgements

The publisher would like to thank the following for permission to reproduce photographs:

p2: NASA/ESA/STSCI/Science Photo Library; **p17t:** Andrey Burmakin/Shutterstock; **p17b:**
James King-Holmes/Science Photo Library; **p24:** Sipa Press/Rex; **p32:** 1_Weblogiq/Shutterstock;
p38: Sebastian Kaulitzki; **p70:** Stephen Rees/Shutterstock; **p71c:** Joe Mercier/Shutterstock;
p71b: Sipa Press/Rex; **p72:** 3D Sculptor/Shutterstock; **p94:** Junne/Shutterstock; **p120:** Jacques
Descloitres/MODIS Land Rapid Response Team/GSFC/NASA; **p146t:** Tim Gainey/Alamy; **p146c:**
Irafael/Shutterstock; **p146b:** NASA; **p147t:** Yury Dmitrienko/Shutterstock; **p148:** ntdanai/
Shutterstock p147b: Gabriel Georgescu/Shutterstock; **p166:** Dinodia Photos/Alamy; **p194:** Nicolas
Boyer; **p226:** Mark Evans/Getty Images; **p252:** Alfred Pasieka/Science Photo Library; **p294tr:**
sfam_photo/Shutterstock; **p294cr:** Zurijeta/Shutterstock; **p294l:** Nerthuz/Shutterstock; **p294cl:**
Peter Dazeley/Getty; **p294br:** Norman Chan/Shutterstock; **p294b:** mamahoohooba/Shutterstock;
p295: KPG_Payless;

MIX
Paper from
responsible sources
FSC® C007785

Contents

Introduction

About this book

This book has been written to cover the **Cambridge AS and A level International Mathematics (9709)** course, and is fully aligned to the syllabus. The first six chapters of the book cover material applicable to both Pure 2 and Pure 3, and the final five chapters cover Pure 3 material only.

In addition to the main curriculum content, you will find:

- 'Maths in real-life', showing how principles learned in this course are used in the real world.
- Chapter openers, which outline how each topic in the Cambridge 9709 syllabus is used in real-life.
- Did you know? boxes (as shown below), which give interesting side-notes beyond the scope of the syllabus.

The book contains the following features:

Note	Did you know?
Advice on calculator use	Pure 2
Pure 3	

Throughout the book, you will encounter worked examples and a host of rigorous exercises. The examples show you the important techniques required to tackle questions. The exercises are carefully graded, starting from a basic level and going up to exam standard, allowing you plenty of opportunities to practise your skills. Together, the examples and exercises put maths in a real-world context, with a truly international focus.

At the start of each chapter, you will see a list of objectives that are covered in the chapter. These objectives are drawn from the Cambridge AS and A level syllabus. Each chapter begins with a *Before you start* section and finishes with a *Summary exercise* and *Chapter summary*, ensuring that you fully understand each topic.

Review exercises, separated into Pure 2 and Pure 3, are placed after chapters 3, 6, 8 and 11. These comprise a host of Cambridge past-paper questions which cover the topics from the previous 2–3 chapters and are in no particular order of difficulty.

The answers given at the back of the book are concise. However, when answering exam-style questions, you should show as many steps in your working as possible. All exam-style questions, as well as *Exam-style papers A, B, C* and *D*, have been written by the authors.

About the authors

Brian Western has almost 40 years of experience in teaching mathematics up to A Level and beyond, and is also a highly experienced examiner. He currently teaches Mathematics and Further Mathematics, and was until recently an Assistant Headteacher in a large state school. Brian has written and consulted on a number of mathematics textbooks.

James Nicholson is an experienced teacher of mathematics at secondary level, having taught for 12 years at Harrow School and spent 13 years as Head of Mathematics in a large Belfast grammar school. He is the author of several A-Level texts, and editor of the *Concise Oxford Dictionary of Mathematics*. He has also contributed to a number of other sets of curriculum and assessment materials, is an experienced examiner and has acted as a consultant for UK government agencies on accreditation of new specifications.

Jean Linsky has been a Mathematics teacher for over 30 years, as well as Head of Mathematics in Watford, Herts, and is also an experienced examiner. Jean has authored and consulted on numerous Mathematics textbooks.

A note from the authors

The aim of this book is to help students prepare for the Pure 2 and Pure 3 units of the Cambridge International AS and A Level Mathematics syllabus, although it may also be useful in providing support material for other AS and A Level courses. The book contains a large number of practice questions, many of which are exam-style, in addition to questions from past Cambridge examination papers.

In writing the book we have drawn on our experiences of teaching Mathematics and Further Mathematics to A Level over many years as well as on our experiences as examiners, and our discussions with mathematics educators from many countries at international conferences.

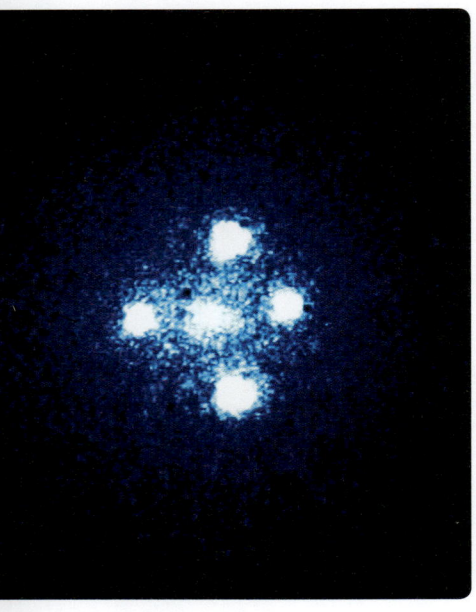

Algebra is used extensively in mathematics, chemistry, physics, economics and social sciences. For example, the study of polynomials in astrophysics has led to our understanding of gravitational lensing.

Gravitational lensing occurs when light from a distant source bends around a massive object (such as a galaxy) between a source and an observer. Multiple images of the same object may be seen. Here, the 'Einstein Cross', four images of a very distant supernova, is seen in a photograph taken by the Hubble telescope. The supernova is at a distance of approximately 8 billion light years, and is twenty times further away than the galaxy, which is at a distance of 400 million light years. The light from the supernova is bent in its path by the gravitational field of the galaxy. This bending produces the four bright outer images. The bright central region of the galaxy is seen as the central object. This phenomena was predicted by Einstein's general theory of relativity published in 1915, but was not observed until 1979.

Objectives

- Understand the meaning of $|x|$, and use relations such as $|a| = |b| \Leftrightarrow a^2 = b^2$ and $|x - a| < b \Leftrightarrow a - b < x < a + b$ in the course of solving equations and inequalities.
- Divide a polynomial, of degree not exceeding 4, by a linear or quadratic polynomial, and identify the quotient and remainder (which may be zero).
- Use the factor theorem and the remainder theorem, e.g. to find factors, solve polynomial equations or evaluate unknown coefficients.

Before you start

You should know how to:

1. Do long division,
   e.g. $357 \div 21$

$$
\begin{array}{r}
17 \\
21\overline{)357} \\
\underline{21} \\
147 \\
\underline{147} \\
0
\end{array}
$$

   Therefore $\dfrac{357}{21} = 17$

2. Find the remainder,
   e.g. $461 \div 37$

$$
\begin{array}{r}
12 \\
37\overline{)461} \\
\underline{37} \\
91 \\
\underline{74} \\
17
\end{array}
$$

   Remainder $= 17$

Skills check:

1. Find the following using long division.
   a) $608 \div 19$
   b) $2774 \div 38$
   c) $1081 \div 23$
   d) $1392 \div 24$

2. Find the remainder of the following after doing long division.
   a) $923 \div 21$
   b) $742 \div 32$
   c) $1527 \div 43$
   d) $4258 \div 26$

1.1 The modulus function

> The **modulus** of a real number is the magnitude of that number.

If we have a real number x, then the modulus of x is written as $|x|$. We say this as 'mod x'.

Thus $|2| = 2$ and $|-2| = 2$, and if we write $|x| < 2$ this means that $-2 < x < 2$.

> The **modulus function** $f(x) = |x|$ is defined as
> $$|x| = x \quad \text{for} \quad x \geq 0$$
> $$|x| = -x \quad \text{for} \quad x < 0$$

Consider the impact that the modulus function has by looking at the graphs of $y = x^2 - 4$ and $y = |x^2 - 4|$.

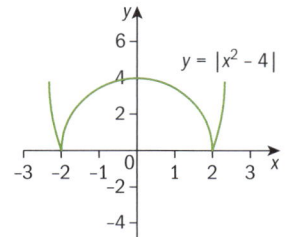

Note: For $f(x) < 0$, $|f(x)| = -f(x)$.

When graphing $y = |f(x)|$, notice that we reflect the graph of $y = f(x)$ in the x-axis whenever $f(x) < 0$.

Example 1

Solve the equation $|x + 2| = |3x|$.

....

Method 1: Using a graph

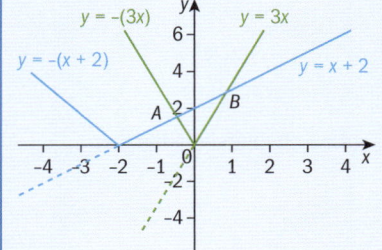

Sketch the graphs and find where they intersect.

The lines cannot be drawn below the x-axis.

For $x < 0$, $|3x| = -(3x)$

For $x < -2$, $|x + 2| = -(x + 2)$

Graphs intersect at A and B.

At A, $x + 2 = -3x$ ◀──────── At A, the line $y = x + 2$ intersects the line $y = -(3x)$.

$\qquad 4x = -2 \Rightarrow x = -\dfrac{1}{2}$

At B, $x + 2 = 3x$ ◀──────── At B, the line $y = x + 2$ intersects the line $y = 3x$.

$\qquad 2x = 2 \Rightarrow x = 1$

Answer: $x = -\dfrac{1}{2}$ or $x = 1$

▶ Continued on the next page

Method 2: Squaring both sides of the equation

$(x + 2)^2 = (3x)^2$

$x^2 + 4x + 4 = 9x^2$ | We do this to ensure both sides of the equation are positive.

$8x^2 - 4x - 4 = 0$

$2x^2 - x - 1 = 0$

$(2x + 1)(x - 1) = 0$

$x = -\dfrac{1}{2}$ or $x = 1$ | **Note:** You can only use this method if the variable (e.g. x) is inside the modulus expression.

Method 3: Remove modulus signs by equating the left-hand side with both 'plus and minus' the right-hand side.

$x + 2 = 3x$ or $x + 2 = -3x$ | We get the same result if we say $3x = \pm(x + 2)$.

$2x = 2$ or $4x = -2$

$x = 1$ or $x = -\dfrac{1}{2}$

Example 2

Solve the inequality $|4x + 3| > |2x - 1|$

Method 1: Using a graph

Sketch the graphs and find where they intersect. The lines cannot be drawn below the x-axis.

For $x < -\dfrac{3}{4}$, $|4x + 3| = -(4x + 3)$

For $x < \dfrac{1}{2}$, $|2x - 1| = -(2x - 1)$

Graphs meet at A and B.

At A, $-(4x + 3) = -(2x - 1)$ | At A, the line $y = -(4x + 3)$ intersects the line $y = -(2x - 1)$.

$\quad 2x = -4, \quad x = -2$

At B, $4x + 3 = -(2x - 1)$ | At B, the line $y = 4x + 3$ intersects the line $y = -(2x - 1)$.

$\quad 6x = -2, \quad x = -\dfrac{1}{3}$

We want the region where $|4x + 3| > |2x - 1|$.

This is where the 'blue' lines are above the 'green' lines.

Answer: $x < -2$ and $x > -\dfrac{1}{3}$

▶ Continued on the next page

Method 2: Squaring both sides of the equation

$(4x + 3)^2 > (2x - 1)^2$

$16x^2 + 24x + 9 > 4x^2 - 4x + 1$

$12x^2 + 28x + 8 > 0$

$3x^2 + 7x + 2 > 0$

$(3x + 1)(x + 2) > 0$

We do this to ensure both sides of the inequality are positive.

$x < -2 \text{ or } x > -\dfrac{1}{3}$

Draw a rough sketch to see where $y > 0$. There are two separate blue regions that satisfy $y > 0$.

There are 2 separate regions that satisfy $y > 0$.

Method 3: Remove modulus signs by equating the left-hand side with both 'plus and minus' the right-hand side.

First find where the graphs of $y = |4x + 3|$ and $y = |2x - 1|$ intersect (the values of x at these points are called the critical values).

$4x + 3 = 2x - 1 \quad \text{or} \quad 4x + 3 = -(2x - 1)$

$\qquad 2x = -4 \qquad \text{or} \quad 6x = -2$

$\qquad x = -2 \qquad \text{or} \quad x = -\dfrac{1}{3}$

This gives us the 2 critical values.

To find the correct inequalities you need to take values of x on either side of the critical values

Take values on either side of $x = -2$ and $x = -\dfrac{1}{3}$.

e.g.

$x = -3, |4x + 3| = 9 \text{ and } |2x - 1| = 7$

Thus $|4x + 3| > |2x - 1|$, so $x < -2$

One of the solutions is less than -2.

is one region where $|4x + 3| > |2x - 1|$

When $x = -1, |4x + 3| = 1 \text{ and } |2x - 1| = 3$

$\qquad\qquad |4x + 3| > |2x - 1|$ is false so there is no solution

between -2 and $-\dfrac{1}{3}$, so we do **not** require this region.

When $x = 0, |4x + 3| = 3 \text{ and } |2x - 1| = 1$

Thus $|4x + 3| > |2x - 1|$ so $x > -\dfrac{1}{3}$

One of the solutions is greater than $-\dfrac{1}{3}$.

$x < -2 \quad \text{or} \quad x > -\dfrac{1}{3}$

Exercise 1.1

1. Solve each of these equations.

   a) $|1 - 2x| = 3$

   b) $|x - 3| = |x + 1|$

   c) $|5x - 2| = |2x|$

   d) $|5 - 4x| = 4$

   e) $|2x - 1| = |x + 2|$

   f) $|x| = |4 - 2x|$

   g) $|3x + 1| = |4 - 2x|$

   h) $|2x - 6| = |3x + 1|$

   i) $|x + 4| = |3x + 1|$

   j) $|1 - 3x| = |5x - 3|$

   k) $3|x - 4| = |x + 2|$

   l) $5|2x - 3| = 4|x - 5|$.

2. Solve each of these inequalities.

   a) $|2x - 3| < |x|$

   b) $|x - 1| \geq 4$

   c) $|x + 3| \geq |2x + 2|$

   d) $|2x + 3| > x + 6$

   e) $|x + 6| \leq 3|x - 2|$

   f) $|3x - 2| < |x + 4|$

   g) $|2x| < |1 - x|$

   h) $5 \leq |2x - 1|$

   i) $2|x - 1| < |x + 3|$

   j) $|2x + 1| \geq |1 - 4x|$

   k) $|x + 2| < 2|x + 1|$

   l) $|3x - 1| \leq |x + 3|$.

1.2 Division of polynomials

We can use **long division** to divide a polynomial by another polynomial.

Example 3

Divide $x^3 - 5x^2 + x + 10$ by $(x - 2)$

$$
\begin{array}{r}
x^2 - 3x - 5 \\
x - 2 \overline{)\, x^3 - 5x^2 + x + 10} \\
\underline{x^3 - 2x^2} \\
-3x^2 + x + 10 \\
\underline{-3x^2 + 6x} \\
-5x + 10 \\
\underline{-5x + 10}
\end{array}
$$

$x^3 \div x = x^2$, so multiply $(x - 2)$ by x^2.

Subtract $(x^3 - 2x^2)$ from $(x^3 - 5x^2)$ and bring down $(+ x + 10)$.

$-3x^2 \div x = 3x$, so multiply $(x - 2)$ by $-3x$.

Subtract $(3x^2 + 6x)$ from $(3x^2 + x)$ and bring down $+ 10$.

We cannot continue the process as $(-5x + 10) - (-5x + 10) = 0$.

Thus $(x^3 - 5x^2 + x + 10) \div (x - 2) = x^2 - 3x - 5$

Note: $(x^3 - 5x^2 + x + 10)$ is called the **dividend**, $(x - 2)$ is called the **divisor**, $(x^2 - 3x - 5)$ is called the **quotient**, and when we subtract $(-5x + 10)$ from $(-5x + 10)$ we are left with nothing, so we say there is no **remainder**. Because there is no remainder, we say that $(x - 2)$ is a **factor** of $x^3 - 5x^2 + x + 10$.

Example 4

Find the remainder when $4x^3 - 7x - 1$ is divided by $(2x + 1)$.

$$
\begin{array}{r}
2x^2 - x - 3 \\
2x + 1 \overline{)\, 4x^3 + 0x^2 - 7x - 1} \\
\underline{4x^3 + 2x^2} \\
-2x^2 - 7x - 1 \\
\underline{-2x^2 - x} \\
-6x - 1 \\
\underline{-6x - 3} \\
+ 2
\end{array}
$$

As there is no term in x^2 in the dividend, it is useful to write $0x^2$ as part of the dividend.

$(4x^3 \div 2x) = 2x^2$, so multiply $(2x + 1)$ by $2x^2$.

Subtract $(4x^3 + 2x^2)$ from $(4x^3 + 0x^2)$ and bring down $-7x - 1$.

$(-2x^2 \div 2x) = -x$, so multiply $(2x + 1)$ by $-x$.

Subtract $(-2x^2 - x)$ from $(-2x^2 - 7x)$ and bring down -1.

Subtract $(-6x - 3)$ from $(-6x - 1)$ to get a remainder of $+2$.

We cannot continue the process as 2 cannot be divided by $(2x + 1)$.
Thus, when $(4x^3 - 7x - 1)$ is divided by $(2x + 1)$, the remainder is 2.

Note: Looking at Example 4 we can write $(4x^3 - 7x - 1) = (2x^2 - x - 3)(2x + 1) + 2$

In general, f(x) = quotient × divisor + remainder.

Exercise 1.2

1. Divide
   a) $x^3 + 3x^2 + 3x + 2$ by $(x + 2)$
   b) $x^3 - 2x^2 + 6x + 9$ by $(x + 1)$
   c) $x^3 - 3x^2 + 6x - 8$ by $(x - 2)$
   d) $x^3 + x^2 - 3x - 2$ by $(x + 2)$
   e) $2x^3 - 6x^2 + 7x - 21$ by $(x - 3)$
   f) $3x^3 - 20x^2 + 10x + 12$ by $(x - 6)$.

2. Find the remainder when

    a) $6x^3 + 28x^2 - 7x + 10$ is divided by $(x + 5)$

    b) $2x^3 + x^2 + 5x - 4$ is divided by $(x - 1)$

    c) $x^3 + 2x^2 - 17x - 2$ is divided by $(x - 3)$

    d) $2x^3 + 3x^2 - 4x + 5$ is divided by $(x^2 + 2)$

    e) $4x^3 - 5x + 4$ is divided by $(2x - 1)$

    f) $3x^3 - x^2 + 1$ is divided by $(x + 2)$.

3. Show that $(2x + 1)$ is a factor of $2x^3 - 3x^2 + 2x + 2$.

4. **a)** Show that $(x - 1)$ is a factor of $x^3 - 6x^2 + 11x - 6$.

    b) Hence factorise $x^3 - 6x^2 + 11x - 6$.

5. Show that when $4x^3 - 6x^2 + 5$ is divided by $(2x - 1)$ the remainder is 4.

6. Divide $x^3 + 1$ by $(x + 1)$.

7. Find the quotient and the remainder when $x^4 + 2x^3 + 3x^2 + 7$ is divided by $(x^2 + x + 1)$.

8. Find the quotient and the remainder when $2x^3 + 3x^2 - 4x + 5$ is divided by $(x + 2)$.

9. **a)** Show that $(2x - 1)$ is a factor of $12x^3 + 16x^2 - 5x - 3$.

    b) Hence factorise $12x^3 + 16x^2 - 5x - 3$.

10. The expression $2x^3 - 5x^2 - 16x + k$ has a remainder of -6 when divided by $(x - 4)$.

    Find the value of k.

1.3 The remainder theorem

You can find the remainder when a polynomial is divided by $(ax - b)$ by using the **Remainder Theorem**.

We know that if $f(x)$ is divided by $(x - a)$ then $f(x) =$ quotient $\times (x - a) +$ remainder.

When $x = a$, $f(x) =$ quotient $\times (a - a) +$ remainder $=$ remainder.

Thus $f(a) =$ remainder

> When a polynomial $f(x)$ is divided by $(x - a)$, the remainder is $f(a)$.
>
> When a polynomial $f(x)$ is divided by $(ax - b)$, the remainder is $f\left(\dfrac{b}{a}\right)$.

Example 5

Find the remainder when $4x^3 + x^2 - 3x + 7$ is divided by $(x + 2)$.

$f(x) = 4x^3 + x^2 - 3x + 7$ ← Write the polynomial as a function.

$f(-2) = 4(-2)^3 + (-2)^2 - 3(-2) + 7$ ← $(x + 2) = 0 \Rightarrow x = -2$ so use $f(-2)$.

$\quad = -32 + 4 + 6 + 7 = -15$

The remainder is -15.

Example 6

When $16x^4 - ax^3 + 8x^2 - 4x - 1$ is divided by $(2x - 1)$, the remainder is 3.

Find the value of a.

$f(x) = 16x^4 - ax^3 + 8x^2 - 4x - 1$ ← Write the polynomial as a function.

$f\left(\frac{1}{2}\right) = 16\left(\frac{1}{2}\right)^4 - a\left(\frac{1}{2}\right)^3 + 8\left(\frac{1}{2}\right)^2 - 4\left(\frac{1}{2}\right) - 1$ ← $(2x - 1) = 0 \Rightarrow x = \frac{1}{2}$ so use $f\left(\frac{1}{2}\right)$.

$\quad = 1 - \frac{1}{8}a + 2 - 2 - 1 = 3$ ← Equate this to 3, since the remainder is 3.

$-\frac{1}{8}a = 3 \Rightarrow a = -24$

Exercise 1.3

1. Find the remainder when

   a) $2x^3 + 8x^2 - x + 4$ is divided by $(x - 3)$

   b) $5x^4 - 3x^3 - 2x^2 + x - 1$ is divided by $(x + 1)$

   c) $x^3 + 4x^2 + 8x - 3$ is divided by $(2x + 1)$

   d) $3x^3 - 2x^2 - 5x - 7$ is divided by $(2 - x)$

   e) $9x^3 - 8x + 3$ is divided by $(1 - x)$

   f) $243x^4 - 27x^3 + 6x + 4$ is divided by $(3x - 2)$.

2. When $x^5 + 4x^4 - 6x^2 + ax + 3$ is divided by $(x + 2)$ the remainder is 5. Find the value of a.

3. When $ax^3 + 16x^2 - 5x - 5$ is divided by $(2x - 1)$ the remainder is -2. Find the value of a.

4. The polynomial $4x^3 - 4x^2 + ax + 1$, where a is a constant, is denoted by p(x). When p(x) is divided by $(2x - 3)$ the remainder is 13. Find the value of a.

5. The polynomial $x^3 + ax^2 + bx + 1$, where a and b are constants, is denoted by p(x). When p(x) is divided by $(x - 2)$ the remainder is 9 and when p(x) is divided by $(x + 3)$ the remainder is 19. Find the value of a and the value of b.

6. When $5x^3 + ax + b$ is divided by $(x - 2)$, the remainder is equal to the remainder obtained when the same expression is divided by $(x + 2)$.

   a) Explain why b can take any value.

   b) Find the value of a.

7. The polynomial $2x^4 + 3x^2 - x + 2$ is denoted by p(x). Show that the remainder when p(x) is divided by $(x + 2)$ is 8 times the remainder when p(x) is divided by $(x - 1)$.

8. The polynomial $x^3 + ax + b$, where a and b are constants, is denoted by p(x). When p(x) is divided by $(x - 1)$ the remainder is 14 and when p(x) is divided by $(x - 4)$ the remainder is 56. Find the values of a and b.

9. The polynomial $x^3 + ax^2 + 2$, where a is a constant, is denoted by p(x). When p(x) is divided by $(x + 1)$ the remainder is one more than when p(x) is divided by $(x + 2)$. Find the value of a.

10. When $6x^2 + x + 7$ is divided by $(x - a)$, the remainder is equal to the remainder obtained when the same expression is divided by $(x + 2a)$, where $a \neq 0$. Find the value of a.

1.4 The factor theorem

We can deduce the **Factor Theorem** directly from the Remainder Theorem.

> **The Factor Theorem:** For any polynomial f(x), if f(a) = 0 then the remainder when f(x) is divided by ($x - a$) is zero. Thus ($x - a$) is a factor of f(x).
>
> For any polynomial f(x), if f$\left(\dfrac{b}{a}\right)$ = 0, then ($ax - b$) is a factor of f(x).

Example 7

The polynomial $x^3 - ax^2 + 2x + 8$, where a is a constant, is denoted by p(x).

It is given that ($x - 2$) is a factor of p(x).

a) Find the value of a.

b) When a has this value, factorise p(x) completely.

· ·

a) p(2) = 8 − 4a + 4 + 8 = 0

4a = 20 ⟹ a = 5

b) We can factorise $x^3 - 5x^2 + 2x + 8$ using either **(i)** long division

or **(ii)** testing other factors using the factor theorem.

i)
$$
\begin{array}{r}
x^2 - 3x - 4 \\
x - 2\ \overline{)x^3 - 5x^2 + 2x + 8} \\
\underline{x^3 - 2x^2} \\
-3x^2 + 2x + 8 \\
\underline{-3x^2 + 6x} \\
-4x + 8 \\
\underline{-4x + 8}
\end{array}
$$

← Put $a = 5$

← You would expect there would be no remainder since $x - 2$ is a factor.

and $x^2 - 3x - 4 = (x - 4)(x + 1)$ ← Factorise the quotient.

So p(x) = ($x - 2$)($x - 4$)($x + 1$)

ii) f(+1) = 1 − 5 + 2 + 8 ≠ 0   ($x - 1$) is not a factor   ← Try a value of x.

f(−1) = −1 − 5 − 2 + 8 = 0   ($x + 1$) is a factor

f(4) = 64 − 80 + 8 + 8 = 0   ($x - 4$) is a factor

So p(x) = ($x - 2$)($x - 4$)($x + 1$)

> **Note:** Instead of performing this last trial we could work out the final factor is ($x - 4$) as we know $x \times x \times x = x^3$ and (−2) × (+1) × (−4) = +8.

Example 8

Solve $x^3 - 3x^2 - 4x + 12 = 0$

Let $f(x) = x^3 - 3x^2 - 4x + 12$

$f(1) = 1 - 3 - 4 + 12 \neq 0$

    so $(x - 1)$ is not a factor.

$f(2) = 8 - 12 - 8 + 12$

    so $(x - 2)$ is a factor.

$f(-2) = -8 - 12 + 8 + 12$

    so $(x + 2)$ is a factor.

We can deduce that the third factor is $(x - 3)$

$f(3) = 27 - 27 - 12 + 12$

    so $(x - 3)$ is a factor.

Thus $(x - 2)(x + 2)(x - 3) = 0$

$x = 2$ or $x = -2$ or $x = 3$

> To solve, we must first factorise $x^3 - 3x^2 - 4x + 12$.

> Trial any value of x that is a factor of 12.

> Alternatively, at this stage you could also do a long division to find the other two factors since you already know one factor.

> $12 \div 2 \div -2 = 3$ (and the coefficient of x^3 is 1).

Example 9

The polynomial $ax^3 + x^2 + bx + 6$, where a and b are constants, is denoted by p(x). It is given that $(2x - 1)$ is a factor of p(x) and that when p(x) is divided by $(x - 1)$ the remainder is -4. Find the values of a and b.

p$(x) = ax^3 + x^2 + bx + 6$

$\text{p}\left(\dfrac{1}{2}\right) = \dfrac{a}{8} + \dfrac{1}{4} + \dfrac{b}{2} + 6 = 0$

    $a + 2 + 4b + 48 = 0$

    $a + 4b = -50$ (1)

p$(1) = a + 1 + b + 6 = -4$

    $a + b = -11$ (2)

(1) - (2) $\Rightarrow 3b = -39$

           $b = -13$

(2) $\Rightarrow$ $a - 13 = -11$

           $a = 2$

$a = 2$ and $b = -13$

> $\text{p}\left(\dfrac{1}{2}\right) = 0$ as $(2x - 1)$ is a factor.

> Multiply each term by 8.

> p$(1) = -4$ as the remainder is -4.

> Solve the two equations simultaneously.

Exercise 1.4

1. Factorise the following as a product of three linear factors.
   In each case, one of the factors has been given.

   a) $2x^3 - 5x^2 - 4x + 3$ One factor is $(x - 3)$

   b) $x^3 - 6x^2 + 11x - 6$ One factor is $(x - 2)$

   c) $5x^3 + 14x^2 + 7x - 2$ One factor is $(5x - 1)$

   d) $2x^3 + 3x^2 - 18x + 8$ One factor is $(x + 4)$

   e) $x^3 + x^2 - 4x - 4$ One factor is $(x + 2)$

   f) $6x^3 + 13x^2 - 4$ One factor is $(3x + 2)$.

2. Solve the following equations.

   a) $2x^3 + 7x^2 - 7x - 12 = 0$ b) $2x^3 - 5x^2 - 14x + 8 = 0$

   c) $x^3 - 6x^2 + 3x + 10 = 0$ d) $x^3 + 3x^2 - 6x - 8 = 0$

   e) $2x^3 - 15x^2 + 13x + 60 = 0$ f) $3x^3 - 2x^2 - 7x - 2 = 0$

3. Show that $(x - 3)$ is a factor of $x^5 - 3x^4 + x^3 - 4x - 15$.

4. Factorise $x^4 + x^3 - 7x^2 - x + 6$ as a product of four linear factors.

5. $(x - 2)$ is a factor of $x^3 - 3x^2 + ax - 10$. Find the value of a.

6. a) Show that $(2x - 5)$ is a factor of $4x^3 - 20x^2 + 19x + 15$.

   b) Hence factorise $4x^3 - 20x^2 + 19x + 15$ as a product of three linear factors.

7. The polynomial $ax^3 - 3x^2 - 5ax - 9$ is denoted by p(x).
   It is given that $(x - a)$ is a factor of p(x). Find the possible values of a.

8. The polynomial $3x^3 + 2x^2 - bx + a$, where a and b are constants, is denoted by p(x). It is given that $(x - 1)$ is a factor of p(x) and that when p(x) is divided by $(x + 1)$ the remainder is 10. Find the values of a and b.

9. The polynomial $ax^3 + bx^2 - 5x + 3$, where a and b are constants, is denoted by p(x). It is given that $(2x - 1)$ is a factor of p(x) and that when p(x) is divided by $(x - 1)$ the remainder is -3. Find the remainder when p(x) is divided by $(x + 3)$.

10. Factorise $2x^4 + 5x^3 - 5x - 2$ as a product of four linear factors.

Summary exercise 1

1. Solve the equation $|5 - 2x| = 7$.

2. Solve the equation $|3x - 4| = |5 - 2x|$.

3. Solve the inequality $\left|\frac{1}{2}x - 2\right| \le \left|\frac{3}{2}x\right|$.

4. Solve the inequality $2|x - 2| < |x|$.

5. Solve the inequality $|2x - 1| < |3x - 4|$.

EXAM-STYLE QUESTION

6. Solve the inequality $|x + 3| \ge 2|x - 3|$.

7. Solve the inequality $|x - 2| \le 3|x + 1|$.

EXAM-STYLE QUESTION

8. Solve the inequality $2|x - a| > |2x + a|$ where a is a constant.

9. Divide $2x^4 - 9x^3 + 13x^2 - 15x + 9$ by $(x - 3)$.

10. Find the quotient and the remainder when $x^3 - 3x^2 + 6x + 1$ is divided by $(x - 2)$.

EXAM-STYLE QUESTIONS

11. a) Show that $(x - 4)$ is a factor of $x^3 - 3x^2 - 10x + 24$.

    b) Hence factorise $x^3 - 3x^2 - 10x + 24$.

12. The expression $x^3 + 3x^2 + 6x + k$ has a remainder of -3 when divided by $(x + 1)$. Find the value of k.

13. The polynomial $ax^4 + bx^3 - 8x^2 + 6$ is denoted by p(x). When p(x) is divided by $(x^2 - 1)$ the remainder is $2x + 1$. Find the value of a and the value of b.

14. The polynomial $x^4 + ax^3 + bx^2 - 16x - 12$ is denoted by p(x).

    $(x + 1)$ and $(x - 2)$ are factors of p(x).

    a) Find the value of a and the value of b.

    b) Hence factorise p(x) fully.

15. The polynomial $x^4 + x^3 - 22x^2 - 16x + 96$ is denoted by p(x).

    a) Find the quotient when p(x) is divided by $x^2 + x - 6$.

    b) Hence solve the equation p$(x) = 0$.

16. The polynomial $6x^3 - 23x^2 + ax + b$ is denoted by p(x). When p(x) is divided by $(x + 1)$ the remainder is -21. When p(x) is divided by $(x - 3)$ the remainder is 11.

    a) Find the value of a and the value of b.

    b) Hence factorise p(x) fully.

17. A polynomial is defined by p$(x) = x^3 + Ax^2 + 49x - 36$, where A is a constant. $(x - 9)$ is a factor of p(x).

    a) Find the value of A.

    b) i) Find all the roots of the equation p$(x) = 0$.
       ii) Find all the roots of the equation p$(x^2) = 0$.

18. The polynomial $x^3 - 15x^2 + Ax + B$, where A and B are constants is denoted by p(x). $(x - 16)$ is a factor of p(x). When p(x) is divided by $(x - 2)$ the remainder is -56.

    a) Find the value of A and the value of B.

    b) i) Find all 3 roots of the equation p$(x) = 0$.
       ii) Find the 4 real roots of the equation p$(x^4) = 0$.

19. Solve the inequality $|2x - 1| > |x|$. [4]

    Cambridge International AS & A Level Mathematics 9709 Paper 2 Q1 October/November 2006

20. The polynomial $4x^3 - 7x + a$, where a is a constant, is denoted by p(x). It is given that $(2x - 3)$ is a factor of p(x).

    a) Show that $a = -3$. [2]

    b) Hence, or otherwise, solve the equation p$(x) = 0$. [4]

    Cambridge International AS & A Level Mathematics 9709 Paper 2 Q3 October/November 2006

Chapter summary

The modulus function

- The **modulus** of a real number is the magnitude of that number.
- The **modulus function** $f(x) = |x|$ is defined as

  $|x| = x$ for $x \geq 0$

  $|x| = -x$ for $x < 0$

Division of polynomials

- When dividing algebraic expressions, for example $(4x^3 - 7x - 3) \div (2x + 1) = (2x^2 - x - 3)$, you need to know the following terms:
  - $(4x^3 - 7x - 3)$ is called the **dividend**
  - $(2x + 1)$ is called the **divisor**
  - $(2x^2 - x - 3)$ is called the **quotient** and there is no **remainder**.
  - $(2x + 1)$ is a **factor** of $(4x^3 - 7x - 3)$.
- $f(x)$ = quotient × divisor + remainder.

The remainder theorem

- When a polynomial $f(x)$ is divided by $(x - a)$, the remainder is $f(a)$.
- When a polynomial $f(x)$ is divided by $(ax - b)$, the remainder is $f\left(\dfrac{b}{a}\right)$.

The factor theorem

- For any polynomial $f(x)$, if $f(a) = 0$ then the remainder when $f(x)$ is divided by $(x - a)$ is zero. Thus $(x - a)$ is a factor of $f(x)$.
- For any polynomial $f(x)$, if $f\left(\dfrac{b}{a}\right) = 0$, then $(ax - b)$ is a factor of $f(x)$.

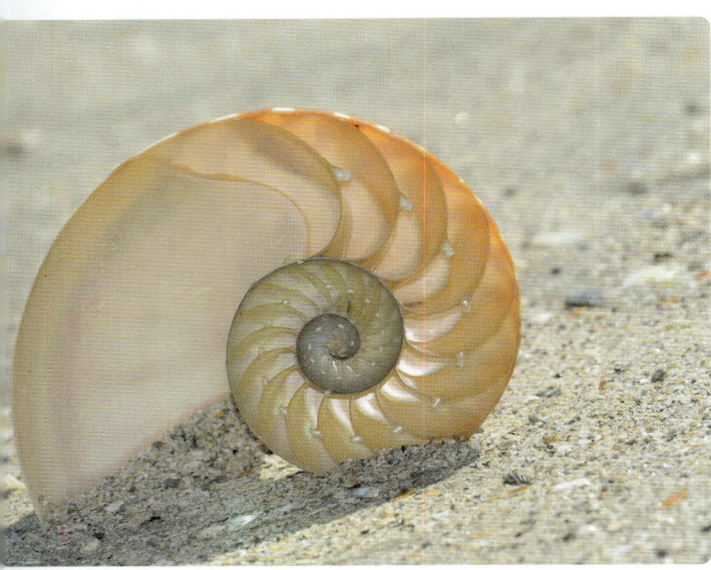

The shell of a Nautilus grows in a manner where each new chamber is an approximate copy of the last one, but is enlarged by a constant factor. This gives rise to the aesthetically pleasing shape known as a *logarithmic spiral*.

Many things in nature grow in the same way; that is, its rate of growth is proportional to its current size. Birth rates, for example, behave in this way: as a population increases in size, its birth rate increases proportionally.

As such, understanding the mathematics which describe the behaviour of entities like birth rates is crucially important if we are to predict how they will behave in future. A lot of important physical quantities, such as the energy released in an earthquake, have very large ranges and thus logarithmic scales (the Richter scale, in the case of the earthquake) are helpful in providing a way of making the numerical values of these measurements manageable.

Objectives

- Understand the relationship between logarithms and indices, and use the laws of logarithms (excluding change of base).
- Understand the definition and properties of e^x and $\ln x$, including their relationship as inverse functions and their graphs.
- Use logarithms to solve equations of the form $a^x = b$, and similar inequalities.
- Use logarithms to transform a given relationship to linear form, and hence determine unknown constants by considering the gradient and/or intercept.

Before you start

You should know how to:

1. Apply the laws of indices,

   e.g. a) Simplify $(x^3)^4$
   $$(x^3)^4 = x^{3 \times 4} = x^{12}$$

   e.g. b) Simplify $x^5 \times x^4$
   $$x^5 \times x^4 = x^{5+4} = x^9$$

2. Know the properties of powers,

   e.g. If $x \neq 0$, state the value of x^0.
   $$x^0 = 1 \text{ if } x \neq 0$$

Skills check:

1. Simplify

   a) $\dfrac{x^8}{x^3}$ **b)** $\sqrt{x^{10}}$ **c)** $(x^7)^{\frac{1}{3}}$

2. Simplify $(2^a)^b$.

2.1 Continuous exponential growth and decay

In P1 you met the idea of exponential growth and decay, modelled by a geometric series of the form $\sum_n ar^{n-1}$, where n was an integer and ratio $r > 1$ indicates growth, whilst $r < 1$ indicates decay. When we model exponential change with a geometric series of this form, we assume that the change takes place in discrete steps. Situations involving exponential growth or decay are not restricted to discrete steps, but often occur in continuous time. We sometimes simplify a phenomenon; modelling it as changing in discrete steps to make the mathematics easier. We might look at the change over a unit of time (an hour, a day, a year) and treat it as though it was a discrete step process – for example, where an investment has the interest added to the principal once a year. Shortening the units of time will give a better approximation to reality. As we consider change over shorter and shorter time divisions, we can consider the limit of that process as a function $f(t) = a \times r^t$, where t can take any real value rather than just integer values (in practice, t is mostly restricted to any non-negative value).

A function $f(t) = a \times r^t$ where t is real, gives exponential growth ($r > 1$) or decay ($r < 1$).

You have defined r^t only for rational values of t so far, but any irrational value of t can be approximated above and below by a pair of rational numbers which differ by as small an amount as you wish – giving a process to define $f(t)$ for all t.

$f(t) = 1.4 \times 1.3^t$

$f(t) = 5 \times 0.8^t$

$f(t) = 1.4 \times 1.3^t$ The initial value of function, a, is 1.4. We also see that $r = 1.3 > 1$ meaning this function describes exponential growth.

$f(t) = 5 \times 0.8^t$ The initial value of the function, a, is now 5 and $r = 0.8 < 1$ meaning this function describes exponential decay.

Exponential cell growth

Did you know?

Examples of exponential growth include the growth of cells, nuclear chain reactions, epidemics, feedback in an amplification signal, and computer processing power.

Examples of exponential decay include radioactive decay (used in carbon dating), rate of some chemical reactions (which depend on the concentration of one or more of the reactants), and light intensity in an absorbent medium.

One of the characteristics of exponential decay is that the graph will get closer to the horizontal axis as time increases, but will never reach it.

Carbon dating relies on the fact that radioactive material decays exponentially.

Logarithms and exponential functions

Example 1

The mass in grams of a decaying radioactive material is given by $M(t) = 200(0.8)^t$ after t days. Draw a graph of M and use your graph to estimate the number of days it takes for M to fall to

a) 100 g **b)** 50 g **c)** 25 g.

a) $M(3.1) = 100 \Rightarrow$ after 3.1 days, $M \approx 100$ g

b) $M(6.2) = 50 \Rightarrow$ after 6.2 days, $M \approx 50$ g

c) $M(9.3) = 25 \Rightarrow$ after 9.3 days, $M \approx 25$ g

The same length of time is taken for the amount remaining to halve each time. The 'half-life' of this radioactive material is approximately 3.1 days.

Example 2

The value of car which is t years old is $V(t) = \$35\,000(0.75)^t$.
a) Find its value after 2 years.
b) Show that the value goes below \$15 000 before it is 3 years old.

a) $V(2) = \$35\,000(0.75)^2 = \$19\,687.50 = \$19\,700$ (3 s.f.)

b) $V(3) = \$35\,000(0.75)^3 = \$14\,765.63 < \$15\,000$ so the value goes below \$15 000 before the car is 3 years old.

Example 3

In a container, the number of cells doubles every six hours. At 10 a.m. on Monday a new culture is started with 20 cells and placed in a container which will hold 10 million cells. In which 6 hour period will the container be filled up?

During the 19th period (10 pm Friday – 4 am Saturday) it goes from 5 242 880 to 10 485 760 cells.

On your calculator if you do $20 \times 2 =$ and then $\times 2 =$ repeatedly, you can count the number of periods until it goes over 10 million.

Exercise 2.1

1. Calculate

   a) $0.6 \times (4)^5$ **b)** $400\,000 \times (0.3)^5$ **c)** $63.2 \times (1.03)^8$ **d)** $5.1 \times (0.08)^4$

2. Air is escaping from a balloon. The volume V (ml) of the balloon after t minutes is given by $V = 1500 \times (0.7)^t$. What is the volume after

   a) 5 minutes   **b)** a quarter of an hour?

3. A company has an accounting policy that equipment is depreciated at 20% of current book value each financial year, and the initial book value is set at the price paid for the equipment. The company buys equipment for $3 million in 2012, and makes a further purchase in 2013 for $2 million. What is the book value of the equipment in 2015?

> **Did you know?**
>
> 'Depreciation' is the loss in value over a period of time for equipment, and 'book value' is the value at any time in the company accounts.

4. The amount of a drug (in mg) in a patient's bloodstream t hours after the first injection is $100 \times (0.8)^t$. If injections are to be repeated at 12-hourly intervals, what should the dose be (to the nearest mg) for the second and subsequent injections to bring the total in the bloodstream back to 100 mg each time?

5. At 4 p.m. one afternoon, Jeremy's car bottoms out when going over a speed bump too quickly and oil starts leaking from an engine sump. The volume, in ml, remaining after t hours is $3000 \times (0.98)^t$. If the oil level falls below 2 litres there is a risk of permanently damaging the engine. The next morning at 9 am Jeremy notices the oil underneath his parked car. Show that there is still enough oil for him to drive the car to the garage for repair.

2.2 The logarithmic function

The graph shows the function $y = 2^x$ in blue, and the inverse of this function in green (recall that the inverse function is the mirror image of the function in the line $y = x$).

This inverse function is known as the logarithmic function.

> The logarithmic function is the inverse of the exponential function to the same base.

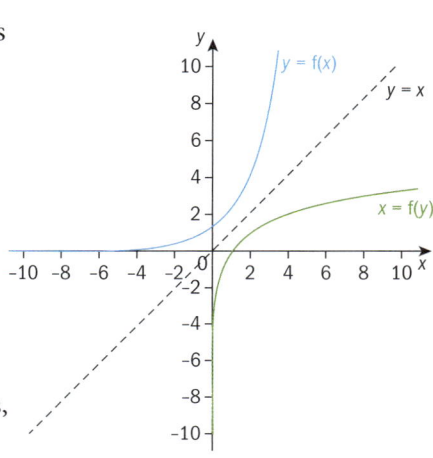

The blue graph shown here is the **logarithm to base 2** because the green graph shows 2^x.

The exponential function has domain the set of all real numbers, with range the positive real numbers, so it follows that the logarithmic function is defined on a domain of the positive real numbers and its range is the set of all real numbers.

Formally, we define the logarithmic function by

$$y = b^x \quad \Leftrightarrow \quad x = \log_b y, \text{ where } x \in \mathbb{R}, \ y \in \mathbb{R}, \ y > 0$$

These two special cases are worth remembering:
For any b, $\log_b 1 = 0$
For any n, $\log_b b^n = n$

Example 4

Find the value of

a) $\log_2 16$ b) $\log_2 \dfrac{1}{8}$ c) $\log_8 2$ d) $\log_{\frac{1}{2}} 4$.

a) $16 = 2^4 \Rightarrow \log_2 16 = 4$

b) $\dfrac{1}{8} = 2^{-3} \Rightarrow \log_2 \dfrac{1}{8} = -3$

c) $2 = \sqrt[3]{8} = (8)^{\frac{1}{3}} \Rightarrow \log_8 2 = \dfrac{1}{3}$

d) $4 = \left(\dfrac{1}{2}\right)^{-2} \Rightarrow \log_{\frac{1}{2}} 4 = -2$

We identify the logarithm by thinking of what power we have to raise the base to.

Example 5

Find the value of

a) $\log_3 243$ b) $\log_{10} 1000$ c) $\log_7 343$.

a) $243 = 3^5 \Rightarrow \log_3 243 = 5$

b) $1000 = 10^3 \Rightarrow \log_{10} 1000 = 3$

c) $343 = 7^3 \Rightarrow \log_7 343 = 3$

Example 6

a) Write these in the form $y = b^x$.

     i) $\log_2 64 = 6$ ii) $\log_k m = p$

b) Write these in the form $x = \log_b y$.

     i) $5^{-3} = 0.008$ ii) $h^5 = R$

a) i) $64 = 2^6$ ii) $m = k^p$

b) i) $\log_5 0.008 = -3$ ii) $\log_h R = 5$

Example 7

On the same axes sketch the graphs of

a) $y = 0.7^x$ **b)** $y = 1.2^x$ **c)** $y = 1.5^x$

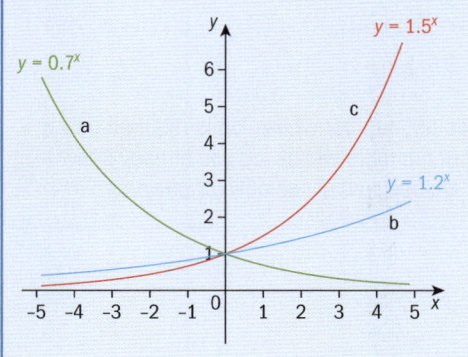

In part (**a**) the base value (0.7) is less than one so the function decreases as x gets larger. In parts (**b**) and (**c**) the base values are larger than 1 (1.2, 1.5) and so these functions increase as x gets larger.

Note also that (**c**) is steeper than (**b**) because (**c**) has a greater base value.

Since the logarithm function is the inverse of the exponential or power function, the rules of indices give rise to corresponding properties of logarithms.

$$\log(xy) = \log x + \log y$$

$$\log\left(\frac{x}{y}\right) = \log x - \log y$$

$$\log(x^n) = n \log x$$

These results are true for logarithms in any base.

Example 8

Express as a single logarithm or number.

a) $\log_5 4 + \log_5 7$ **b)** $\log_6 4 + \log_6 9$.

a) $\log_5 4 + \log_5 7 = \log_5 28$ **b)** $\log_6 4 + \log_6 9 = \log_6 36 = \log_6 (6^2) = 2$

Example 9

Express in terms of $\log x$, $\log y$, and $\log z$.

a) $\log xy$ **b)** $\log\left(\dfrac{xy^2}{z}\right)$ **c)** $\log\sqrt{\left(\dfrac{x^3 y^2}{z^4}\right)}$

a) $\log xy = \log x + \log y$ **b)** $\log\left(\dfrac{xy^2}{z}\right) = \log x + 2\log y - \log z$

c) $\log\sqrt{\left(\dfrac{x^3 y^2}{z^4}\right)} = \log\left(x^{\frac{3}{2}} y z^{-2}\right) = \dfrac{3}{2}\log x + \log y - 2\log z$

Identify the power of each variable first.

Example 10

It is given that $\log_a 3 = x$, $\log_a 5 = y$, and $\log_a 10 = z$.

Express the following in terms of x, y, and z.

a) $\log_a 50$ **b)** $\log_a 75$ **c)** $\log_a 0.75$

....

a) $\log_a 50 = \log_a (5 \times 10) = y + z$ **b)** $\log_a 75 = \log_a (3 \times 5 \times 5) = x + 2y$

c) $\log_a 0.75 = \log_a \left(\dfrac{3 \times 5 \times 5}{10 \times 10} \right) = x + 2y - 2z$

> Express the number whose log is required only in terms of numbers whose log you already know – using multiply and divide only (or powers).

Exercise 2.2

1. Evaluate the following logarithms.

   a) $\log_3 \left(\dfrac{1}{81} \right)$ **b)** $\log_9 27$ **c)** $\log_4 128$ **d)** $\log_3 6561$

   e) $\log_3 216$ **f)** $\log_8 \dfrac{1}{32}$ **g)** $\log_{16} 32$ **h)** $\log_4 \left(\dfrac{1}{16} \right)$

   i) $\log_9 \left(\dfrac{1}{3} \right)$ **j)** $\log_{\frac{1}{2}} \left(\dfrac{1}{32} \right)$ **k)** $\log_{\frac{1}{3}} 27$ **l)** $\log_{\frac{1}{4}} 8$

2. Write in the form $y = b^x$.

   a) $\log_2 256 = 8$ **b)** $\log_3 \left(\dfrac{1}{27} \right) = -3$ **c)** $\log_4 32 = 2.5$

   d) $\log_3 (9\sqrt{3}) = 2.5$ **e)** $\log_a x = q$ **f)** $\log_s t = u$

3. Write in the form $x = \log_b y$.

   a) $7^{-3} = \left(\dfrac{1}{343} \right)$ **b)** $10^9 = 1\,000\,000\,000$ **c)** $2^{-4} = \left(\dfrac{1}{16} \right)$

   d) $5^{2.5} = (25\sqrt{5})$ **e)** $t^{-3} = v$ **f)** $p^x = m$

4. Evaluate the following logarithms.

   a) $\log_{10} 100$ **b)** $\log_{10} \sqrt{10}$ **c)** $\log_{10} \dfrac{10}{\sqrt[3]{10}}$ **d)** $\log_{10} \left(\dfrac{1}{\sqrt{10}} \right)$

   e) $\log_{10} (100\,000\sqrt{10})$ **f)** $\log_{10} \left(\dfrac{1}{100\sqrt[3]{10}} \right)$ **g)** $\log_{10} 1$

5. On the same axes, sketch the graphs of

   a) $y = 0.5^x$ **b)** $y = 1.3^x$ **c)** $y = 1^x$

6. Express the following as a single logarithm or number.

   a) $\log_5 3 + \log_5 8$
   b) $\log_6 7 + \log_6 10$
   c) $\log_5 4 + \log_5 6.25$
   d) $2\log_5 3 + \log_5 4$
   e) $2\log_5 6 - \log_5 8 + \log_5 12$
   f) $2\log_{10} 4 - 4\log_{10} 2$
   g) $2\log_p 4 - 4\log_p 2$
   h) $\log_{10} 12 - \frac{1}{2}\log_{10} 9 + \log_{10} 25$

7. Express as a single logarithm.

   a) $\log_5 3 + 2$
   b) $3 + \log_6 10$
   c) $2\log_{10} x - \log_{10} 8 + \log_{10} (x + 3)$
   d) $3 + 2\log_5 x$
   e) $\log_{10} x + 3$
   f) $2 + \log_{10} 8 - \log_{10} x$

8. Express in terms of $\log x$, $\log y$, and $\log z$.

   a) $\log\left(\dfrac{xy}{z^2}\right)$
   b) $\log\left(x^2\left(xy^2z\right)^3\right)$
   c) $\log\sqrt{\left(\dfrac{x^4 y^2}{z^4}\right)}$

9. Given that $\log_a 2 = x$, $\log_a 5 = y$, and $\log_a 6 = z$, express the following in terms of x, y, and z.

   a) $\log_a 50$
   b) $\log_a 75$
   c) $\log_a 0.75$
   d) $\log_a 0.001$

2.3 e^x and logarithms to base e

You have seen that you can use any positive real number as a base for a power or logarithmic function, but in practice two bases are very commonly used, and they are available on your calculator.

- The button marked $\boxed{\log}$ uses base 10 since our number system is based on powers of 10.

- $\boxed{\ln}$ uses base e (= 2.71828......). These are sometimes called natural logarithms because of some special properties the number represented by 'e' has when used in an exponential function.

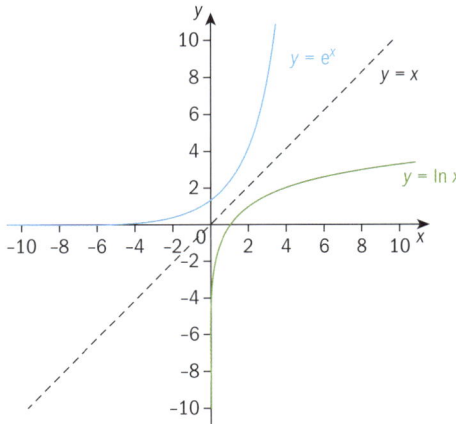

The blue graph is $y = e^x$ and the green graph is $y = \log_e x$ (also known as $y = \ln x$). These are inverses and are similar to the graphs you saw earlier at the beginning of section **2.2**. One feature to notice about $y = e^x$ is that as x increases, not only does e^x also increase, but its rate of increase (the gradient of the graph) gets larger.

$\ln x$ or $\log_e x$ is the inverse function to e^x.

Did you know?

Consider the following results, which you encountered in section **2.2**.

$$\log(xy) = \log x + \log y \quad (1)$$

$$\log\left(\frac{x}{y}\right) = \log x - \log y \quad (2)$$

$$\log(x^n) = n \log x \quad (3)$$

Before hand-held electronic calculators became commonly available in the 1980s, scientists and students used these results to simplify numerical calculations. Using them, we are able to transform a calculation and so merely add instead of multiply (using (1)), to subtract instead of divide (using (2)), and to multiply instead of raising something to a power (using (3)).

Using base 10 made those new operations very simple, since (for example)

$$257.2 = 2.572 \times 100 \Rightarrow \log_{10} 257.2 = \log_{10} 2.572 + \log_{10} 100$$
$$= \log_{10} 2.572 + 2$$
$$= 2.4102...$$

Check on your calculator that $\log_{10} 257.2 = 2 + \log_{10} 2.572$

When they did not have calculators, scientists and students used books of logarithm values. However, as they were able to transform the logarithm of a big number into the logarithm of a small one, as shown above, the tables only needed to show the logarithms of numbers x satisfying $1 \leq x < 10$.

An earthquake much under 3 will go unnoticed by most people going about their normal business, but 3 will be noticed and 4 will cause anxiety amongst those unfamiliar with the feeling of earthquakes. Once it gets up to 5, buildings may be damaged and 6 is likely to cause injuries, so the effects increase enormously for small increases in the reported measurement.

An earthquake of magnitude 8.0 occurred in Beichuan, Sichuan Province, China on May 12, 2009

When logarithmic scales are used with natural phenomena, it is because the range of measurement is so large that the numbers are not easily comprehensible, but it is then very important to remember that the scale is logarithmic when making comparisons. An earthquake which measures 4.0 on the Richter scale creates waves recorded on a seismograph which have ten times the amplitude of the waves an earthquake measuring 3.0 creates, and releases around 31.7 times as much energy

Other commonly used logarithmic scales include the ph scale for measuring acidity, noise levels measured in decibels and the stellar magnitude scale for the brightness of stars.

Example 11

Use your calculator to find the values of

a) $\log_{10} 16$ b) $\log_{10} 0.01457$ c) $\ln 10$

d) $\ln 2$ e) $\ln 1$ f) $\ln 0.8$.

Give your answers to 3 d.p.

....

a) 1.204   b) −1.837   c) 2.303

d) 0.693   e) 0   f) −0.223

> Remember that
> - 1 has logarithm 0 in any base
> - numbers below 1 will have a negative logarithm
> - anything between 1 and the base will have a positive logarithm less than 1.

Example 12

Find the values of

a) $\dfrac{\log_{10} 7}{\log_{10} 4}$ b) $\dfrac{\log_{10} 3.52}{\log_{10} 7.94}$ c) $\dfrac{\ln 8.2}{\ln 6.5}$

   Give your answers to 3 d.p.

....

a) 1.404   b) 0.607   c) 1.124

> Often when you solve logarithmic equations your solution will be in a form involving a quotient. Make sure you can enter these correctly in your calculator – often you need to remember to insert parentheses separately around both the numerator and the denominator.

Example 13

An earthquake of magnitude 3.0 on the Richter scale releases the energy equivalent of 480 kg of TNT. How much energy will be released by an earthquake of the following magnitudes?

> Remember that an increase of 0.2 on the Richter scale roughly doubles the amount of energy released by an earthquake. An increase of 1.0 gives multiplying factor of 31.7 (very close to 2^5).

a) 3.2   b) 4.0   c) 7.5

....

a) approximately 1 tonne of TNT   ⟵ $0.48 \times 31.7^{0.2} = 0.958$

b) approximately 15 tonnes of TNT   ⟵ $0.48 \times 31.7 = 15.2$

c) approximately $0.48 \times (31.7)^{4.5} = 2.7$ million tonnes (2.7 megatonnes) of TNT

Example 14

A radioactive isotope has a half-life of 20 days. Initially it has a mass of 60 grams. How much will there be after

a) 20 days   b) 30 days?

....

a) 30 grams

b) 21.2 grams ($= 60 \times (0.5)^{1.5}$) since 30 days = 1.5 half-lives

Exercise 2.3

1. Use your calculator to find the value of these. Give your answers to 3 d.p.

   a) $\log_{10} 23.2$ b) $\log_{10} 0.0232$

   c) $\log_{10} (2.7183)$ (which is e)   d) $\ln 6.5$

   e) $\ln 0.0065$

2. Find the value of these. Give your answers to 3 d.p.

   a) $\dfrac{\log_{10} 70}{\log_{10} 40}$ b) $\dfrac{\log_{10} 352}{\log_{10} 794}$ c) $\dfrac{\ln 0.82}{\ln 0.65}$

   > Compare your answers to those in Example 12. Make sure you never think of cancelling in an expression like this.

3. An earthquake of magnitude 3.0 on the Richter scale releases the energy equivalent of 480 kg of TNT.

   a) How much energy will be released by an earthquake of magnitude

      i) 2   ii) 6?

   b) How much energy will be released by six earthquakes of magnitude 5?

   > Remember from Example 13, an increase of 1.0 gives a multiplying factor of 31.7.

4. The radioactive isotope carbon-14 has a half-life of 5730 years. Once a living organism dies, the decay process starts and can be used to date how old a fossil is.

   An archeologist finds a skull which he thinks is from an animal which became extinct 4000 years ago. For the size of the skull, he knows it would have had a mass of 28 mg of carbon-14 when the creature died. If the skull is 4000 years old, how much carbon-14 should be in the skull now?

2.4 Equations and inequalities using logarithms

You can use the laws of logarithms to transform equations in order to make their solutions accessible.

Example 15

Solve the equation $4^x = 16$ by taking logarithms of both sides.

$4^x = 16 \Rightarrow x \log 4 = \log 16 \Rightarrow x = \dfrac{\log 16}{\log 4} = 2$ (using any base)

Example 16

Solve the equation $5^x = 3^{2x+1}$

$5^x = 3^{2x+1} \Rightarrow x \log 5 = (2x+1) \log 3 \Rightarrow x(\log 5 - 2 \log 3) = \log 3$

$\Rightarrow x = \dfrac{\log 3}{\log 5 - 2 \log 3} = -1.869$

Since x appears in more than one power, you need to collect x terms together to solve for x.

One important feature of the power (and logarithmic) functions is that they are strictly monotonic – that is to say they are always increasing (if the base is > 1) or always decreasing (if the base is < 1). So if the logarithms are equal it follows that the expressions are equal. Examples 17 and 21 express both sides as a single logarithm and use this feature to remove the logarithm to enable the equation to be solved.

Example 17

Solve the equation $\log_{10} (2 + x) = 2 + \log_{10} x$

$\log_{10} (2 + x) = 2 + \log_{10} x \Rightarrow \log_{10} (2 + x) = \log_{10} 100 + \log_{10} x$

$\Rightarrow \log_{10}(2 + x) = \log_{10} 100x \Rightarrow 2 + x = 100x \Rightarrow x = \dfrac{2}{99}$

We need to express the right-hand side as a single logarithm.

Another important consequence of the strictly monotonic property is that inequalities can be solved by considering the case of equality and then selecting the values above or below that critical value as appropriate.

Example 18

Solve the inequalities

a) $5^x \leq 13.3$ **b)** $(0.4)^x < 0.0001$

a) $5^x = 13.3 \Rightarrow x\log 5 = \log 13.3 \Rightarrow x = \dfrac{\log 13.3}{\log 5} = 1.608$

so $5^x \leq 13.3$ when $x \leq 1.608$

> Powers of 5 give a monotonically increasing function, since $5 > 1$.

b) $(0.4)^x = 0.0001 \Rightarrow x\log 0.4 = \log 0.0001$

$\Rightarrow x = \dfrac{\log 0.0001}{\log 0.4} = 10.1$

so $(0.4)^x < 0.0001$ when $x > 10.1$

> Powers of 0.4 give a monotonically decreasing function, since $0.4 < 1$ so the direction of the inequality must be reversed.

Example 19

How many terms of the geometric series $2 + 2 \times (1.1) + 2 \times (1.1)^2 + 2 \times (1.1)^3 + ...$ must be taken for the sum to exceed one thousand?

If you solve for the sum of n terms of this GP to be 1000 it will (almost certainly) give a non-integer value. The solution will be the next integer above that value.

$1000 = \dfrac{2(1.1^n - 1)}{1.1 - 1} \Rightarrow 50 = 1.1^n - 1 \Rightarrow 1.1^n = 51$

Then taking logs gives $n = \dfrac{\log 51}{\log 1.1} = 41.25...$

So 42 terms are needed.

> Remember $S_n = \dfrac{a(r^n - 1)}{r - 1}$ is the sum of n terms of a GP with first term a and common ratio r.

Example 20

Solve the equation $\log_a (2 - x) - 2\log_a x = \log_a 3$

$\log_a (2 - x) - 2\log_a x = \log_a 3 \Rightarrow \log_a \dfrac{(2 - x)}{x^2} = \log_a 3$

$\Rightarrow \dfrac{(2 - x)}{x^2} = 3 \Rightarrow 2 - x = 3x^2 \Rightarrow 3x^2 + x - 2 = 0$

This is now a standard quadratic form, with solutions $x = \dfrac{2}{3}, -1$.

However, only $x = \dfrac{2}{3}$ is valid because the logarithm of a negative number is undefined.

> Having equated the expressions which were logarithms you need to check the 'solutions' actually gave a valid equation in log form.

Example 21

Solve the equation $\log_{10}(2 - x) - 2\log_{10} x = 1$

$\log_{10}(2 - x) - 2\log_{10} x = 1 \Rightarrow \log_{10}\dfrac{(2-x)}{x^2} = \log_{10} 10$

$\Rightarrow \dfrac{(2-x)}{x^2} = 10 \Rightarrow 2 - x = 10x^2 \Rightarrow 10x^2 + x - 2 = 0$

This is now a standard quadratic form, with solutions $x = \dfrac{2}{5}, -\dfrac{1}{2}$

but only $x = \dfrac{2}{5}$ is valid.

> You need to express both sides as a single logarithm in order to use this technique so writing 1 as $\log_{10} 10$ is a critical step.

Example 22

A culture contains 300 000 bacteria at 9 am on Monday and increases at a rate of 20% every hour. At what time will there be

a) 600 000 bacteria   **b)** 2 million bacteria

The number of bacteria t hours after 9 am on Monday is given by $300\,000 \times (1.2)^t$

a) $300\,000 \times (1.2)^t = 600\,000 \Rightarrow (1.2)^t = 2 \Rightarrow t \log 1.2 = \log 2$

$\Rightarrow t = 3.8017....$ so at 12:48 pm.

b) $300\,000 \times (1.2)^t = 2\,000\,000 \Rightarrow (1.2)^t = \dfrac{20}{3} \Rightarrow t\log 1.2 = \log\left(\dfrac{20}{3}\right)$

$\Rightarrow t = 10.405....$ so at 7:24 pm.

Did you know?

The time for a bacteria population to double in size is known as the **generation time** and typically varies from 10 minutes to about 20 hours, though bacteria deep below the earth's surface may have a generation time of several thousand years.

Exercise 2.4

For this exercise, give your answers to 3 significant figures (unless they are exact).

1. Solve for x.

a) $5^x = 100$ b) $6^x = 3$ c) $0.4^x = 0.1$

d) $0.4^x = 2.5$ e) $7^{2x} = 12$ f) $7^{3x} = 64$

g) $4^{2x-1} = 7$ h) $7^{\frac{1}{2}x} = 10$ i) $2^{4x-1} = 128$

j) $3^{2x+1} = \dfrac{1}{27}$ k) $3^{\frac{1}{2}x} = 27$ l) $\left(\dfrac{1}{2}\right)^{2x-1} = 128$

2. Solve for x.

a) $2^{2x} = 3^{x+2}$ b) $5^{x+1} = 4^{2x-3}$ c) $3^{2x-1} = 7^{x+1}$

d) $4^{3x-1} = 2^{2x-6}$ e) $3^{x+4} = 5^{x+1}$ f) $7^{2x} \times 4^x = 3^{x+5}$

3. Solve for x.

   a) $3^x \geq 40$

   b) $6^x \leq 0.8$

   c) $0.4^x < 0.2$

   d) $0.4^x \leq 2$

   e) $7^{2x} \leq 5$

   f) $7^{\frac{1}{3}x} < 25$

4. Solve for x, v, z.

   a) $7^x = 5$

   b) $7^v = 12$

   c) $7^z = 60$

Since $5 \times 12 = 60$ you should find that $x + y = z$.

5. Solve for x, y.

   a) $5^x = 30$

   b) $5^{2y} = 60$

Since this shows us that $x \neq y$, we conclude that in general, $5^{2x} \neq 2 \times 5^x$.

6. How many terms of the geometric series $1 + 3 + 9 + 27 + 81 + \ldots$ must be taken for the sum to exceed one million?

7. How many terms of the geometric series $1 + 3 + 9 + 27 + 81 + \ldots$ must be taken for the sum to exceed one billion? (1 billion = 1000 million)

8. How many terms of the geometric series $1 + 5 + 25 + 125 + 625 + \ldots$ must be taken for the sum to exceed one million?

9. How many terms of the geometric series $3 + 6 + 12 + 24 + 48 + \ldots$ must be taken for the sum to exceed ten thousand?

10. How many terms of the geometric series $10 + 5 + 2.5 + 1.25 + 0.625 + \ldots$ must be taken for the sum to exceed 19.999?

11. How many terms of the geometric series $8 + 4 + 2 + 1 + 0.5 + \ldots$ must be taken for the sum to get within 10^{-4} of its sum to infinity?

12. How many terms of the geometric series $1 + \dfrac{1}{3} + \dfrac{1}{9} + \dfrac{1}{27} + \dfrac{1}{81} + \ldots$ must be taken for the sum to get to within 0.01% of its sum to infinity?

13. Solve for x.

    a) $\log_a (2 - x) = \log_a x + \log_a 5$

    b) $\log_{10} (4 + x) = \log_{10} x + \log_{10} 7$

    c) $\ln(8x - 1) = \ln x + \ln 6$

    d) $\log_{10} (4 + x) = \log_{10} x + 2$

    e) $\ln(1 - x) = \ln x + 1$ (Give your answer in exact form, i.e. in terms of e)

14. Solve for x.

    a) $\ln (6 - 13x) = 2 \ln x + \ln 5$

    b) $\log_{10} (13x - 6) = 2 \log_{10} x + \log_{10} 6$

    c) $\ln (17x - 6) = \ln x + \ln 12$

    d) $\log_{10} (13x - 4) = 2 \log_{10} x + 1$

15. There were 80 mg of a radioactive material stored at the start of 1950. The material has a half-life of 12 years.

    a) How much radioactive material will there be at the start of 2050?

    b) When will there be 1 mg left?

16. An accident occurs while transporting radioactive waste, and results in the area being contaminated by material with a half-life of 8 years. Experts say that the area should be quarantined until the radioactive material has reduced to 10% of its original level. How long are the experts recommending the quarantine should be in place for?

17. A bacteria has a generation time (the time in which the population doubles) of 30 minutes under ideal conditions. Sunil conducts an experiment in which he grows one culture of the bacteria under ideal conditions and another under conditions he thinks will have a generation time of 40 minutes. If he starts with equal sizes of culture in both cases, after how long would he expect one to be three times the size of the other?

2.5 Using logarithms to reduce equations to linear form

Many scientific, economic, and social science quantities can be described (at least approximately) by relationships which follow either an exponential growth or decay law, or else a power law. If we take logarithms of an equation which follows either of these two laws, we transform it into a linear expression. This allows values of unknown constants to be estimated from observational data.

Exponential growth / decay:

$y = ab^t \Rightarrow \log y = \log a + t \log b$

The graph of '$\log y$' against 't' has intercept '$\log a$', and gradient '$\log b$'.

Power law:

$y = ax^n \Rightarrow \log y = \log a + n \log x$

The graph is of '$\log y$' against '$\log x$' and has intercept '$\log a$' and gradient n.

> You know some examples of power laws in geometrical formulae for areas and volumes.

Example 23

By taking logarithms, transform these relationships between the two stated variables into a linear relationship between two new variables, and state the new variables.

a) y and t are related by $y = 7b^t$

b) y and x are related by $y = ax^3$.

- -

a) $\log y = \log 7 + t \log b$ is linear with variables '$\log y$' and 't'.

b) $\log y = \log a + 3 \log x$ is linear with variables '$\log y$' and '$\log x$'.

Example 24

For the following linear equations involving logarithms, find the relationship between the unknown variables, giving your answer in a form not involving logarithms.

a) $\log_{10} y = 2 + 3\log_{10} x$ b) $\log_{10} A = \log_{10} \pi + 2\log_{10} r$ c) $\log_{10} y = 0.3 + 0.7x$

a) $\log_{10} y = 2 + 3\log_{10} x \Rightarrow \log_{10} y = \log_{10} 100 + \log_{10} x^3 = \log_{10} 100\, x^3$
$\Rightarrow y = 100\, x^3$

b) $\log_{10} A = \log_{10} \pi + 2\log_{10} r = \log_{10} \pi r^2 \Rightarrow A = \pi r^2$

c) $\log_{10} y = 0.3 + 0.7\, x = \log_{10} (10^{0.3}) + \log_{10} (10^{0.7x})$
$= \log_{10} (10^{0.3}) + \log_{10} ((10^{0.7})^x)$
$= \log_{10} (10^{0.3} \times (10^{0.7})^x)$
$\Rightarrow y = 1.995 \times 5.012^x$

This step, using one of the laws of indices, is the vital one in transforming this into a power law relationship between y and x.

Example 25

The metabolic rate of mammals (M) and their body mass (B) are thought to obey a power law in the form $M = aB^k$. Data of 4 species of mammals are given in the table. Plot a graph of $\log M$ against $\log B$ and use it to estimate the values of a and k.

B	1.6	8	700	9000
M	1.9	6.4	143	1100

$M = aB^k \Rightarrow \log M = \log a + k \log B$

$\log B$	0.204	0.903	2.845	3.954
$\log M$	0.279	0.806	2.155	3.041

Taking logarithms of the equation shows that the gradient is k and the intercept is $\log a$.

Plotting the points on the graph and drawing a line of best fit by eye gives $k = 0.72$, and $a = 10^{0.13} = 1.3$.

Example 26

The variables x and y satisfy the relation $5^y = 3^{2x+1}$. By taking logarithms, show that the graph of y against x is a straight line, and find the exact values of the gradient and the intercept.

$$5^y = 3^{2x+1} \Rightarrow y\log 5 = (2x+1)\log 3$$

$$\Rightarrow y = \left(\frac{2\log 3}{\log 5}\right)x + \left(\frac{\log 3}{\log 5}\right)$$

This equation describes a straight line with gradient $\dfrac{2\log 3}{\log 5}$ and intercept $\dfrac{\log 3}{\log 5}$.

Exercise 2.5

1. By taking logarithms, transform these relationships between the two stated variables into a linear relationship between two new variables, and state the new variables.

    a) p and t are related by $p = 3b^t$

    b) y and x are related by $y = Kx^{-2}$

    c) y and x are related by $y = \alpha\sqrt{x}$.

2. For the following linear equations involving logarithms find the relationship between the unknown variables in a form which does not involve logarithms.

    a) $\log_{10} y = 1 + 2\log_{10} x$

    b) $\log_{10} V = \log_{10}\left(\dfrac{4\pi}{3}\right) + 3\log_{10} r$

    c) $\log_{10} y = 0.1 + 1.3x$.

3. Two variable quantities are related by the equation $y = Ax^n$ where A and n are constants. The graph shows the results of plotting $\log_{10} y$ against $\log_{10} x$ for four pairs of values of x and y. Use the diagram to estimate the values of A and n.

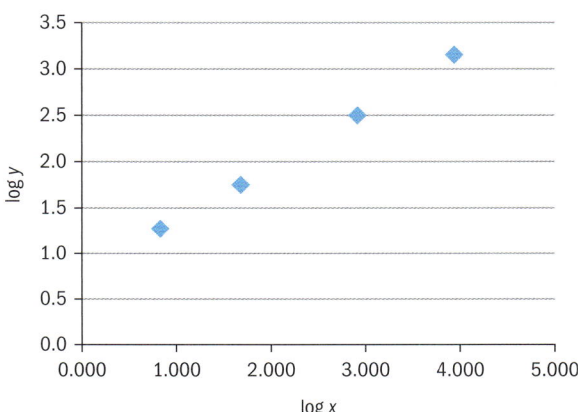

Logarithms and exponential functions   33

4. Two variable quantities are related by the equation
    $p = Ak^{-q}$ where A and k are constants. The graph
    shows the results of plotting $\ln p$ against q for four
    pairs of values of p and q. Use the diagram to estimate
    the values of A and k.

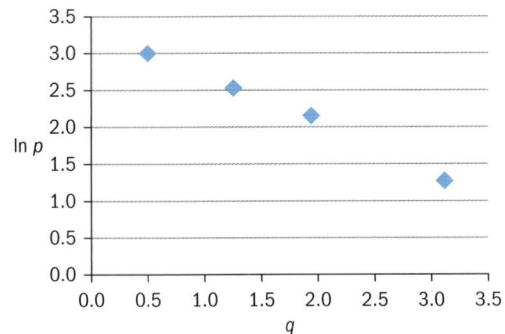

5. Two variable quantities are related by the equation $y = Ax^n$
    where A and n are constants. The table shows four pairs of
    values of x and y. Estimate the values of A and n.

x	62	150	400	921
y	352	232	141	88.1

6. The resistance R (in Newtons) to the motion of a train is measured
    at a number of speeds v (in ms^{-1}) and is shown in the table. If the
    resistance follows the relationship $R = kv^n$, find the values of k and n.

v	25.7	45.3	71	102.3
R	3562	4635	6037	7324

7. The distance D (in millions of km) from the Sun of the first 4 planets
    in the solar system, and the time T (in days) for the planet to orbit the
    Sun are given in the table. If the orbit time follows the relationship
    $T = kD^n$, find the values of k and n.

	Mercury	Venus	Earth	Mars
D	58	108	150	228
T	88	225	365	687

8. The population P of a colony of bacteria is thought to obey an exponential
    growth law in the form $P = kr^t$ where P is thousands of bacteria, and t is
    hours since the experiment started. Plot an appropriate graph and use it
    to estimate the values of k and r.

t	2	5	8	12
P	919	2049	4475	12817

9. The variables x and y satisfy the relation $7^y = 4^{3x-2}$. By taking logarithms
    show that the graph of y against x is a straight line, and find the exact
    values of the gradient and the intercept.

Summary exercise 2

1. Find the values of

   a) $\log_2\left(\dfrac{1}{32}\right)$ b) $\log_4 8$ c) $\log_{81} 3$.

2. Write in the form $y = b^x$.

   a) $\log_4 256 = 4$ b) $\log_2\left(\dfrac{1}{32}\right) = -5$

   c) $\log_2\left(32\sqrt{2}\right) = 5.5$.

3. Write in the form $x = \log_b y$.

   a) $6^{-3} = \left(\dfrac{1}{216}\right)$ b) $10^7 = 10\,000\,000$.

4. Find the values of

   a) $\log_{10} 10\,000$ b) $\log_{10} \sqrt{1000}$

   c) $\log_{10} \dfrac{100}{\sqrt[4]{10}}$

5. Express as a single logarithm or number.

   a) $\log_4 2 + \log_4 7$

   b) $\log_{10} 17 + \log_{10} 12$

   c) $\ln 3 + \ln 4.5$

   d) $2\log_6 3 + \log_6 4$

   e) $\log_{10} 24 - \log_{10} 3 + \log_{10} 12.5$

   f) $2 + 3\log_5 x$

   g) $3\ln(2x) - \ln 8 + \ln(2x + 3)$

   h) $2 + \ln 4 - 2\ln x$.

6. Express $\log\left(\dfrac{\left(xy^2 z\right)^5}{x^2}\right)$ in terms of $\log x$, $\log y$, and $\log z$.

7. Given $\log_a 2 = x$, $\log_a 5 = y$, and $\log_a 6 = z$, express $\log_a 0.15$ in terms of x, y, and z.

8. Solve

   a) $7^x = 83$ b) $5^x = 0.3$

   c) $0.2^x = 0.8$.

9. Solve

   a) $4^{2x} = 8^{x+2}$ b) $3^{2x+1} = 5^{3x-1}$

   c) $7^{2x-1} = 3^{x+1}$.

10. Solve

    a) $5^x \geq 43.2$ b) $5^x \leq 0.7$

    c) $0.7^x < 0.3$.

11. How many terms of the geometric series $2 + 10 + 50 + 250 + 1250 + \ldots$ must be taken for the sum to exceed one million?

12. How many terms of the geometric series $5 + 2.5 + 1.25 + 0.625 + \ldots$ must be taken for the sum to get within 10^{-5} of its sum to infinity?

13. Solve

    a) $\log_a(3 - 2x) = \log_a x + \log_a 7$

    b) $\log_{10}(3 + x) = \log_{10} x + 2$

    c) $\ln(5x - 2) = \ln x + \ln 4$

    d) $\ln(3 - 2x) = \ln x + 2$ (leave your answer in terms of e).

⋮ EXAM-STYLE QUESTION

14. Solve

    a) $\ln(2 - 5x) = 2\ln x + \ln 3$

    b) $\log_{10}(1 - 3x) = 2\log_{10} x + 1$.

15. There were $150\,\text{mg}$ of a radioactive material stored at the start of the year 2000. The material has a half-life of 15 years.

    a) How much radioactive material will there be at the start of 2040?

    b) When will there be 1 mg of radioactive material left?

16. A bacteria has a generation time (the time in which the population doubles) of 30 minutes. Sunil conducts an experiment in which he grows a culture of the bacteria starting with a colony of $10\,000$ bacteria, in a container which will hold 10 million bacteria. When will the container be filled?

17. Warren is an investor. He puts $A into a fund. Its value, $V, T years later is expected to follow the equation $V = Ar^T$. The value of the fund on four occasions is given in the table. Find estimates of A and r and hence estimate the average annual return Warren gets on his investment in the fund.

T	2	4	6	9
V	3572	4000	4440	5106

18. The population P of a colony of bacteria is thought to obey an exponential growth law in the form $P = kr^t$ where P is thousands of bacteria, and t is the number of hours since the experiment started. Plot an appropriate graph and use it to estimate the values of k and r.

t	1.2	2.5	4.2	6.2
P	12.3	17.2	26.9	45.9

19. The variables x and y satisfy the relation $5^y = 6^{2x-5}$. By taking logarithms show that the graph of y against x is a straight line, and find the exact values of the gradient and the intercept.

20. Two variable quantities are related by the equation $y = Ak^x$ where A and k are constants. The graph shows the results of plotting ln y against x for four pairs of values of x and y. Use the diagram to estimate the values of A and k.

21.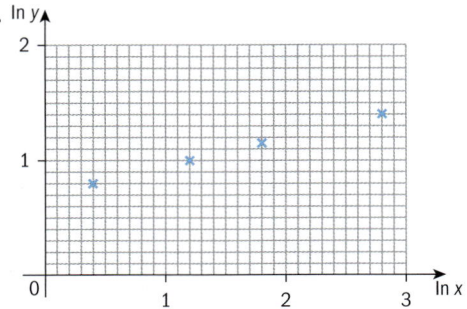

Two variable quantities x and y are related by the equation $y = Ax^n$, where A and n are constants. The diagram shows the result of plotting ln y against ln x for four pairs of values of x and y. Use the diagram to estimate the values of A and n. [5]

Cambridge International AS and A Level Mathematics 9709 Paper 3 Q2 October/November 2005

22. Given that $2^x = 5^y$, use logarithms to find the value of $\frac{x}{y}$ correct to 3 significant figures. [3]

Cambridge International AS and A Level Mathematics 9709 Paper 2 Q1 May/June 2004

Chapter summary

Exponential growth & decay and logarithmic functions

- A function $f(t) = a \times r^t$ where t is real, describes exponential growth when $r > 1$, and describes exponential decay when $r < 1$.
- $y = b^x \Leftrightarrow x = log_b\, y$, where $x \in \mathbb{R}$, $y \in \mathbb{R}$, $y > 0$.
- The logarithmic function is the inverse of the exponential function to the same base.

Properties of logarithms

$$\log(xy) = \log x + \log y$$

$$\log\left(\frac{x}{y}\right) = \log x - \log y$$

$$\log(x^n) = n \log x$$

Special bases

In practice, two bases are very commonly used for logarithms and both are available on your calculator.

- The button marked $\boxed{\log}$ uses base 10 since our number system is based on powers of 10.
- $\boxed{\ln}$ uses base e ($= 2.71828\ldots$). These are sometimes called natural logarithms because of some special properties the number represented by 'e' has when used in an exponential function.

Equations and inequalities using logarithms

- You can use the laws of logarithms to transform equations to make their solutions accessible.
- Power (and logarithmic) functions are strictly monotonic, so if the logarithms are equal it follows that the expressions are equal. This also allows inequalities to be solved by solving for the critical value at which equality holds.

Using logarithms to reduce equations to linear form

- Exponential growth/decay: $y = ab^t \Rightarrow \log y = \log a + t \log b$. The graph of '$\log y$' against '$t$' has intercept '$\log a$' and gradient '$\log b$'.
- Power law: $y = ax^n \Rightarrow \log y = \log a + n \log x$. The graph of '$\log y$' against '$\log x$' has intercept '$\log a$' and gradient '$n$'.

3 Trigonometry

Trigonometry plays a major role in musical theory and production. A musical note can be represented by a sine curve, and a chord can be represented by multiple sine curves. A graphical representation of music allows computers to create and understand sounds. It also allows sound engineers to visualize sound waves so that they can adjust volume, pitch and other elements to create the desired sound. Trigonometry also plays an important role in speaker placement, since the angles of sound waves hitting our ears can influence the quality of the sound.

Objectives

- Understand the relationship of the secant, cosecant, and cotangent functions to cosine, sine, and tangent, and use properties and graphs of all six trigonometric functions for angles of any magnitude.
- Use trigonometric identities for the simplification and exact evaluation of expressions in the course of solving equations, and select an identity or identities appropriate to the context, showing familiarity in particular with the use of
  - $\sec^2 \theta \equiv 1 + \tan^2 \theta$ and $\operatorname{cosec}^2 \theta \equiv 1 + \cot^2 \theta$
  - the expansions of $\sin(A \pm B)$, $\cos(A \pm B)$, and $\tan(A \pm B)$
  - the formulae for $\sin 2A$, $\cos 2A$, and $\tan 2A$,
  - the expressions of $a \sin \theta + b \cos \theta$ in the forms $R \sin(\theta \pm \alpha)$ and $R \cos(\theta \pm \alpha)$.

Before you start

You should know how to:

1. Use exact trigonometric ratios,

   e.g. $\sin 30° = \dfrac{1}{2}$, $\cos 45° = \dfrac{1}{\sqrt{2}} \left(= \dfrac{\sqrt{2}}{2} \right)$,

   $\tan 60° = \sqrt{3}$, $\sin \dfrac{\pi}{4} = \dfrac{1}{\sqrt{2}} \left(= \dfrac{\sqrt{2}}{2} \right)$,

   $\cos \dfrac{\pi}{3} = \dfrac{1}{2}$, $\tan \dfrac{3\pi}{4} = -1$.

Skills check:

1. State the exact values of

   a) $\sin 0°$ b) $\cos 30°$

   c) $\tan 150°$ d) $\sin \dfrac{2\pi}{3}$

   e) $\cos \dfrac{\pi}{6}$ f) $\tan \dfrac{7\pi}{4}$

2. Use inverse trigonometric functions,
    e.g. write down the principal values of
    $\tan^{-1}\left(\dfrac{1}{\sqrt{3}}\right)$, $\cos^{-1} 0$, and $\sin^{-1}\left(-\dfrac{\sqrt{3}}{2}\right)$.

    $$\tan^{-1}\left(\dfrac{1}{\sqrt{3}}\right) = \dfrac{\pi}{6} \; (=30°)$$

    $$\cos^{-1} 0 = \dfrac{\pi}{2} \; (=90°)$$

    $$\sin^{-1}\left(-\dfrac{\sqrt{3}}{2}\right) = -\dfrac{\pi}{3} \; (=-60°).$$

2. **a)** State the principal value, in radians, of
    i) $\sin^{-1}\left(\dfrac{1}{2}\right)$, **ii)** $\tan^{-1}(-1)$
    b) State the principal value, in degrees, to 1 decimal place, of
    i) $\cos^{-1}(0.85)$, **ii)** $\sin^{-1}(-0.2)$

3. Solve trigonometric equations using the
    identities $\tan \theta = \dfrac{\sin \theta}{\cos \theta}$ and $\sin^2 \theta + \cos^2 \theta = 1$.

    e.g. Solve $\sqrt{2} \sin x - \cos x = 0$
    for $0 \le x \le 360°$.

    $$\sqrt{2} \sin x = \cos x$$

    $$\tan x = \dfrac{1}{\sqrt{2}}$$

    $$x = 35.3°, 215.3°$$

3. Solve
    a) $4 \sin \theta = 2 \cos \theta$ for $0° \le \theta < 360°$
    b) $2(\sin^2 \theta - \cos^2 \theta) = 1$ for $0 \le \theta \le 2\pi$.

3.1 Secant, cosecant, and cotangent

We are familiar with the trigonometric functions sine, cosine, and tangent.
In P1, we studied these functions together with their graphs, their inverse
functions, and the trigonometric identities $\tan \theta = \dfrac{\sin \theta}{\cos \theta}$ and $\sin^2 \theta + \cos^2 \theta = 1$.
We also studied how to use these identities to solve trigonometric equations.

We now introduce three other trigonometric functions: **secant**, **cosecant** and
cotangent. These functions, usually abbreviated to sec, cosec, and cot, are
closely related to sin, cos, and tan. They are defined as follows:

$$\sec \theta = \dfrac{1}{\cos \theta}$$ $$\operatorname{cosec} \theta = \dfrac{1}{\sin \theta}$$ $$\cot \theta = \dfrac{1}{\tan \theta} = \left(\dfrac{\cos \theta}{\sin \theta}\right)$$

These functions are the reciprocals of cos θ, sin θ and tan θ. Remember, for example,
that $\operatorname{cosec} \theta = \dfrac{1}{\sin \theta}$. Be careful not to confuse these functions with the inverse functions
$\sin^{-1} \theta$, $\cos^{-1} \theta$, and $\tan^{-1} \theta$.

In order to sketch the graphs of sec θ, cosec θ, and cot θ we can consider the reciprocals of the functions cos θ, sin θ, and tan θ.

For example, to sketch the graph of $y = \sec\theta$, we consider the graph of $y = \cos\theta$.

As $y = \cos\theta$ is zero at $x = 90°, 270°, 450°, ...$, we know that sec θ will not be defined at each of these points, and instead there will be vertical asymptotes on the graph of $y = \sec\theta$ at these points.

The graph can then be sketched by considering the value of $\dfrac{1}{\cos\theta}$ at different points.

We see that, just as the graph of $y = \cos\theta$ is periodic with period 360°, then so is the graph of $y = \sec\theta$.

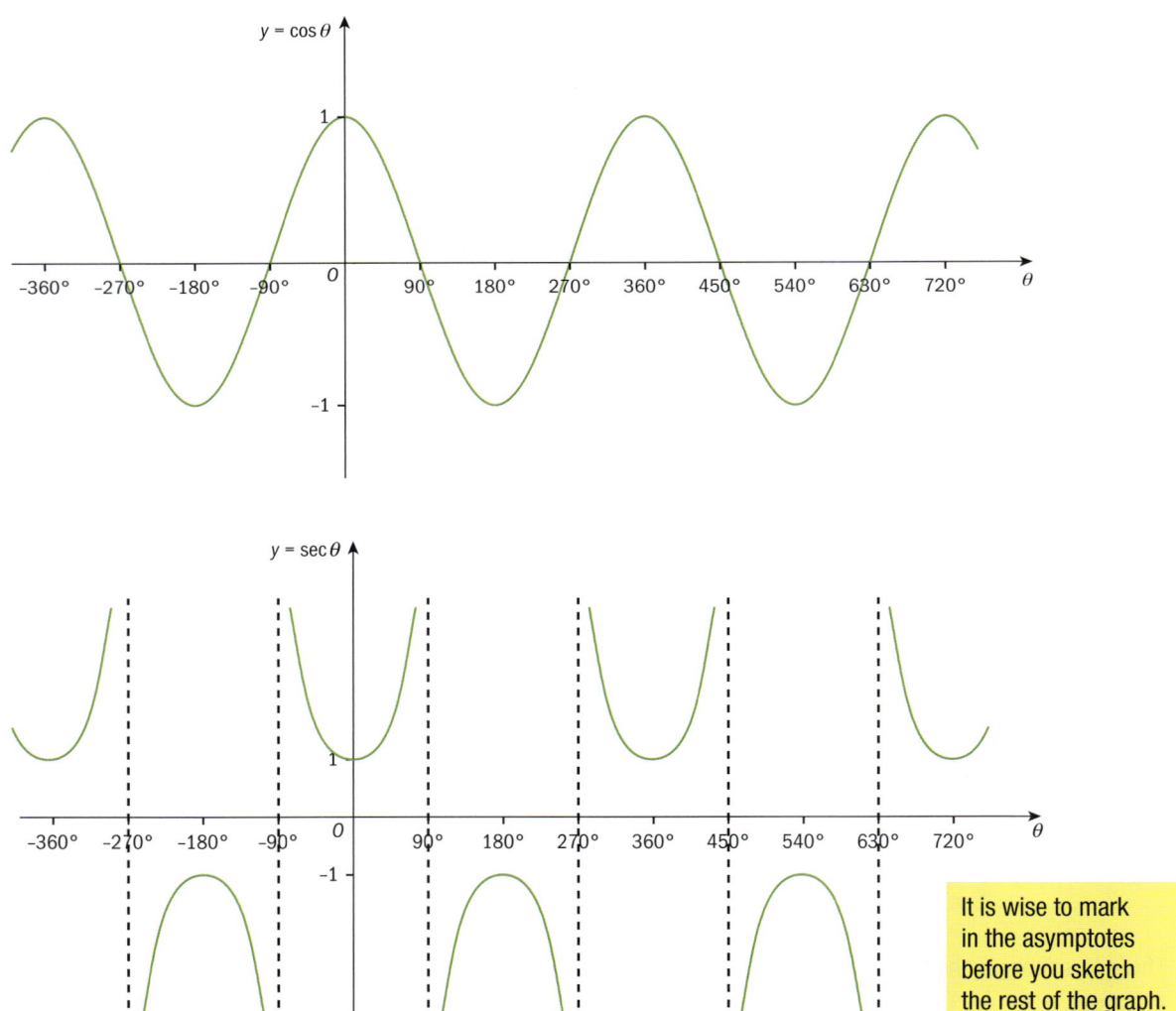

It is wise to mark in the asymptotes before you sketch the rest of the graph.

Secant, cosecant, and cotangent

The graph of $y = \text{cosec } \theta$ can be obtained from the graph of $y = \sin \theta$ in a similar way.

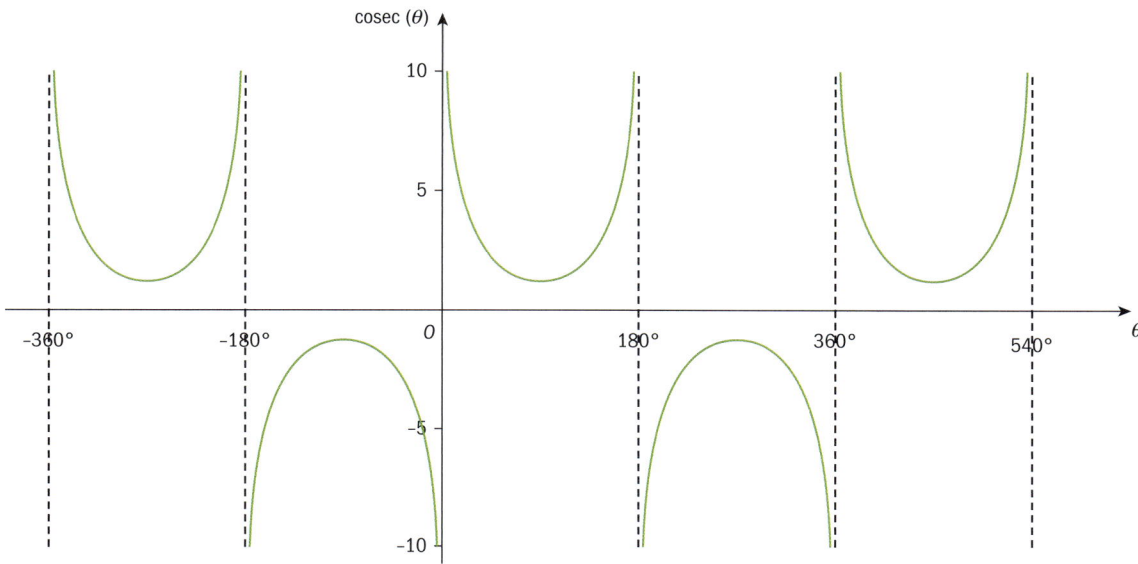

The graph of $y = \cot \theta$ can be obtained from the graph of $y = \tan \theta$

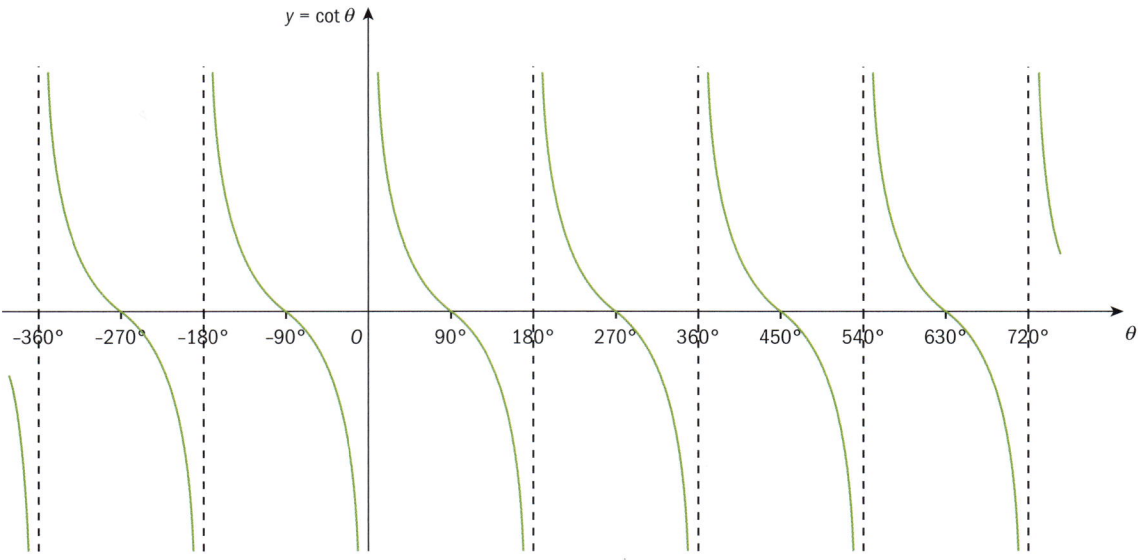

Example 1

Without using a calculator, write down the values of

a) $\sec 0°$ **b)** $\sec 60°$ **c)** $\operatorname{cosec} 120°$ **d)** $\cot 270°$ **e)** $\operatorname{cosec}\left(-\dfrac{\pi}{4}\right)$ **f)** $\cot\left(\dfrac{5\pi}{4}\right)$.

a) $\sec 0° = \dfrac{1}{\cos 0°} = 1$ ⟵ Use $\cos 0° = 1$.

b) $\sec 60° = \dfrac{1}{\cos 60°} = 2$ ⟵ Use $\cos 60° = \dfrac{1}{2}$.

c) $\operatorname{cosec} 240° = \dfrac{1}{\sin 240°} = -\dfrac{2}{\sqrt{3}}\left(\text{or } -\dfrac{2\sqrt{3}}{3}\right)$ ⟵ Use $\sin 240° = -\dfrac{\sqrt{3}}{2}$ and take the reciprocal.

d) $\cot 270° = \dfrac{1}{\tan 270°} = \dfrac{1}{\infty} = 0$ ⟵ $\tan 270°$ is infinitely large so $\cot 270° = 0$.

e) $\operatorname{cosec}\left(-\dfrac{\pi}{4}\right) = \dfrac{1}{\sin\left(-\dfrac{\pi}{4}\right)} = -\dfrac{2}{\sqrt{2}}\ (\text{or } -\sqrt{2})$ ⟵ $\sin\left(-\dfrac{\pi}{4}\right) = -\sin\left(\dfrac{\pi}{4}\right) = -\dfrac{\sqrt{2}}{2}$

f) $\cot\left(\dfrac{5\pi}{4}\right) = \dfrac{\cos\left(\dfrac{5\pi}{4}\right)}{\sin\left(\dfrac{5\pi}{4}\right)} = \left(-\dfrac{\sqrt{2}}{2}\right) \div \left(-\dfrac{\sqrt{2}}{2}\right) = 1$ ⟵ You can use the reciprocal of $\tan\left(\dfrac{5\pi}{4}\right)$ or use $\cot\theta = \dfrac{\cos\theta}{\sin\theta}$

We can use the definitions of $\sec\theta$, $\operatorname{cosec}\theta$, and $\cot\theta$ to help us solve some simple trigonometric equations.

Example 2

Solve for $0° \le \theta \le 360°$.

a) $\sec\theta = 2$ **b)** $\cot^2\theta = 3$ **c)** $11 + 3\operatorname{cosec} 2\theta = 1$

a) $\sec\theta = \dfrac{1}{\cos\theta} = 2$ ⟵ Replace $\sec\theta$ with $\dfrac{1}{\cos\theta°}$.

$\cos\theta = \dfrac{1}{2}$ ⟵ Rearrange to obtain the value of $\cos\theta$.

$\theta = 60°, 300°$ ⟵ Solve the equation for $0° \le \theta \le 360°$.

▶ Continued on the next page

b) $\cot \theta = \pm \sqrt{3}$ ← Write down the two values, $\pm \sqrt{3}$.

$\tan \theta = \pm \dfrac{1}{\sqrt{3}} \left(\text{or } \pm \dfrac{\sqrt{3}}{3} \right)$ ← Use $\cot \theta = \dfrac{1}{\tan \theta}$ to write down the values of $\tan \theta$.

$\theta = 30°, 210°$

or $\theta = 150°, 330°$ ← Solve each of the equations $\tan \theta = \dfrac{1}{\sqrt{3}}$ and $\tan \theta = -\dfrac{1}{\sqrt{3}}$.

$\theta = 30°, 150°, 210°, 330°$

c) $3 \operatorname{cosec} 2\theta = -10$ ← Rearrange to obtain the value of $\operatorname{cosec} 2\theta$.

$\operatorname{cosec} 2\theta = -\dfrac{10}{3}$

$\sin 2\theta = -0.3$ ← Use $\operatorname{cosec} 2\theta = \dfrac{1}{\sin 2\theta°}$ find the value of $\sin 2\theta$.

$2\theta = 180° + 17.45°, 360° - 17.45°,$
$\quad 540° + 17.45°, 720° - 17.45°$ ← Find all values of 2θ in the range $0°$ to $720°$.

$\theta = 98.8°, 171.3°, 278.8°, 351.3°$ ← Give each solution for θ correct to one decimal place.

We can also sketch the graphs of composite functions involving $\sec \theta$, $\operatorname{cosec} \theta$, and $\cot \theta$ in the same way that we sketched the graphs of composite functions involving $\sin \theta$, $\cos \theta$, and $\tan \theta$ in P1. You can practise this by answering questions **13** and **14** in Exercise 4.1.

Exercise 3.1

1. Find the exact values of
   - **a)** $\sec 30°$
   - **b)** $\operatorname{cosec} 30°$
   - **c)** $\cot 60°$
   - **d)** $\sec 45°$
   - **e)** $\sec 210°$
   - **f)** $\operatorname{cosec} 135°$
   - **g)** $\cot 300°$
   - **h)** $\sec 150°$.

2. Find the exact values of
   - **a)** $\sec \dfrac{3\pi}{4}$
   - **b)** $\sec \dfrac{\pi}{3}$
   - **c)** $\cot \dfrac{\pi}{6}$
   - **d)** $\operatorname{cosec} \dfrac{\pi}{2}$
   - **e)** $\operatorname{cosec} \dfrac{4\pi}{3}$
   - **f)** $\operatorname{cosec} \dfrac{3\pi}{2}$
   - **g)** $\sec \dfrac{5\pi}{6}$
   - **h)** $\cot \dfrac{7\pi}{4}$.

3. State whether each of the following values are defined or undefined.
   - **a)** $\operatorname{cosec} 90°$
   - **b)** $\operatorname{cosec} 130°$
   - **c)** $\sec 180°$
   - **d)** $\cot (-180°)$
   - **e)** $\sec \dfrac{3\pi}{2}$
   - **f)** $\cot \dfrac{7\pi}{6}$
   - **g)** $\operatorname{cosec} (-\pi)$
   - **h)** $\cot \dfrac{5\pi}{2}$.

4. Use your calculator to find these values correct to 3 significant figures.

    a) $\cot 32°$ **b)** $\operatorname{cosec} 70°$ **c)** $\sec 215°$ **d)** $\sec(-22°)$

    e) $\operatorname{cosec} \dfrac{\pi}{5}$ **f)** $\cot \dfrac{7\pi}{10}$ **g)** $\sec(0.6\pi)$ **h)** $\sec \dfrac{7\pi}{5}$

5. Solve each of these equations for $0° < \theta < 360°$.

    a) $\cot \theta° = 1$ **b)** $\sec \theta° = \sqrt{2}$ **c)** $\operatorname{cosec} \theta° = \dfrac{2}{\sqrt{3}}$

    d) $\operatorname{cosec} \theta° = 2$ **e)** $\cot \theta° = -\sqrt{3}$ **f)** $\sec \theta° = -1$

6. Solve each of these equations for $0 \le \theta \le 2\pi$. Give your answers in exact form.

    a) $\operatorname{cosec} \theta = 1$ **b)** $\sec \theta = 1$ **c)** $\cot \theta = -1$

    d) $\sec \theta = \dfrac{2}{\sqrt{3}}$ **e)** $\cot \theta = 0$ **f)** $\operatorname{cosec} \theta = -\dfrac{2}{\sqrt{3}}$

7. Solve each of these equations for $0° \le x < 360°$. Give your answers correct to 1 decimal place.

    a) $\sec x° = 3$ **b)** $\operatorname{cosec} x° = 4$ **c)** $\cot x° = 0.9$

    d) $\cot x° = 5$ **e)** $\operatorname{cosec} x° = -\dfrac{9}{7}$ **f)** $\sec x° = -\sqrt{6}$

8. **a)** Solve the equation $\sec x° = 7$ for $0° < x < 360°$.

    b) Solve the equation $\sec^2 x° = 49$ for $0° < x < 360°$.

9. Solve the equation $8 \cot x° - 5 = 7$ for $0° < x < 360°$.

10. Solve the equation $7 + 2 \operatorname{cosec} \theta = 4$ for $0 < \theta < 2\pi$.

11. Solve each of these equations for $0° \le x \le 360°$.

    a) $\sec 2x° = 1$ **b)** $\operatorname{cosec} 2x° = 4$ **c)** $6 \cot 3x° + 5 = 13$

12. Solve each of these equations for $0° < x < 360°$.

    a) $\operatorname{cosec}(2x + 90)° = 1$ **b)** $4 - 3\cot \dfrac{1}{2}x° = 3$ **c)** $2\sec 3x° - 5 = \sqrt{2}$

13. Sketch these graphs on separate axes.

    a) $y = \operatorname{cosec} 3x°$ for $0° \le x \le 360°$ **b)** $y = \sec 2x°$ for $0° \le x \le 360°$

    c) $y = \cot \dfrac{1}{2}x°$ for $-360° \le x \le 360°$ **d)** $y = 2\operatorname{cosec} x°$ for $0° \le x \le 180°$

    e) $y = \cot(x + 90)°$ for $0° \le x \le 360°$ **f)** $y = 3\sec 2x°$ for $0° \le x \le 360°$

14. Sketch, on separate sets of axes, the graphs of the following functions for $-2\pi \le x \le 2\pi$.

    a) $y = 6\operatorname{cosec} x$ **b)** $y = \cot\left(x - \dfrac{\pi}{4}\right)$

    c) $y = \dfrac{1}{2}\sec\left(x + \dfrac{\pi}{2}\right)$ **d)** $y = \operatorname{cosec}(2x - \pi)$

3.2 Further trigonometric identities

In P1, we used the identities $\tan\theta \equiv \dfrac{\sin\theta}{\cos\theta}$ and $\sin^2\theta + \cos^2\theta \equiv 1$. We now turn our

attention to finding and using two new identities involving sec θ, cosec θ, and cot θ.

Using the identity

$$\sin^2\theta + \cos^2\theta \equiv 1$$

We divide through by $\cos^2\theta$ to give $\dfrac{\sin^2\theta}{\cos^2\theta} + \dfrac{\cos^2\theta}{\cos^2\theta} \equiv \dfrac{1}{\cos^2\theta}$

$$\Rightarrow \tan^2\theta + 1 \equiv \sec^2\theta$$

Similarly, dividing $\sin^2\theta + \cos^2\theta \equiv 1$ by $\sin^2\theta$ gives

$$\dfrac{\sin^2\theta}{\sin^2\theta} + \dfrac{\cos^2\theta}{\sin^2\theta} \equiv \dfrac{1}{\sin^2\theta}$$

$$\Rightarrow 1 + \cot^2\theta \equiv \operatorname{cosec}^2\theta$$

$$\boxed{\begin{aligned} 1 + \tan^2\theta &\equiv \sec^2\theta \\ 1 + \cot^2\theta &\equiv \operatorname{cosec}^2\theta \end{aligned}}$$

> Remember that we should use '≡' rather than '=' in identities to show that the statement is true for all values of θ. Although, don't worry if you use the '=' sign.

These identities are useful in helping us to simplify expressions, prove identities, and solve trigonometric equations.

Example 3

Prove the identity $(\tan\theta + \cot\theta)^2 \equiv \sec^2\theta + \operatorname{cosec}^2\theta$

$(\tan\theta + \cot\theta)^2 \equiv \tan^2\theta + 2\tan\theta\cot\theta + \cot^2\theta$ ← Start with the left-hand side and expand the brackets.

$\equiv \tan^2\theta + 2\tan\theta\,\dfrac{1}{\tan\theta} + \cot^2\theta$ ← Replace cot θ with $\dfrac{1}{\tan\theta}$.

$\equiv \tan^2\theta + 2 + \cot^2\theta$

$\equiv \sec^2\theta - 1 + 2 + \operatorname{cosec}^2\theta - 1$ ← Use the identities to replace $\tan^2\theta$ with $\sec^2\theta - 1$ and $\cot^2\theta$ with $\operatorname{cosec}^2\theta - 1$ and then simplify.

$\equiv \sec^2\theta + \operatorname{cosec}^2\theta.$

Example 4

Solve the equation $\sec^2 \theta + \tan \theta - 1 = 0$ for $0° \le \theta \le 360°$.

$1 + \tan^2 \theta + \tan \theta - 1 = 0$ ← Substitute $1 + \tan^2 \theta$ for $\sec^2 \theta$ to get a quadratic equation in $\tan \theta$.

$\tan^2 \theta + \tan \theta = 0$

$\tan \theta (\tan \theta + 1) = 0$ ← Factorise the equation and solve for $\tan \theta$.

$\tan \theta = 0$ or $\tan \theta + 1 = 0$

$\tan \theta = 0$ or $\tan \theta = -1$

$\theta = 0°, 135°, 180°, 315°, 360°$

Exercise 3.2

1. Simplify the following expressions.

   a) $\sec^2 \theta - \tan^2 \theta$

   b) $\dfrac{\sin \theta}{\csc \theta - \cos \theta \cot \theta}$

   c) $\dfrac{\tan \theta}{\sqrt{1 + \tan^2 \theta}}$

   d) $\tan \theta \cot \theta$

   e) $\dfrac{\sin \theta}{1 + \cot^2 \theta}$

   f) $\dfrac{1 + \tan^2 \theta}{1 + \cot^2 \theta}$

2. Solve the equation $2 \sec^2 \theta + 3 \tan \theta - 4 = 0$ for $0° \le \theta \le 360°$.

3. Solve the equation $\sec \theta \tan \theta = 1$ for $0 < \theta < 2\pi$.

4. Solve the equation $\tan^2 \theta - \sec \theta - 5 = 0$ for $0° \le x \le 360°$.

5. Prove the identity $\csc \theta + \cot \theta \equiv \dfrac{1}{\csc \theta - \cot \theta}$.

   Hint: Start with multiplying the left hand side by $\csc \theta - \cot \theta$ and then consider the right hand side.

6. Prove the identity $\dfrac{\tan \theta}{1 + \sec \theta} + \dfrac{1 + \sec \theta}{\tan \theta} \equiv \dfrac{2}{\sin \theta}$.

7. Prove the identity $(\sec \theta - \tan \theta)^2 \equiv \dfrac{1 - \sin \theta}{1 + \sin \theta}$.

8. Prove the identity $\dfrac{\cot \theta}{1 + \cot^2 \theta} \equiv \sin \theta \cos \theta$.

9. Prove the identity $\sec^2 \theta + \csc^2 \theta \equiv \sec^2 \theta \csc^2 \theta$.

10. Solve the equation $4 \cot^2 \theta - 2 \cot \theta = 3 \csc^2 \theta$ for $0° \le \theta \le 360°$.

11. Solve the equation $\sec \theta = 3 \cos \theta + 1$ for $0° \le \theta \le 360°$.

12. Solve the equation $\operatorname{cosec}^2 \theta + 2 \cot \theta = 0$ for $0 \le \theta \le 2\pi$.

13. Prove the identity $\tan \theta + \cot \theta \equiv \sec \theta \operatorname{cosec} \theta$.

14. Solve the equation $3 \cot \theta + 2 \tan \theta = 5$ for $0° \le \theta \le 360°$.

15. Solve the equation $\tan^2 \theta = \sec \theta$ for $0° \le \theta \le 360°$.

16. Prove the identity $\cot^2 \theta \cos^2 \theta - \sin^2 \theta = \cot^2 \theta - 1$.

A geometrical interpretation of the trigonometric functions

All of the trigonometric functions of an angle can be constructed geometrically, in terms of a unit circle centre O, as shown in the diagram.

> **Remember**: In P1 we saw that the term 'cosine(of θ)' originates from 'the sine of the complementary angle (to θ)'.

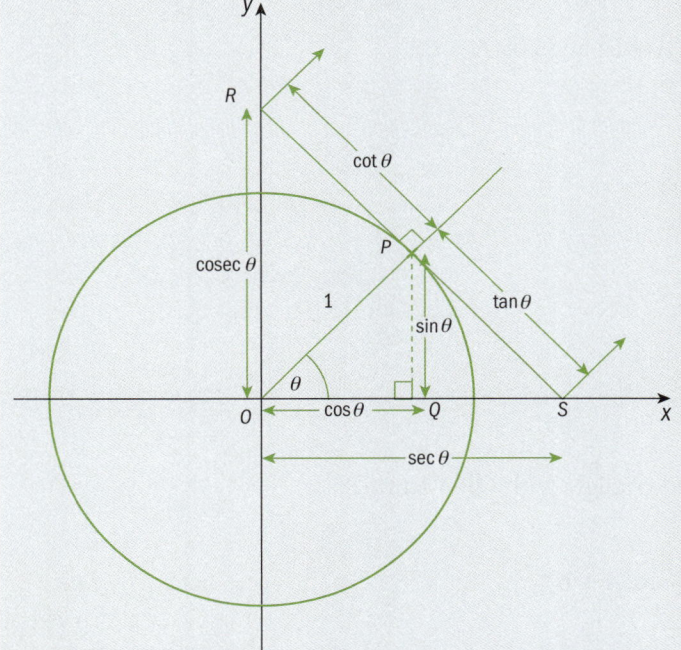

Can you show how the diagram may be used to help explain the following identities?

$\cos \theta \equiv \sin(90° - \theta)$ and $\sin \theta \equiv \cos(90° - \theta)$

$\cot \theta \equiv \tan(90° - \theta)$ and $\tan \theta \equiv \cot(90° - \theta)$

$\operatorname{cosec} \theta \equiv \sec(90° - \theta)$ and $\sec \theta \equiv \operatorname{cosec}(90° - \theta)$

$\tan \theta \equiv \dfrac{\sin \theta}{\cos \theta}$

$\sin^2 \theta + \cos^2 \theta \equiv 1$

$1 + \tan^2 \theta \equiv \sec^2 \theta$

$1 + \cot^2 \theta \equiv \operatorname{cosec}^2 \theta$

3.3 Addition formulae

In this section we will learn how to use formulae for $\sin(A \pm B)$, $\cos(A \pm B)$ and $\tan(A \pm B)$.

It is important to recognise that $\sin(A + B) \neq \sin A + \sin B$.

We can see this if we let $A = 30°$ and $B = 60°$, for example.

If the identity $\sin(A + B) \neq \sin A + \sin B$ were true, then,

$$\sin(30 + 60)° = \sin 90° = 1$$

However, we know that really $\sin 30° + \sin 60° = \frac{1}{2} + \frac{\sqrt{3}}{2} \neq 1$.

So $\sin(30 + 60)° \neq \sin 30° + \sin 60°$.

We will now establish the formulae for $\sin(A + B)$ and $\cos(A + B)$.

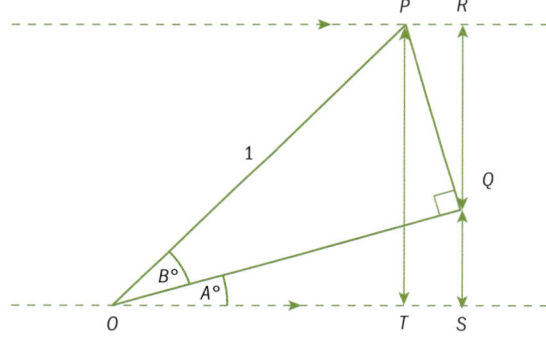

In the diagram above, OPQ is a right-angled triangle with $OP = 1$ unit.

From triangle OPT,

$$PT = 1 \times \sin(A + B) = \sin(A + B) \qquad (1)$$

From triangle OPQ,

$$PQ = 1 \times \sin B = \sin B \text{ and } OQ = 1 \times \cos B = \cos B$$

We know that angle $OQS = 90° - A$
and angle $PQR = A$

We also know that $RQ = PQ \cos A = \sin B \cos A \qquad (2)$
and $QS = OQ \sin A = \cos B \sin A \qquad (3)$

Now

$$PT = RQ + QS$$

Alternatively, $\sin(A + B) = \sin B \cos A + \cos B \sin A$ <mark>Using equations (1), (2), and (3).</mark>

This is usually written as

$$\sin(A + B) \equiv \sin A \cos B + \cos A \sin B.$$

This result is the first of six formulae we are going to find.

Similarly, from triangle OPT,

$$OT = 1 \times \cos(A + B) = \cos(A + B)$$

but

$$OT = OS - TS = OS - PR$$

and

$$OS = \cos B \cos A$$
$$PR = \sin B \sin A$$

so

$$\cos(A + B) = \cos B \cos A - \sin A \sin B$$

or

$$\cos(A + B) \equiv \cos A \cos B - \sin A \sin B.$$

This is the second of our six formulae.

Replacing B with $-B$ in $\sin(A + B) = \sin A \cos B + \cos A \sin B$, we obtain

$$\sin(A - B) = \sin A \cos(-B) + \cos A \sin(-B)$$

or $\quad \sin(A - B) \equiv \sin A \cos B - \cos A \sin B \quad$ as $\quad \cos(-B) = \cos B$
$$\sin(-B) = -\sin B.$$

Similarly, replacing B with $-B$ in $\cos(A + B) = \cos A \cos B - \sin A \sin B$, we obtain
$$\cos(A - B) = \cos A \cos(-B) \sin A \sin(-B)$$
or $\quad \cos(A - B) \equiv \cos A \cos B + \sin A \sin B.$

We now turn our attention to obtaining expressions for $\tan(A + B)$ and $\tan(A - B)$ by using the identity $\tan \theta \equiv \dfrac{\sin \theta}{\cos \theta}$.

$$\tan(A + B) = \frac{\sin(A + B)}{\cos(A + B)} = \frac{\sin A \cos B + \cos A \sin B}{\cos A \cos B - \sin A \sin B}$$

Dividing the numerator and denominator by $\cos A \cos B$,

$$\tan(A + B) = \frac{\dfrac{\sin A \cos B}{\cos A \cos B} + \dfrac{\cos A \sin B}{\cos A \cos B}}{\dfrac{\cos A \cos B}{\cos A \cos B} - \dfrac{\sin A \sin B}{\cos A \cos B}}$$

$$\equiv \frac{\tan A + \tan B}{1 - \tan A \tan B}$$

Replacing B with $-B$, we obtain

$$\tan(A - B) \equiv \frac{\tan A - \tan B}{1 + \tan A \tan B}.$$

$$\sin(A + B) \equiv \sin A \cos B + \cos A \sin B$$
$$\sin(A - B) \equiv \sin A \cos B - \cos A \sin B$$
$$\cos(A + B) \equiv \cos A \cos B - \sin A \sin B$$
$$\cos(A - B) \equiv \cos A \cos B + \sin A \sin B$$
$$\tan(A + B) \equiv \frac{\tan A + \tan B}{1 - \tan A \tan B}$$
$$\tan(A - B) \equiv \frac{\tan A - \tan B}{1 + \tan A \tan B}$$

These formulae complete our set of six **Addition Formulae** (or Compound Angle Formulae).

Example 5

By considering $\sin(45° + 30°)$, prove that $\sin 75° = \dfrac{1+\sqrt{3}}{2\sqrt{2}}$.

$$\sin(A + B) = \sin A \cos B + \cos A \sin B$$
$$\sin(45° + 30°) = \sin 45° \cos 30° + \cos 45° \sin 30°$$

Put $A = 45°$ and $B = 30°$ into the addition formula for $\sin(A + B)$.

$$= \frac{1}{\sqrt{2}} \times \frac{\sqrt{3}}{2} + \frac{1}{\sqrt{2}} \times \frac{1}{2}$$

Substitute exact values for $\sin 30°$, $\sin 45°$, $\cos 30°$, and $\cos 45°$.

$$\sin 75° = \frac{1+\sqrt{3}}{2\sqrt{2}}$$

Example 6

Prove that $2 \cos\left(\theta - \dfrac{\pi}{3}\right) \equiv \cos\theta + \sqrt{3} \sin\theta$.

$$2 \cos\left(\theta - \frac{\pi}{3}\right) \equiv 2\left(\cos\theta \cos\frac{\pi}{3} + \sin\theta \sin\frac{\pi}{3}\right)$$

Use the formula
$\cos(A - B) = \cos A \cos B + \sin A \sin B$.

$$\equiv 2\left(\cos\theta \times \frac{1}{2} + \sin\theta \times \frac{\sqrt{3}}{2}\right)$$

Substitute $\cos\dfrac{\pi}{3} = \dfrac{1}{2}$ and $\sin\dfrac{\pi}{3} = -\dfrac{\sqrt{3}}{2}$.

$$\equiv \cos\theta + \sqrt{3} \sin\theta$$

Example 7

Solve the equation $\cos(\theta + 60°) = 2\sin(\theta - 45°)$ for $0° \leq \theta \leq 360°$.

$\cos\theta\cos 60° - \sin\theta\sin 60° = 2(\sin\theta\cos 45° - \cos\theta\sin 45°)$ ← Expand both expressions.

$\dfrac{1}{2}\cos\theta - \dfrac{\sqrt{3}}{2}\sin\theta = 2\left(\dfrac{1}{\sqrt{2}}\sin\theta - \dfrac{1}{\sqrt{2}}\cos\theta\right)$ ← Substitute values for $\sin 45°$, $\cos 45°$, $\sin 60°$, $\cos 60°$.

$\left(\dfrac{1}{2} + \dfrac{2}{\sqrt{2}}\right)\cos\theta = \left(\dfrac{2}{\sqrt{2}} + \dfrac{\sqrt{3}}{2}\right)\sin\theta$ ← Collect terms in $\sin\theta$ and $\cos\theta$.

$\dfrac{\sqrt{2}+4}{2\sqrt{2}}\cos\theta = \dfrac{4+\sqrt{3}\sqrt{2}}{2\sqrt{2}}\sin\theta$

$\dfrac{\sqrt{2}+4}{4+\sqrt{3}\sqrt{2}} = \tan\theta$ ← Rearrange to find $\tan\theta$.

$\tan\theta = 0.8394...$ ← Find all the angles.
$\theta = 40.0°, 220.0°$

We have worked in exact terms until the final step, but you may wish to use a calculator at an earlier stage in the solution.

Exercise 3.3

1. Use the addition formulae to find, in exact form, the values of the following expressions.

   a) $\cos 75°$ **b)** $\cos 15°$ **c)** $\tan 105°$ **d)** $\dfrac{1 + \sin 15°}{1 - \sin 15°}$

2. **a)** Prove that $\tan(\theta + 45°) \equiv \dfrac{1 + \tan\theta}{1 - \tan\theta}$

   b) Show that $\tan 75° \equiv \dfrac{\sqrt{3} + 1}{\sqrt{3} - 1}$

3. Given that $\sin\alpha = \dfrac{4}{5}$ and $\sin\beta = \dfrac{5}{13}$ where α and β are acute angles, find the exact values of

   a) $\cos\alpha$ and $\cos\beta$

   b) $\sin(\alpha + \beta)$, $\sin(\alpha - \beta)$, $\cos(\alpha + \beta)$, and $\cos(\alpha - \beta)$

   c) $\tan(\alpha + \beta)$ and $\tan(\alpha - \beta)$.

4. Find, in exact form, the values of

   a) $\sin\dfrac{5\pi}{12}$ **b)** $\cos\dfrac{7\pi}{12}$ **c)** $\tan\dfrac{\pi}{12}$ **d)** $\sin\left(-\dfrac{\pi}{12}\right)$.

5. Find, in exact form, the values of

   a) $\cos 10°\cos 20° - \sin 10°\sin 20°$ **b)** $\sin 75°\cos 45° + \cos 75°\sin 45°$

   c) $\dfrac{\tan 103° - \tan 58°}{1 + \tan 103°\tan 58°}$ **d)** $\dfrac{\tan 75° + 1}{\tan 75° - 1}$.

6. Simplify

   a) $\sin(\alpha + \beta) + \sin(\alpha - \beta)$

   b) $\cos(\alpha + \beta) + \cos(\alpha - \beta)$

   c) $\sin(\alpha - \beta)\cos\alpha - \cos(\alpha - \beta)\sin\alpha.$

7. Use the addition formulae to show that

   a) $\sin\left(\theta + \dfrac{\pi}{2}\right) \equiv \cos\theta$ **b)** $\sin(90° - \theta) \equiv \cos\theta$

   c) $\cos(90° - \theta) \equiv \sin\theta$ **d)** $\cos(180° + \theta) \equiv -\cos\theta$

   e) $\tan(\theta + 180°) \equiv \tan\theta$ **f)** $\tan(\pi - \theta) \equiv -\tan\theta.$

8. Prove the identity $\operatorname{cosec}(\theta + \varphi) \equiv \dfrac{\operatorname{cosec}\theta \ \operatorname{cosec}\varphi}{\cot\theta + \cot\varphi}.$

9. Express $\cot(\theta + \varphi)$ in terms of $\cot\theta$ and $\cot\varphi$.

10. Prove that $\tan(\alpha + \beta + \gamma) = \dfrac{\tan\alpha + \tan\beta + \tan\gamma - \tan\alpha\tan\beta\tan\gamma}{1 - \tan\alpha\tan\beta - \tan\beta\tan\gamma - \tan\alpha\tan\gamma}.$

    Hint: Consider $\tan\{\alpha + (\beta + \gamma)\}$

11. Find the exact values of

    a) the sine of angle *PSR*

    b) the tangent of angle *PSR*

    c) the secant of angle *PQR*.

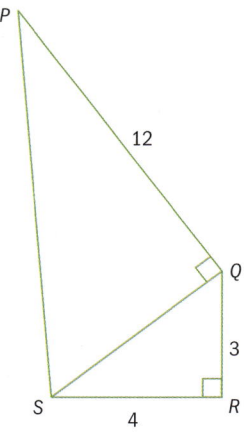

12. Solve the equation $\tan\theta = 2\tan(45° - \theta)$ for $0° \leq \theta \leq 360°$.

13. Solve the equation $\cos\left(\theta - \dfrac{\pi}{6}\right) = \cos\left(\theta + \dfrac{\pi}{6}\right)$ for $0 \leq \theta \leq 2\pi$.

14. Solve the equation $5\cos(\theta + 45°) = \sin(\theta - 45°)$ for $0° \leq \theta \leq 360°$.

3.4 Double angle formulae

We can substitute $B = A$ into the Addition Formulae to obtain three new identities.

$\sin(A + B) \equiv \sin A \cos B + \cos A \sin B$

$\Rightarrow \sin(A + A) \equiv \sin A \cos A + \cos A \sin A$

$\Rightarrow \sin 2A \equiv 2 \sin A \cos A.$

$\cos(A + B) \equiv \cos A \cos B - \sin A \sin B$

$\Rightarrow \cos(A + A) \equiv \cos A \cos A - \sin A \sin A$

$\Rightarrow \cos 2A \equiv \cos^2 A - \sin^2 A.$

$\tan(A + B) \equiv \dfrac{\tan A + \tan B}{1 - \tan A \tan B}$

$\Rightarrow \tan(A + A) \equiv \dfrac{\tan A + \tan A}{1 - \tan A \tan A}$

$\Rightarrow \tan 2A \equiv \dfrac{2 \tan A}{1 - \tan^2 A}.$

Further, using the identity $\cos^2 A + \sin^2 A \equiv 1$ to rewrite the identity for $\cos 2A$ in terms of $\sin A$ or $\cos A$ only, we obtain

$\cos 2A \equiv \cos^2 A - (1 - \cos^2 A) = 2\cos^2 A - 1$

and $\cos 2A \equiv (1 - \sin^2 A) - \sin^2 A = 1 - 2\sin^2 A$

$\sin 2A \equiv 2 \sin A \cos A$	$\cos 2A \equiv \cos^2 A - \sin^2 A$ $\equiv 2\cos^2 A - 1$ $\equiv 1 - 2\sin^2 A$	$\tan 2A \equiv \dfrac{2\tan A}{1 - \tan^2 A}$

These results are called the Double Angle Formulae.

It is important to know all three versions of the formula for $\cos 2A$.

Example 8

Solve the following equations for $0° \leq \theta \leq 360°$.

a) $4 \sin 2\theta = \sin \theta$ **b)** $\cos 2\theta = \cos \theta$

a) $4 \times 2\sin \theta \cos \theta = \sin \theta$ ← Write the equation in terms of $\sin \theta$ and $\cos \theta$ only.

$8 \sin \theta \cos \theta - \sin \theta = 0$ ← Write in the form $f(\theta) = 0$.

$\sin \theta (8 \cos \theta - 1) = 0$ ← Factorise the LHS and solve the equation.

$\sin \theta = 0$ or $\cos \theta = \dfrac{1}{8}$

$\theta = 0°, 180°, 360°$ or $\theta = 82.8°, 360° - 82.8°$

$\theta = 0°, 82.8°, 180°, 277.2°, 360°$

b) $2 \cos^2 \theta - 1 = \cos \theta$ ← Write the equation in terms of $\cos \theta$ only.

$2 \cos^2 \theta - \cos \theta - 1 = 0$

$(2 \cos \theta + 1)(\cos \theta - 1) = 0$ ← Factorise and solve the resulting quadratic equation in $\cos \theta$.

$\cos \theta = -\dfrac{1}{2}$ or $\cos \theta = 1$

$\theta = 120°, 240°$ or $\theta = 0°, 360°$

$\theta = 0°, 120°, 240°, 360°$.

Example 9

Express $\cos 3\theta$ in terms of $\cos \theta$.

$\cos 3\theta \equiv \cos(2\theta + \theta)$ ← Express $\cos 3\theta$ in a way that the addition formulae may be used.

$\equiv \cos 2\theta \cos \theta - \sin 2\theta \sin \theta$ ← Use the double angle formulae.

$\equiv (2 \cos^2 \theta - 1) \times \cos \theta - 2 \sin \theta \cos \theta \times \sin \theta$

$\equiv 2 \cos^3 \theta - \cos \theta - 2 \sin^2 \theta \cos \theta$

$\equiv 2 \cos^3 \theta - \cos \theta - 2(1 - \cos^2 \theta)\cos \theta$ ← Ensure that the expression is expressed in terms of $\cos \theta$ only.

$\equiv 4 \cos^3 \theta - 3 \cos \theta$.

Exercise 3.4

1. Solve these equations for $0° \le \theta \le 360°$.

   a) $\sin 2\theta = \cos \theta$ b) $\sin 2\theta - \sqrt{3} \cos \theta = 0$

   c) $\cos 2\theta + 3 \cos \theta - 1 = 0$ d) $4 \cos 2\theta + 2 \sin \theta - 1 = 0$

   e) $\tan 2\theta + \tan \theta = 0$ f) $\sin 2\theta = \tan \theta$

2. Solve these equations for $0 \le \theta \le 2\pi$.

   a) $2 \tan 2\theta = 5 \tan \theta$ b) $5 \sin 2\theta = 2 \sin \theta$ c) $2 \cos 2\theta = 1 - 4 \cos \theta$

3. An acute angle A is such that $\sin \theta = \dfrac{3}{5}$. Find the exact value of

   a) $\cos 2\theta$ b) $\tan 2\theta$ c) $\tan 4\theta$

4. For the triangle ABC, calculate the exact values of

   a) $\sin 2A$ b) $\tan 2A$ c) $\sec 2A$.

5. If $\tan \theta = 3$, where $180° \le \theta \le 270°$, find the exact values of

   a) $\sin \theta$ b) $\cos \theta$ c) $\tan 2\theta$ d) $\cos 2\theta$.

6. a) Express $(\sin x - \cos x)^2$ in terms of $\sin 2x$.

   b) Express $(\cos^4 x - \sin^4 x)(\cos^2 x - \sin^2 x)$ in terms of $\cos 2x$.

7. Express $\sin 4\theta \sin \theta$ in terms of $\cos \theta$.

8. a) i) By considering $\sin(2\theta + \theta)$, express $\sin 3\theta$ in terms of $\sin \theta$.

        ii) Solve the equation $\sin 3\theta = \sin \theta$ for $0° \le \theta \le 360°$.

   b) Express $\tan 3\theta$ in terms of $\tan \theta$.

9. Express $\cot 2\theta$ in terms of $\cot \theta$.

10. Prove that   a) $\dfrac{\sin 2\theta}{1 + \cos 2\theta} = \tan \theta$ b) $\dfrac{\sin 2\theta}{1 - \cos 2\theta} = \cot \theta$.

11. Prove that $\dfrac{\sin \alpha}{\cos \beta} + \dfrac{\cos \alpha}{\sin \beta} = \dfrac{2 \cos(\alpha - \beta)}{\sin 2\beta}$.

12. Prove that $\sec 2\theta - \tan 2\theta = \dfrac{\cos \theta - \sin \theta}{\cos \theta + \sin \theta}$.

3.5 Expressing $a \sin \theta + b \cos \theta$ in the form $R \sin(\theta \pm \alpha)$ or $R \cos(\theta \pm \alpha)$

We saw in P1 how to solve equations such as $2 \sin(\theta + 30°) = 1$.
In this section, we will learn how to write expressions of the form
$a \sin \theta + b \cos \theta$ as a single sine or cosine and hence be able to solve
equations such as $3 \sin \theta + 4 \cos \theta = 2$.

If we draw the graph of $y = 3 \sin \theta + 4 \cos \theta$, we obtain a single wave
(see the graph on the right). This wave is the same as the graph we obtain
if we draw $y = 5 \sin(\theta + 53.1°)$, so the expressions $3 \sin \theta + 4 \cos \theta$ and
$5 \sin(\theta + 53.1°)$ are equivalent.

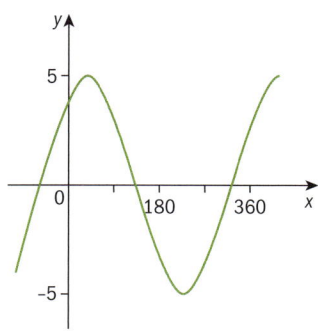

In order to write $a \sin \theta + b \cos \theta$, $(a, b > 0)$, in the form $R \sin(\theta + \alpha)$,
let $a \sin \theta + b \cos \theta \equiv R \sin(\theta + \alpha)$ where R and α are constants to be determined.
Using addition formulae,
$$a \sin \theta + b \cos \theta \equiv R(\sin \theta \cos \alpha + \cos \theta \sin \alpha)$$
$$\equiv R \sin \theta \cos \alpha + R \cos \theta \sin \alpha$$

We require the identity to be true for all values of θ, so we must equate the
coefficients of $\sin \theta$ and $\cos \theta$.
$$R \cos \alpha = a$$
$$R \sin \alpha = b$$

Squaring and adding these equations gives
$$R^2\cos^2 \alpha + R^2\sin^2 \alpha = a^2 + b^2$$
$$R^2(\cos^2 \alpha + \sin^2 \alpha) = a^2 + b^2$$
$$R^2 = a^2 + b^2$$
$$R = \sqrt{a^2 + b^2}, \text{ taking } R > 0$$

> Remember $\cos^2 \alpha + \sin^2 \alpha = 1$.

Also,
$$\frac{R \sin\alpha}{R \cos\alpha} = \frac{b}{a}$$
$$\tan \alpha = \frac{b}{a}$$
$$\alpha = \tan^{-1}\frac{b}{a}$$

> Remember $\dfrac{\sin \alpha}{\cos \alpha} \equiv \tan \alpha$.

$$a \sin \theta + b \cos \theta \equiv R \sin(\theta + \alpha) \quad \text{where } R = \sqrt{a^2 + b^2}$$
$$\text{and } \alpha = \tan^{-1}\frac{b}{a}$$

Examples 10 and 11 demonstrate how we can use this technique and how we
can extend it to writing expressions in one of the other three possible forms:
$R \sin(\theta - \alpha)$, $R \cos(\theta + \alpha)$, or $R \cos(\theta - \alpha)$.

$a \sin \theta + b \cos \theta \equiv R \cos(\theta - \alpha)$ where $R = \sqrt{a^2 + b^2}$

and $\alpha = \tan^{-1} \dfrac{a}{b}$

$a \sin \theta - b \cos \theta \equiv R \sin(\theta - \alpha)$ where $R = \sqrt{a^2 + b^2}$

and $\alpha = \tan^{-1} \dfrac{b}{a}$

$a \cos \theta - b \sin \theta \equiv R \cos(\theta + \alpha)$ where $R = \sqrt{a^2 + b^2}$

and $\alpha = \tan^{-1} \dfrac{b}{a}$

Note: It is important that you learn the process of how to find the values of R and α as outlined in Examples 10 and 11, rather than relying on learning these formulae.

Example 10

a) Express $2 \sin \theta + \cos \theta$ in the form $R \sin(\theta + \alpha)$ where $R > 0$ and $0° \leq \alpha \leq 90°$, giving the exact value of R and the value of α correct to 2 decimal places.

b) Hence solve the equation $2 \sin \theta + \cos \theta = 2$ for $0° \leq \theta \leq 360°$.

a) $2 \sin \theta + \cos \theta \equiv R \sin(\theta + \alpha)$

$\equiv R \sin \theta \cos \alpha + R \cos \theta \sin \alpha$ ⟵ Expand the expression.

$R \cos \alpha = 2$ ⟵ Compare the coefficients of $\sin \theta$ and $\cos \theta$.
$R \sin \alpha = 1$
$R^2 \cos^2 \alpha + R^2 \sin^2 \alpha = 2^2 + 1^2$ ⟵ Square and add to find the exact value of R.
$R^2 (\cos^2 \alpha + \sin^2 \alpha) = 5$
$R = \sqrt{5}$

Also, ⟵ Divide to find $\tan \alpha$ and then find α to the required accuracy.

$\dfrac{\cancel{R} \sin \alpha}{\cancel{R} \cos \alpha} = \dfrac{1}{2}$

$\tan \alpha = \dfrac{1}{2}$

$\alpha = \tan^{-1} \dfrac{1}{2} = 26.57°$

Thus, $2 \sin \theta + \cos \theta = \sqrt{5} \sin(\theta + 26.57°)$

b) $2 \sin \theta + \cos \theta = 2$

$\sqrt{5} \sin(\theta + 26.57°) = 2$ ⟵ Substitute $\sqrt{5} \sin(\theta + 26.57°)$ for $2 \sin \theta + \cos \theta$.

$\sin(\theta + 26.57°) = \dfrac{2}{\sqrt{5}}$

$\theta + 26.57° = \sin^{-1} \dfrac{2}{\sqrt{5}}$ ⟵ Divide both sides by $\sqrt{5}$.

$\theta + 26.57° = 63.43°,\ 180° - 63.43°,\ 360° + 63.43°$ ⟵ Find all the values for $\theta + 26.57°$.

$\theta = 36.9°,\ 90°,\ 396.9°$

$\theta = 36.9°,\ 90°$ for $0° \leq \theta \leq 360°$. ⟵ Subtract 26.57° from each value and check you only include values in the required interval.

Example 11

a) Express $3 \cos \theta + \sqrt{3} \sin \theta$ in the form $R \cos(\theta - \alpha)$ where $R > 0$ and $0° \leq \alpha \leq 90°$, stating the exact values of R and α.

b) Find the greatest and least possible values of $[(3 \cos \theta + \sqrt{3} \sin \theta)^2 - 5]$ as θ varies.

....

a) $3 \cos \theta + \sqrt{3} \sin \theta = R \cos(\theta - \alpha)$

$$\equiv R \cos \theta \cos \alpha + R \sin \theta \sin \alpha \qquad \longleftarrow \boxed{\text{Expand the expression.}}$$

$$R \cos \alpha = 3$$
$$R \sin \alpha = \sqrt{3} \qquad \longleftarrow \boxed{\text{Equate coefficients of } \sin \theta \text{ and } \cos \theta.}$$

$$R^2\cos^2 \alpha + R^2\sin^2 \alpha = 3^2 + (\sqrt{3})^2$$
$$R = \sqrt{12} \text{ (or } 2\sqrt{3})$$

Also,

$$\tan\alpha = \frac{\sqrt{3}}{3}$$
$$\alpha = 30°$$
$$3 \cos \theta + \sqrt{3} \sin \theta = \sqrt{12}\cos(\theta - 30°) \qquad \longleftarrow \boxed{\text{Write out the alternative form of the expression.}}$$

b) $3 \cos \theta + \sqrt{3} \sin \theta = \sqrt{12} \cos(\theta - 30°) \qquad \longleftarrow \boxed{\text{Use the expression found in part (a).}}$

$$\text{but} \quad -1 \leq \cos(\theta - 30°) \leq 1$$

$$\qquad\qquad\qquad\qquad\qquad\qquad \longleftarrow \boxed{\begin{array}{l}\text{Use the fact that the cosine of an angle of any} \\ \text{size is between } -1 \text{ and } 1.\end{array}}$$

$$\text{so} \quad -\sqrt{12} \leq \sqrt{12} \cos(\theta - 30°) \leq \sqrt{12}$$

$$-\sqrt{12} \leq (3 \cos \theta + \sqrt{3} \sin \theta) \leq \sqrt{12}$$

$$0 \leq (3 \cos \theta + \sqrt{3} \sin \theta)^2 \leq 12 \qquad \longleftarrow \boxed{\begin{array}{l}\text{Remember than the square of an expression} \\ \text{cannot be negative.}\end{array}}$$

$$-5 \leq (3 \cos \theta + \sqrt{3} \sin \theta)^2 - 5 \leq 7$$

Greatest value of $[(3 \cos \theta + 3 \sin \theta)^2 - 5] = 7$

Least value of $[(3 \cos \theta + 3 \sin \theta)^2 - 5] = -5$

Exercise 3.5

1. Express each of the following in the form $R \sin(\theta + \alpha)$, where $R > 0$ and $0° \leq \alpha \leq 90°$, giving the exact value of R and the value of α correct to 2 decimal places.

    a) $3 \sin \theta + 2 \cos \theta$ b) $10 \sin \theta + 7 \cos \theta$

    c) $2 \sin \theta + \cos \theta$ d) $40 \sin \theta + 9 \cos \theta$

Expressing $a \sin \theta + b \cos \theta$ in the form $R \sin (\theta \pm \alpha)$ or $R \cos(\theta \pm \alpha)$

2. a) Express each of the following in the form $R\sin(\theta - \alpha)$, where $R > 0$ and $0° \le \alpha \le 90°$, giving the exact value of R and the value of α correct to 2 decimal places.

i) $5\sin\theta - 8\cos\theta$ **ii)** $7\sin\theta - 24\cos\theta$

iii) $2\sin\theta - \cos\theta$ **iv)** $\sqrt{3}\sin\theta - 4\cos\theta$

b) Find the greatest and least possible values of each of the expressions in part **(a)**.

3. Express each of the following in the form $R\cos(\theta + \alpha)$, where $R > 0$ and $0° \le \alpha \le 90°$, giving the exact value of R and the value of α correct to 2 decimal places.

a) $\cos\theta - \sin\theta$ **b)** $2\cos\theta - \sqrt{2}\sin\theta$

c) $4\cos\theta - 3\sin\theta$ **d)** $\sqrt{5}\cos\theta - \sin\theta$

4. a) Express each of the following in the form $R\cos(\theta - \alpha)$, where $R > 0$ and $0° \le \alpha \le 90°$, giving the exact value of R and the value of α correct to 2 decimal places.

i) $3\cos\theta + 4\sin\theta$ **ii)** $5\cos\theta + 12\sin\theta$

iii) $\sqrt{2}\cos\theta + \sin\theta$ **iv)** $\sin\theta + \cos\theta$

b) Find the greatest and least possible values of each of the expressions in part **(a)**.

c) For each of the expressions in part **(a)** find a value of θ for which the expression has its greatest value.

5. a) Express $\sin\theta + 2\cos\theta$ in the form $R\sin(\theta + \alpha)$, where $R > 0$ and $0° \le \alpha \le 90°$, giving the exact value of R and the value of α correct to 2 decimal places.

b) Hence solve the equation $\sin\theta + 2\cos\theta = 1$ for $0° \le x \le 360°$.

6. a) Express $\sqrt{3}\sin\theta - \cos\theta$ in the form $R\sin(\theta - \alpha)$, where $R > 0$ and $0° \le \alpha \le 90°$, giving the exact values of R and α.

b) Hence solve the equation $\sqrt{3}\sin\theta - \cos\theta = 1$ for $0° \le x \le 360°$.

c) Find the greatest and least possible values of $(\sqrt{3}\sin\theta - \cos\theta)^2$ as θ varies.

7. a) Express $2\cos\theta - 2\sin\theta$ in the form $R\cos(\theta + \alpha)$, where $R > 0$ and $0° \le \alpha \le 90°$, giving the exact values of R and α.

b) Hence solve the equation $\cos\theta - \sin\theta = \dfrac{1}{2}$ for $0° \le x \le 360°$.

8. a) Express $4 \cos \theta + 6 \sin \theta$ in the form $R \cos(\theta - \alpha)$, where $R > 0$ and $0° \leq \alpha \leq 90°$, giving the exact value of R and the value of α correct to 2 decimal places.

b) Hence solve the equation $4 \cos \theta + 6 \sin \theta = 5$ for $0° \leq x \leq 360°$.

c) Find the greatest and least possible values of $[(4 \cos \theta + 6 \sin \theta)^2 + 5]$ as θ varies.

9. Solve the following equations for $0° \leq \theta \leq 360°$.

a) $4 \cos \theta - 6 \sin \theta = 5$ **b)** $7 \sin \theta - 24 \cos \theta = 25$

c) $7 \cos \theta = 5 - \sin \theta$ **d)** $4 \cos \theta = 2 + 3 \sin \theta$

Hint: You may use any appropriate form of a single sine or single cosine to help you.

10. a) Express $2 \sin 2x + \cos 2x$ in the form $R \sin(2x + \alpha)$, where $R > 0$ and $0° \leq \alpha \leq 90°$, giving the exact value of R and the value of α correct to 2 decimal places.

b) Hence solve the equation $2 \sin 2x + \cos 2x = 1$ for $0° \leq x \leq 360°$.

c) Find the greatest and least possible values of $10 - (2 \sin 2x + \cos 2x)$ as θ varies.

11. a) Express $\sin x + 4 \cos x$ in the form $R \sin(x + \alpha)$, where $R > 0$ and $0 \leq \alpha \leq \frac{\pi}{2}$, giving the exact value of R and the value of α correct to 3 decimal places.

b) Hence solve the equation $\sin x + 4 \cos x = 3$ for $0 \leq x \leq 2\pi$.

c) Find the greatest and least possible values of $[(\sin x + 4 \cos x)^2 - 1]$ as x varies.

12. Solve the equation $15 \cos 2\theta + 20 \sin 2\theta + 7 = 0$ for $0° \leq \theta \leq 180°$.

Expressing $a \sin \theta + b \cos \theta$ in the form $R \sin (\theta \pm \alpha)$ or $R \cos(\theta \pm \alpha)$

Summary exercise 3

1. Solve the equation
   $\cos(45° − \theta) = \sin(30° + \theta)$ for $0 \le \theta \le 360°$.

2. Solve the equation $6 \cos \theta − 2 \sec \theta = 1$ for
   $0 \le \theta \le 2\pi$.

3. Solve the equation $5 \cos 2\theta − 11 \sin \theta + 1 = 0$
   for $0° \le x \le 360°$.

4. Prove the identity
   $\tan\left(\dfrac{\pi}{\theta} + \theta\right) − \tan\left(\dfrac{\pi}{4} − \theta\right) \equiv 2 \tan 2\theta$.

5. Prove the identity $\operatorname{cosec} 2\theta + \cot 2\theta \equiv \cot \theta$.

6. a) Prove that $\dfrac{1 + \sin 2\theta − \cos 2\theta}{1 + \sin 2\theta + \cos 2\theta} \equiv \tan \theta$.
   b) Hence find the exact value of $\tan 22.5°$.

7. Prove that $(\sec \theta − \tan \theta)^2 \equiv \dfrac{1 − \sin \theta}{1 + \sin \theta}$.

8. Acute angles θ and α are such that $\tan \theta = \dfrac{1}{7}$
   and $\tan \alpha = \dfrac{1}{3}$.
   Find the exact value of $\tan(\theta + 2\alpha)$.

9. Show that
   $\cos \theta + \cos\left(\theta + \dfrac{2\pi}{3}\right) + \cos\left(\theta + \dfrac{4\pi}{3}\right) = 0$.

10. a) Express $5 \sin \theta + 12 \cos \theta$ in the form
    $R \sin(\theta + \alpha)$, where $R > 0$ and
    $0 \le \alpha \le 90°$, giving the exact value
    of R and the value of α correct to
    2 decimal places.
    b) Hence
    i) Write down the greatest value of
       $5 \sin \theta + 12 \cos \theta$ as θ varies.
    ii) Find a value of θ at which this
       greatest value occurs.

11. a) Express $4 \cos \theta − 5 \sin \theta$ in the form
    $R \cos(\theta + \alpha)$, where $R > 0$ and
    $0 \le \alpha \le 90°$, giving the exact value of R
    and the value of α correct to 2 decimal
    places.
    b) Hence solve the equation
    $4 \cos \theta − 5 \sin \theta = −2$ for $0 \le x \le 360°$.
    c) Find the greatest possible value of
    $20 − (4 \cos \theta − 5 \sin \theta)^2$ as θ varies.

12. a) Express $7 \sin \theta + 24 \cos \theta$ in the form
    $R \sin(\theta + \alpha)$, where $R > 0$ and $0 \le \alpha \le 90°$,
    giving the exact value of R and the value
    of α correct to 2 decimal places.
    b) Hence solve the inequality
    $7 \sin \theta + 24 \cos \theta < 15$ for $0 \le \theta \le 360°$.

13. If $\sin(\theta + \alpha) = 2 \sin(\theta − \alpha)$, prove that
    $\tan \theta = 3 \tan \alpha$.

14. Solve the equation $\tan(\theta + 45°) = 1 − 4 \tan \theta$
    for $0° \le \theta \le 360°$.

15. Solve the equation
    $\sin 60° + \sin(60° + x) + \sin(60° + 2x) = 0$
    for $0 \le x \le 360°$.

16. If $p = \operatorname{cosec} \theta − \sin \theta$ and $q = \sec \theta − \cos \theta$,
    show that $p^2 q^2 (p^2 + q^2 + 3) = 1$.

17. a) Given that $\tan \dfrac{\theta}{2} = t$, find expressions for
    $\sin \dfrac{\theta}{2}$ and $\cos \dfrac{\theta}{2}$ in terms of t.
    b) Show that $\sin \theta = \dfrac{2t}{1 + t^2}$ and $\cos \theta = \dfrac{1 − t^2}{1 + t^2}$.
    c) Hence solve the equation
    $4 \sin \theta = 3 + 2 \cos \theta$ for $0° \le \theta \le 360°$.

18. a) Show that the equation
$\tan(45° + x) - \tan x = 2$
can be written in the form
$\tan^2 x + 2 \tan x - 1 = 0$. [3]

b) Hence solve the equation
$\tan(45° + x) - \tan x = 2$,
giving all solutions in the interval
$0° \le x \le 180°$. [4]

Cambridge International AS and
A Level Mathematics 9709 Paper 3 Q5
October/November 2007

19. Solve the equation
$\tan x \tan 2x = 1$,
giving all solutions in the interval
$0° < x < 180°$. [4]

Cambridge International AS and
A Level Mathematics 9709 Paper 3 Q2
October/November 2006

20. a) Express $7 \cos \theta + 24 \sin \theta$ in the
form $R \cos(\theta - \alpha)$, where $R > 0$ and
$0° < \alpha < 90°$, giving the exact value of
R and the value of α correct to
2 decimal places. [3]

b) Hence solve the equation
$7 \cos \theta + 24 \sin \theta = 15$,
giving all solutions in the interval
$0° \le \theta \le 360°$. [4]

Cambridge International AS and
A Level Mathematics 9709 Paper 3 Q4
May/June 2006

21. a) Prove the identity
$\cos 4\theta + 4 \cos 2\theta \equiv 8 \cos^4 \theta - 3$. [4]

b) Hence solve the equation
$\cos 4\theta + 4 \cos 2\theta = 2$,
for $0° \le \theta \le 360°$. [4]

Cambridge International AS and
A Level Mathematics 9709 Paper 3 Q6
May/June 2005

Chapter summary

Secant, cosecant, and cotangent

- $\sec \theta = \dfrac{1}{\cos \theta}$ $\operatorname{cosec} \theta = \dfrac{1}{\sin \theta}$ $\cot \theta = \dfrac{1}{\tan \theta} \left(= \dfrac{\cos \theta}{\sin \theta} \right)$

- $1 + \tan^2 \theta \equiv \sec^2 \theta$

- $1 + \cot^2 \theta \equiv \operatorname{cosec}^2 \theta$

Graphs of sec θ, cosec θ and cot θ

Complementary angle identities

- $\sin(90 - \theta) = \cos \theta$
- $\cos(90 - \theta) = \sin \theta$
- $\tan(90 - \theta) = \cot \theta$
- $\cot(90 - \theta) = \tan \theta$
- $\sec(90 - \theta) = \csc \theta$
- $\csc(90 - \theta) = \sec \theta$

Addition formulae

- $\sin(A + B) \equiv \sin A \cos B + \cos A \sin B$
- $\sin(A - B) \equiv \sin A \cos B - \cos A \sin B$
- $\cos(A + B) \equiv \cos A \cos B - \sin A \sin B$
- $\cos(A - B) \equiv \cos A \cos B + \sin A \sin B$
- $\tan(A + B) \equiv \dfrac{\tan A + \tan B}{1 - \tan A \tan B}$
- $\tan(A - B) \equiv \dfrac{\tan A - \tan B}{1 + \tan A \tan B}$

Double angle formulae

- $\sin 2A \equiv 2 \sin A \cos A$
- $\cos 2A \equiv \cos^2 A - \sin^2 A$
  $\equiv 2 \cos^2 A - 1$
  $\equiv 1 - 2 \sin^2 A$
- $\tan 2A \equiv \dfrac{2 \tan A}{1 - \tan^2 A}$

Expressing $a \sin \theta + b \cos \theta$ in the form $R \sin(\theta \pm \alpha)$ or $R \cos(\theta \pm \alpha)$

- $a \sin \theta + b \cos \theta \equiv R \sin(\theta + \alpha)$ where $R = \sqrt{a^2 + b^2}$ and $\alpha = \tan^{-1} \dfrac{b}{a}$
- $a \sin \theta + b \cos \theta \equiv R \cos(\theta - \alpha)$ where $R = \sqrt{a^2 + b^2}$ and $\alpha = \tan^{-1} \dfrac{a}{b}$
- $a \sin \theta - b \cos \theta \equiv R \sin(\theta - \alpha)$ where $R = \sqrt{a^2 + b^2}$ and $\alpha = \tan^{-1} \dfrac{b}{a}$
- $a \cos \theta - b \sin \theta \equiv R \cos(\theta + \alpha)$ where $R = \sqrt{a^2 + b^2}$ and $\alpha = \tan^{-1} \dfrac{b}{a}$

Review exercise A – Pure 2

1. Solve the equation $\ln(3 - x^2) = 2 \ln x$, giving your answer correct to 3 significant figures. **[4]**

Cambridge International AS & A Level Mathematics 9709 Paper 21 Q2 October/November 2009

2. a) Express $3 \cos\theta + \sin\theta$ in the form $R\cos(\theta - \alpha)$, where $R > 0$ and $0° < \alpha < 90°$, giving the exact value of R and the value of α correct to 2 decimal places. **[3]**

   b) Hence solve the equation $3 \cos 2x + \sin 2x = 2$, giving all solutions in the interval $0° \le x \le 360°$. **[5]**

Cambridge International AS & A Level Mathematics 9709 Paper 21 Q7 October/November 2013

3. a) Show that the equation $\sin(x + 30°) = 2\cos(x + 60°)$ can be written in the form $(3\sqrt{3})\sin x = \cos x$. **[3]**

   b) Hence solve the equation $\sin(x + 30°) = 2\cos(x + 60°)$, for $-180° \le x \le 180°$. **[3]**

Cambridge International AS & A Level Mathematics 9709 Paper 2 Q4 October/November 2008

4. Solve the equation $\left|2^x - 7\right| = 1$, giving answers correct to 2 decimal places where appropriate. **[5]**

Cambridge International AS & A Level Mathematics 9709 Paper 21 Q1 May/June 2013

5. a) Given that $y = 2^x$, show that the equation $2^x + 3(2^{-x}) = 4$ can be written in the form $y^2 - 4y + 3 = 0$. **[3]**

   b) Hence solve the equation $2^x + 3(2^{-x}) = 4$, giving the values of x correct to 3 significant figures where appropriate. **[3]**

Cambridge International AS & A Level Mathematics 9709 Paper 21 Q5 May/June 2010

6. The polynomial $x^3 + ax^2 + bx + 6$, where a and b are constants, is denoted by p(x). It is given that $(x - 2)$ is a factor of p(x), and that when p(x) is divided by $(x - 1)$ the remainder is 4.

   a) Find the values of a and b. **[5]**

   b) When a and b have these values, find the other two linear factors of p(x). **[3]**

Cambridge International AS & A Level Mathematics 9709 Paper 2 Q6 May/June 2009

7. The polynomial $ax^3 - 5x^2 + bx + 9$, where a and b are constants, is denoted by p(x). It is given that $(2x + 3)$ is a factor of p(x), and that when p(x) is divided by $(x + 1)$ the remainder is 8.

   a) Find the values of a and b. **[5]**

   b) When a and b have these values, factorise p(x) completely. **[3]**

Cambridge International AS & A Level Mathematics 9709 Paper 21 Q4 May/June 2013

8. Use logarithms to solve the equation $5^x = 3^{2x-1}$, giving your answer correct to 3 significant figures. **[4]**

Cambridge International AS & A Level Mathematics 9709 Paper 21 Q2 October/November 2012

9. a) Show that the equation $\sin(60° - x) = 2\sin x$ can be written in the form $\tan x = k$, where k is a constant. **[4]**

   b) Hence solve the equation $\sin(60° - x) = 2\sin x$, for $0° < x < 360°$. **[2]**

Cambridge International AS & A Level Mathematics 9709 Paper 21 Q4 October/November 2009

10. Solve the equation

$2 \cos 2\theta = 4 \cos \theta - 3$,

for $0° \leq \theta \leq 180°$. [4]

**Cambridge International AS &
A Level Mathematics 9709 Paper 21
Q3 October/November 2012**

11. Use logarithms to solve the equation $4^x = 2(3^x)$, giving your answer correct to 3 significant figures. [4]

**Cambridge International AS &
A Level Mathematics 9709 Paper 2
Q2 May/June 2008**

12. The polynomial $3x^3 + 2x^2 + ax + b$, where a and b are constants, is denoted by $p(x)$. It is given that $(x - 1)$ is a factor of $p(x)$, and that when $p(x)$ is divided by $(x - 2)$ the remainder is 10.

a) Find the values of a and b. [5]

b) When a and b have these values, solve the equation $p(x) = 0$. [4]

**Cambridge International AS &
A Level Mathematics 9709 Paper 21
Q7 October/November 2010**

13.

The variables x and y satisfy the equation $y = A(b^x)$, where A and b are constants. The graph of ln y against x is a straight line passing through the points $(0, 2.14)$ and $(5, 4.49)$, as shown in the diagram.

Find the values of A and b, correct to 1 decimal place. [5]

**Cambridge International AS &
A Level Mathematics 9709 Paper 21
Q2 May/June 2012**

14. The polynomial $p(x)$ is defined by

$p(x) = ax^3 - 3x^2 - 5x + a + 4$,

where a is a constant.

a) Given that $(x - 2)$ is a factor of $p(x)$, find the value of a. [2]

b) When a has this value,

i) factorise $p(x)$ completely, [3]

ii) find the remainder when $p(x)$ is divided by $(x + 1)$. [2]

**Cambridge International AS &
A Level Mathematics 9709 Paper 21
Q3 May/June 2012**

15. Solve the inequality $|2x - 3| > 5$. [3]

**Cambridge International AS &
A Level Mathematics 9709 Paper 21
Q1 May/June 2010**

16. a) Express $4 \sin \theta - 6 \cos \theta$ in the form $R \sin(\theta - \alpha)$, where $R > 0$ and $0° < \alpha < 90°$. Give the exact value of R and the value of α correct to 2 decimal places. [3]

b) Solve the equation $4 \sin \theta - 6 \cos \theta = 3$ for $0° \leq \theta \leq 360°$. [4]

c) Find the greatest and least possible values of $(4 \sin \theta - 6 \cos \theta)^2 + 8$ as θ varies. [2]

**Cambridge International AS &
A Level Mathematics 9709 Paper 21
Q8 May/June 2011**

17. Solve the equation $\ln(3 - 2x) - 2 \ln x = \ln 5$. [5]

**Cambridge International AS &
A Level Mathematics 9709 Paper 21
Q2 May/June 2013**

18. Use logarithms to solve the equation $5^x = 2^{2x+1}$, giving your answer correct to 3 significant figures. [4]

**Cambridge International AS &
A Level Mathematics 9709 Paper 21
Q2 October/November 2010**

19. a) Given that $35 + \sec^2\theta = 12\tan\theta$, find the value of $\tan\theta$. **[3]**

b) Hence, showing the use of an appropriate formula in each case, find the exact value of

i) $\tan(\theta - 45°)$, **[2]**

ii) $\tan 2\theta$. **[2]**

Cambridge International AS & A Level Mathematics 9709 Paper 21 Q4 May/June 2012

20. Solve the inequality $|x + 1| < |3x + 5|$. **[4]**

Cambridge International AS & A Level Mathematics 9709 Paper 21 Q1 October/November 2013

21. Solve the inequality $|4 - 5x| < 3$. **[3]**

Cambridge International AS & A Level Mathematics 9709 Paper 21 Q1 October/November 2011

22. Solve the inequality $|x - 3| > |2x|$. **[4]**

Cambridge International AS & A Level Mathematics 9709 Paper 2 Q1 October/November 2008

23. Solve the equation $3^{2x} - 7(3^x) + 10 = 0$, giving your answers correct to 3 significant figures. **[5]**

Cambridge International AS & A Level Mathematics 9709 Paper 21 Q4 October/November 2011

24. The polynomial f(x) is defined by

$$f(x) = 3x^3 + ax^2 + ax + a,$$

where a is a constant.

a) Given that $(x + 2)$ is a factor of f(x), find the value of a. **[2]**

b) When a has the value found in part **(a)**, find the quotient when f(x) is divided by $(x + 2)$. **[3]**

Cambridge International AS & A Level Mathematics 9709 Paper 21 Q4 May/June 2011

25. Solve the equation $8 + \cot\theta = 2\operatorname{cosec}^2\theta$, giving all solutions in the interval $0° \le \theta \le 360°$. **[6]**

Cambridge International AS & A Level Mathematics 9709 Paper 21 Q5 October/November 2010

26. Given that $(1.25)^x = (2.5)^y$, use logarithms to find the value of $\frac{x}{y}$ correct to 3 significant figures. **[3]**

Cambridge International AS & A Level Mathematics 9709 Paper 2 Q1 May/June 2009

27. Solve the inequality $|3x - 1| < 2$. **[3]**

Cambridge International AS & A Level Mathematics 9709 Paper 2 Q1 May/June 2008

28. a) Express $5\cos\theta - \sin\theta$ in the form $R\cos(\theta + \alpha)$, where $R > 0$ and $0° < \alpha < 90°$, giving the exact value of R and the value of α correct to 2 decimal places. **[3]**

b) Hence solve the equation $5\cos\theta - \sin\theta = 4$, giving all solutions in the interval $0° \le \theta \le 360°$. **[4]**

Cambridge International AS & A Level Mathematics 9709 Paper 2 Q5 May/June 2008

Review exercise A – Pure 3

1. Solve the equation
   $\sin \theta = 2 \cos 2\theta + 1$,
   giving all solutions in the interval
   $0° \le \theta \le 360°$. [6]

   **Cambridge International AS and
   A Level Mathematics 9709 Paper 31 Q2
   May/June 2010**

2. Prove that $\cot \theta + \tan \theta = 2 \csc 2\theta$. [3]

   **Cambridge International AS and
   A Level Mathematics 9709 Paper 31 Q7
   October/November 2013**

3. The polynomial $x^4 + 5x + a$ is denoted by
   p(x). It is given that $x^2 - x + 3$ is a factor
   p(x).

   a) Find the value of a and factorise
      p(x) completely. [6]

   b) Hence state the number of real roots
      of the equation p(x) = 0, justifying
      your answer. [2]

   **Cambridge International AS and
   A Level Mathematics 9709 Paper 3 Q5
   May/June 2005**

4. Let $f(x) = \dfrac{2x^2 - 7x - 1}{(x - 2)(x^2 + 3)}$.

   a) Express f(x) in partial fractions. [5]

   b) Hence obtain the expansion of f(x)
      in ascending powers of x, up to and
      including the term in x^2. [5]

   **Cambridge International AS and
   A Level Mathematics 9709 Paper 31 Q7
   October/November 2013**

5. Solve, correct to 3 significant figures,
   the equation
   $e^x + e^{2x} = e^{3x}$. [5]

   **Cambridge International AS and
   A Level Mathematics 9709 Paper 3 Q2
   May/June 2008**

6. a) Prove the identity
      $\csc 2\theta + \cot 2\theta = \cot \theta$. [3]

   b) Hence solve the equation
      $\csc 2\theta + \cot 2\theta = 2$,
      for $0° \le \theta \le 360°$. [2]

   **Cambridge International AS and
   A Level Mathematics 9709 Paper 3 Q3
   May/June 2009**

7. Solve the equation
   $\ln (x + 2) = 2 + \ln x$,
   giving your answer correct to
   3 decimal places. [3]

   **Cambridge International AS and
   A Level Mathematics 9709 Paper 3 Q1
   October/November 2008**

8. The polynomial $x^4 + 3x^2 + a$, where a is a
   constant, is denoted by p(x). It is given that
   $x^2 + x + 2$ is a factor of p(x). Find the value of
   a and the other quadratic factor of p(x). [4]

   **Cambridge International AS and
   A Level Mathematics 9709 Paper 3 Q2
   October/November 2007**

9. a) Solve the equation $|4x - 1| = |x - 3|$. [3]

   b) Hence solve the equation
      $|4^{y+1} - 1| = |4^y - 3|$ correct to
      3 significant figures. [3]

   **Cambridge International AS and
   A Level Mathematics 9709 Paper 31 Q4
   May/June 2013**

10. Solve the equation
    $\sin (\theta + 45°) = 2 \cos(\theta - 30°)$,
    giving all soultions in the
    interval $0° < \theta < 180°$. [5]

    **Cambridge International AS and
    A Level Mathematics 9709 Paper 31 Q3
    October/November 2012**

11. Given that a is a positive constant, solve the inequality
$$|x - 3a| > |x - a|.$$ [4]

Cambridge International AS and A Level Mathematics 9709 Paper 3 Q1 October/November 2005

12. The polynominal $x^4 + 3x^3 + ax + 3$ is denoted by p(x). It is given that p(x) is divisible by $x^2 - x + 1$.

a) Find the value of a. [4]

b) When a has this value, find the real roots of the equation p$(x) = 0$. [2]

Cambridge International AS and A Level Mathematics 9709 Paper 31 Q3 October/November 2011

13. The polynomial f(x) is defined by
$$f(x) = 12x^3 + 25x^2 - 4x - 12.$$

a) Show that f$(-2) = 0$ and factorise f(x) completely. [4]

b) Given that
$$12 \times 27^y + 25 \times 9^y - 4 \times 3^y - 12 = 0,$$
state the value of 3^y and hence find y correct to 3 significant figures. [3]

Cambridge International AS and A Level Mathematics 9709 Paper 31 Q4 May/June 2011

14. Solve the inequality $|x - 2| > 3|2x + 1|$. [4]

Cambridge International AS and A Level Mathematics 9709 Paper 3 Q1 May/June 2008

15.

The diagram shows a circle with centre O and radius 10 cm. The chord AB divides the circle into two regions whose areas are in the ratio 1 : 4 and it is required to find the length of AB. The angle AOB is θ radians.

Show that $\theta = \frac{2}{5}\pi + \sin\theta$. [3]

Cambridge International AS and A Level Mathematics 9709 Paper 31 Q6i May/June 2011

16. Solve the equation $2|3^x - 1| = 3^x$, giving your answers correct to 3 significiant figures. [4]

Cambridge International AS and A Level Mathematics 9709 Paper 31 Q2 October/November 2013

17. Find the set of values of x satisfying the inequality $|3^x - 8| < 0.5$, Giving 3 significant figures in your answer. [4]

Cambridge International AS and A Level Mathematics 9709 Paper 3 Q1 October/November 2006

Maths in real-life

Predicting tidal behaviour

The surface level of water in oceans all over the world is seen to rise and fall periodically – known as high and low tides. To develop a good mathematical model of what is happening you need to understand the causes of the tidal motion.

Water is fluid and so it can move in response to different gravitational pulls from other bodies in the solar system acting at different points. While the sun has a huge gravitational pull, it is very far away and so the tidal effects of the sun's gravity are minimal. Most of the tidal effects we see are caused by the moon, as the distances on the earth's surface are a much smaller proportion of the total distance between the moon and the earth, and so the gravitational field difference across the earth is very small.

The earth rotates on its axis every 24 hours and the moon orbits around the earth over a period of 27.3 days. Both of these cause points on the earth's surface to move further away from (and closer to) the moon; however, the earth's rotation happens over a much shorter time-period than the lunar cycle, and so the effect is much greater.

The first model would be a 24 hour cycle with two high and low tides – as shown here, a wave function in the form $h = a + b \sin 30t$ gives a starting point.

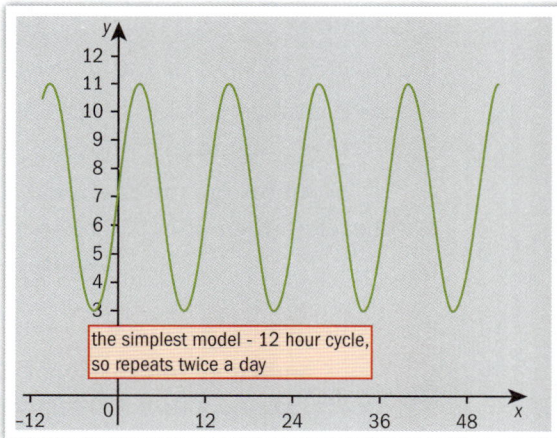

the simplest model - 12 hour cycle, so repeats twice a day

In fact, because of the relative motion of the moon, the tidal cycle is actually about 24 hours and 50 minutes. In this time, two high and low tides of not quite equal sizes are seen. The side of the earth closest to the moon experiences the biggest pull so there is a difference in the size of the two high tides in a day – however, this is hard to model. Changing the period to 12 hours 25 minutes is easy ($h = a + b \sin 29t$).

Once the period is adjusted we can add in the effect of the lunar cycle. When the sun, moon and earth are directly in line with one another, the smaller effect of the sun is added to that of the moon – this happens twice in the cycle: once when the moon is closer to the sun than the earth is (at the new moon), and once when it is further away (at the full moon) and the highest tides occur. These are known as spring tides and the corresponding lowest high tides are known as neap tides.

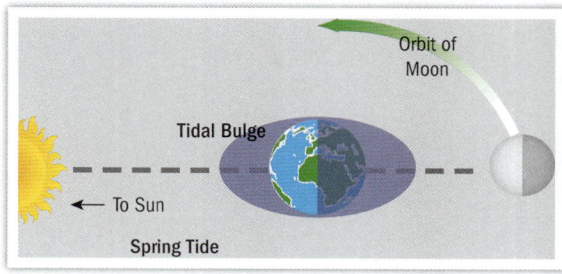

In this better model, the red line (much of it is hidden) is the effect of the earth's rotation and the blue line at the bottom is the effect of the moon's orbit. The purple line shows the combination of the two effects, and the phenomenon of the spring and neap tides can be seen. The model for this is:

$$y = 7 + 4\,(\sin(29x)) + 0.4\sin(1.062x)$$

Very high tides can cause untold damage when floods occur. A further regular perturbation occurs at the two equinoxes – where the axis of the earth's rotation is perpendicular to the line between the earth and the sun. This is when the spring tides have their greatest variation between high and low tides. This is often a time at which stormy meteorological conditions occur (perhaps because of the way the bodies are moving relative to one another) so the risk of abnormally high tides and flooding is raised.

By understanding tides we can gain important insight into other problems, such as how oil spills will behave – whether as the result of an accident in a drilling operation or some mishap to a tanker transporting oil. The picture here shows an oil-drenched pelican being cleaned after the Deepwater Horizon Oil Spill in the Gulf of Mexico in 2010. Experts suggested that the spill was being pulled southeast by the Loop Current and would be extremely difficult to contain. Being able to predict the behavior of pollutants – where they will go, how much they will become diluted, etc. – is a crucial component of making good decisions about what environmental measures are needed in order to contain the damage caused by toxic materials.

4 Differentiation

In real-world applications, differentiation is often complex, or involves trigonometric, exponential or logarithmic functions. For example, differentiation is used to maximise the electrical power produced by a certain source, to minimise the stress on a hollow tube, and to find the acceleration of a rocket which moves such that the only force on it is due to gravity and where its mass is decreasing at a constant rate.

Objectives

- Use the derivatives of e^x, $\ln x$, $\sin x$, $\cos x$, $\tan x$, together with constant multiples, sums, differences, and composites.
- Differentiate products and quotients.
- Find and use the first derivative of a function which is defined parametrically or implicitly.

Before you start

You should know how to:

1. Differentiate expressions that can be simplified to single terms in x,

    e.g. find $\dfrac{dy}{dx}$ when

    a) $y = \dfrac{x^6 - 12x}{3x^3}$ **b)** $y = 2x\left(4\sqrt{x} + \dfrac{1}{\sqrt{x}}\right)$.

    a) $y = \dfrac{1}{3}x^3 - 4x^{-2}$, $\dfrac{dy}{dx} = x^2 + 8x^{-3}$

    b) $y = 8x^{\frac{3}{2}} + 2x^{\frac{1}{2}}$, $\dfrac{dy}{dx} = 12^{\frac{1}{2}} + x^{-\frac{1}{2}}$

2. Differentiate composite functions using the Chain Rule:

    e.g. find the rate of change of y with respect to x when

    a) $y = \dfrac{6}{2x - 3}$ **b)** $y = \dfrac{8}{\sqrt{5x + 1}}$.

    a) $y = 6(2x - 3)^{-1}$,
    $\dfrac{dy}{dx} = -6(2x - 3)^{-2}(2) = -12(2x - 3)^{-2}$

    b) $y = 8(5x + 1)^{-\frac{1}{2}}$
    $\dfrac{dy}{dx} = -4(5x + 1)^{-\frac{3}{2}}(5) = -20(5x + 1)^{-\frac{3}{2}}$

Skills check:

1. Find $\dfrac{dy}{dx}$ when

    a) $y = \dfrac{6x^8 + x^2}{2x^4}$

    b) $y = (4\sqrt{x} - 1)(2\sqrt{x} + 5)$

    c) $y = \dfrac{x^{10} - 2x^5 + 10x}{5x^5}$

    d) $y = \dfrac{1}{3}\sqrt{x}\left(\dfrac{6}{\sqrt{x}} - \sqrt[3]{x^6}\right)$.

2. Find the rate of change of y with respect to x when

    a) $y = \dfrac{5}{4 - x}$ **b)** $y = \dfrac{7}{\sqrt{2x + 3}}$

    c) $y = \dfrac{1}{x^2 - 3x + 7}$ **d)** $y = \dfrac{4}{\sqrt{(5x + 1)^3}}$.

4.1 Differentiating the exponential function

In Chapter 2, you learned about the natural exponential function e^x. This function is the only function whose derivative is the same as the function itself. Consider the graph of $y = e^x$. Let δx be a small increase in x.

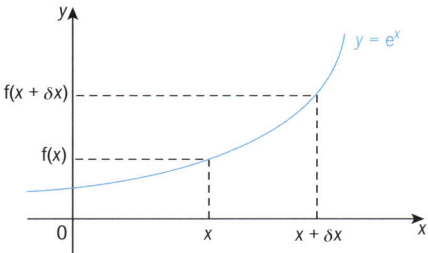

$$\frac{dy}{dx} = \lim_{\delta x \to 0} \frac{f(x + \delta x) - f(x)}{(x + \delta x) - x} = \lim_{\delta x \to 0} \frac{e^{x + \delta x} - e^x}{(x + \delta x) - x} = \lim_{\delta x \to 0} \frac{e^x e^{\delta x} - e^x}{\delta x} = \lim_{\delta x \to 0} \frac{e^x(e^{\delta x} - 1)}{\delta x}$$

Consider $\dfrac{(e^{\delta x} - 1)}{\delta x}$ with values of δx getting smaller and smaller.

δx	0.1	0.01	0.001	0.0001
$\dfrac{(e^{\delta x} - 1)}{\delta x}$	1.05170925	1.005016722	1.00050023	1.00005006

We can see that as $\delta x \longrightarrow 0$, $\dfrac{(e^{\delta x} - 1)}{\delta x} \longrightarrow 1$

Thus $\dfrac{dy}{dx} \longrightarrow e^x(1) = e^x$

> If $y = e^x$ then $\dfrac{dy}{dx} = e^x$

If we want to differentiate $y = e^{kx}$ where k is a constant, we can use the chain rule:
Let $u = kx \Rightarrow y = e^u$.

Thus $\dfrac{du}{dx} = k$ and $\dfrac{dy}{du} = e^u$.

$$\frac{dy}{dx} = \frac{dy}{du} \times \frac{du}{dx} = e^u \times k = e^{kx} \times k = ke^{kx}$$

Note: You are not expected to know this proof for the examination.

> If $y = e^{kx}$ then $\dfrac{dy}{dx} = ke^{kx}$

In general we can say:

If $y = e^{f(x)}$ then $\dfrac{dy}{dx} = f'(x)\, e^{f(x)}$

Example 1

Differentiate

a) e^{7x} **b)** e^{5x-4} **c)** e^{x^3}.

---

a) $\dfrac{d}{dx}\, e^{7x} = 7e^{7x}$ ← Substitute for $k = 7$ in ke^{kx}.

b) $\dfrac{d}{dx}\, e^{5x-4} = 5e^{5x-4}$ ← Use $\dfrac{dy}{dx} = f'(x)\, e^{f(x)}$ where $f'(x) = 5$.

c) $\dfrac{d}{dx}\, e^{x^3} = 3x^2\, e^{x^3}$ ← Use $\dfrac{dy}{dx} = f'(x)\, e^{f(x)}$ where $f'(x) = 3x^2$.

Example 2

Find $\dfrac{dy}{dx}$ when

a) $y = 4e^{2x}$ **b)** $y = -2e^{\sqrt{x}}$ **c)** $y = 3e^{\frac{1}{x}}$.

---

a) $\dfrac{dy}{dx} = 4e^{2x}\,(2) = 8e^{2x}$ ← Write down $4e^{2x}$ and then multiply by $\dfrac{d}{dx}\, 2x$.

b) $\dfrac{dy}{dx} = -2e^{\sqrt{x}}\left(\dfrac{1}{2}x^{-\frac{1}{2}}\right)$ ← Write down $-2e^{\sqrt{x}}$ and then multiply by $\dfrac{d}{dx}\, x^{\frac{1}{2}}$.

$\phantom{\dfrac{dy}{dx}} = -\dfrac{1}{\sqrt{x}}e^{\sqrt{x}}$

c) $\dfrac{dy}{dx} = 3\,e^{\frac{1}{x}}(-1x^{-2})$ ← Write down $3e^{\frac{1}{x}}$ and then multiply by $\dfrac{d}{dx}\, x^{-1}$.

$\phantom{\dfrac{dy}{dx}} = -\dfrac{3e^{\frac{1}{x}}}{x^2}$

Exercise 4.1

1. Differentiate

   a) e^{2x}

   b) e^{-5x}

   c) e^{3x+9}

   d) e^{8-x}

   e) $-7e^x$

   f) $5e^{x^2}$

   g) $3e^{2x-1}$

   h) $-4e^{6+x^2}$

   i) e^{5x^2}

   j) $e^{\frac{3}{x}}$

   k) $6e^{\sqrt{x}}$

   l) $\dfrac{1}{e^{2x}}$.

2. Find $\dfrac{dy}{dx}$ when

   a) $y = e^{-2x}$

   b) $y = e^{5+x}$

   c) $y = 9e^{3x}$

   d) $y = -8e^{\sqrt{x}}$

   e) $y = 3e^x - 2e^{4x}$

   f) $y = \dfrac{1}{2}e^{-8x}$

   g) $y = e^{4x} + e^{-x}$

   h) $y = 6e^{-x} - 4e^{3x}$

   i) $y = (e^x + 2)(e^x - 1)$ **Hint for part (i):** First expand the brackets.

   j) $y = (2e^{3x} - 1)^2$

   k) $y = \dfrac{1+e^x}{e^x}$

   l) $y = e^{-4x+2} + 7e^{5-2x}$.

3. Find the gradient of the tangent to the curve $y = 5e^{2x}$ at the point $(0, 5)$.

4. Find the exact value of the gradient of the tangent to the curve $y = e^x - 6\sqrt{x}$ when $x = 1$.

5. Find the coordinates of the minimum point on the curve with equation $y = e^x - x$.

6. Find the coordinates of the turning point on the curve $y = x - e^x$ and determine whether this point is a maximum or a minimum point.

4.2 Differentiating the natural logarithmic function

In Chapter 2, you learned about the natural logarithmic function $\log_e x$. This is also known as $\ln x$.

If $y = \log_e x$ (or $y = \ln x$) then $e^y = x$.

Since $x = e^y$, then $\dfrac{dx}{dy} = e^y$.

As $\dfrac{dy}{dx} = \dfrac{1}{\frac{dx}{dy}}$, we can say $\dfrac{dy}{dx} = \dfrac{1}{e^x} = \dfrac{1}{x}$.

> If $y = \ln x$ then $\dfrac{dy}{dx} = \dfrac{1}{x}$

Example 3

Find $\dfrac{dy}{dx}$ when

a) $y = \ln 5x$ **b)** $y = \ln(3x^2 - 2)$.

a) Let $u = 5x$, then $\dfrac{du}{dx} = 5$ ← Use the chain rule by letting $u = 5x$.

$y = \ln u$, so $\dfrac{dy}{du} = \dfrac{1}{u}$

$\dfrac{dy}{dx} = \dfrac{dy}{du} \times \dfrac{du}{dx} = \dfrac{1}{u} \times 5 = \dfrac{5}{u}$ ← Use the chain rule.

Thus $\dfrac{dy}{dx} = \dfrac{5}{5x} = \dfrac{1}{x}$ ← Substitute $u = 5x$ to have your answer in terms of x.

b) Let $u = 3x^2 - 2$, then $\dfrac{du}{dx} = 6x$ ← Use the chain rule by letting $u = 3x^2 - 2$.

$y = \ln u$, so $\dfrac{dy}{du} = \dfrac{1}{u}$

$\dfrac{dy}{dx} = \dfrac{dy}{du} \times \dfrac{du}{dx} = \dfrac{1}{u} \times 6x = \dfrac{6x}{u}$ ← Use the chain rule.

Thus $\dfrac{dy}{dx} = \dfrac{6x}{3x^2 - 2}$ ← Eliminate u to present your answer in terms of x.

If $y = \ln[f(x)]$ then $\dfrac{dy}{dx} = \dfrac{f'(x)}{f(x)}$

Example 4

Differentiate

a) $\ln(9x + 2)$ **b)** $\ln(e^x - 3x)$ **c)** $4\ln 8x$ **d)** $\ln(5x - 1)^7$.

a) $\dfrac{dy}{dx} = \dfrac{f'(x)}{f(x)} = \dfrac{9}{9x + 2}$ ← $f(x) = (9x + 2)$, $f'(x) = 9$.

b) $\dfrac{dy}{dx} = \dfrac{e^x - 3}{e^x - 3x}$ ← $f(x) = (e^x - 3x)$, $f'(x) = e^x - 3$.

c) $\dfrac{dy}{dx} = 4 \times \dfrac{f'(x)}{f(x)} = 4 \times \dfrac{8}{8x} = \dfrac{4}{x}$ ← $f(x) = 8x$, $f'(x) = 8$.

d) $\ln(5x - 1)^7 = 7\ln(5x - 1)$

$\dfrac{dy}{dx} = 7 \times \dfrac{f'(x)}{f(x)} = 7 \times \dfrac{5}{5x - 1} = \dfrac{35}{5x - 1}$ ← It is best to use laws of logarithms (covered in Chapter 2) to first simplify the expression.

Exercise 4.2

1. Differentiate

   a) $\ln 4x$

   b) $\ln(1 - x^2)$

   c) $\ln(3 + 5x)$

   d) $\ln(x^3 + 2)$

   e) $\ln(e^x - 7)$

   f) $\ln\sqrt{x}$

   g) $\ln(6x - 3)^4$

   h) $\ln(4x^2 + 2x)$

   i) $\ln\dfrac{1}{x^2}$

   j) $3\ln 4x$

   k) $2\ln\dfrac{1}{x}$

   l) $6\ln\dfrac{9x - 2}{3}$.

2. Find $\dfrac{dy}{dx}$ when

   a) $y = 8\ln 3x$

   b) $y = \ln 2x^3$

   c) $y = \ln(2 - 4x)$

   d) $y = 6\ln\dfrac{x}{2}$

   e) $y = \ln\sqrt{(2x + 1)}$

   f) $y = \ln(5x)^{10}$

   g) $y = \ln(3e^x - 2x)$

   h) $y = 9\ln(x^2 + 3x)^2$

   i) $y = 5\ln\left(7 - \dfrac{2}{x}\right)$

   j) $y = 4\ln\sqrt{(x^2 - 8x)}$

   k) $y = \ln e^{5x}$

   l) $y = \ln\dfrac{7}{(x + 3)^6}$.

3. Find the gradient of the tangent to the curve $y = 4 + \ln 3x$ at the point $(1, 4)$.

4. Find the gradient of the tangent to the curve $y = \ln(x^2 + 1)$ when $x = 3$.

5. Find the exact coordinates of the point on the curve $y = \ln 3x$ where the gradient is $\dfrac{1}{2}$.

6. Find the coordinates of the turning point on the curve $y = x - \ln x$, and determine whether this point is a maximum or a minimum point.

4.3 Differentiating products

Consider $y = uv$ where u and v are functions of x.
Let δx be a small increase in x. Let δu, δv, and δy be the corresponding increases in u, v, and y.

If $\quad\quad y = uv$

$$y + \delta y = (u + \delta u)(v + \delta v)$$

$$uv + \delta y = uv + u\,\delta v + v\,\delta u + \delta u\,\delta v$$

$$\delta y = u\,\delta v + v\,\delta u + \delta u\,\delta v$$

$$\frac{\delta y}{\delta x} = u\frac{\delta v}{\delta x} + v\frac{\delta u}{\delta x} + \delta u\frac{\delta v}{\delta x}$$

As $\delta x \longrightarrow 0$, $\delta u \longrightarrow 0$, $\dfrac{\delta y}{\delta x} \longrightarrow \dfrac{dy}{dx}$, $\dfrac{\delta u}{\delta x} \longrightarrow \dfrac{du}{dx}$, $\dfrac{\delta v}{\delta x} \longrightarrow \dfrac{dv}{dx}$,

Thus, $\dfrac{dy}{dx} = u\dfrac{dv}{dx} + v\dfrac{du}{dx} + 0\dfrac{dv}{dx}$ $\boxed{\text{If } y = uv, \text{ then } \dfrac{dy}{dx} = u\dfrac{dv}{dx} + v\dfrac{du}{dx}}$

This is known as **The Product Rule.**

Example 5

Find $\dfrac{dy}{dx}$ when $y = e^{2x}(x^3 - 3)$.

Let $y = uv$ where $u = e^{2x}$ and $v = (x^3 - 3)$. ← y is a product of two functions.

$\dfrac{du}{dx} = 2e^{2x}$ and $\dfrac{dv}{dx} = 3x^2$ ← Differentiate u and differentiate v.

$\dfrac{dy}{dx} = u\dfrac{dv}{dx} + v\dfrac{du}{dx}$ ← Use the product rule.

$\dfrac{dy}{dx} = (e^{2x})(3x^2) + (x^3 - 3)(2e^{2x})$ ← Substitute for each of the 4 unknowns.

$= e^{2x}(2x^3 + 3x^2 - 6)$ ← Factorise by taking out the common e^{2x}.

Example 6

Find $\dfrac{dy}{dx}$ when $y = (2x + 1)(x - 5)^4$

Let $y = uv$ where $u = (2x + 1)$ and $v = (x - 5)^4$ ← y is a product of two functions.

$\dfrac{du}{dx} = 2$ and $\dfrac{dv}{dx} = 4(x - 5)^3(1)$ ← Differentiate u and differentiate v.

$\dfrac{dy}{dx} = u\dfrac{dv}{dx} + v\dfrac{du}{dx}$ ← Use the product rule.

$\dfrac{dy}{dx} = (2x + 1)[4(x - 5)^3] + [(x - 5)^4](2)$ ← Substitute for each of the 4 unknowns.

$= 2(x - 5)^3(4x + 2 + x - 5)$

$= 2(x - 5)^3(5x + 3)$ ← Factorise by taking out the common $2(x - 5)^3$.

Example 7

Show that $\dfrac{dy}{dx} = \dfrac{4x(5x+2)}{\sqrt{2x+1}}$ when $y = 4x^2\sqrt{2x+1}$.

Let $y = uv$ where $u = 4x^2$ and $v = \sqrt{2x+1}$ ← y is a product of two functions.

$\dfrac{du}{dx} = 8x$ ←   Differentiate u.

$\dfrac{dv}{dx} = \dfrac{1}{2}(2x+1)^{-\frac{1}{2}}(2) = \dfrac{1}{\sqrt{2x+1}}$ ←   Differentiate $v = (2x+1)^{\frac{1}{2}}$.

$\dfrac{dy}{dx} = u\dfrac{dv}{dx} + v\dfrac{du}{dx}$ ←   Use the product rule.

$\dfrac{dy}{dx} = (4x^2)\left(\dfrac{1}{\sqrt{2x+1}}\right) + (\sqrt{2x+1})(8x)$ ←   Substitute for each of the 4 unknowns.

$= \dfrac{4x^2 + (2x+1)(8x)}{\sqrt{2x+1}}$ ← $(2x+1) \div \sqrt{2x+1} = \sqrt{2x+1}$.

$= \dfrac{4x^2 + 16x^2 + 8x}{\sqrt{2x+1}}$

$= \dfrac{20x^2 + 8x}{\sqrt{2x+1}}$

$= \dfrac{4x(5x+2)}{\sqrt{2x+1}}$

Thus $\dfrac{dy}{dx} = \dfrac{4x(5x+2)}{\sqrt{2x+1}}$

Exercise 4.3

1. Use the product rule to find $\dfrac{dy}{dx}$ when

a) $y = x\,e^x$

b) $y = (4x-1)(x^2-5)$

c) $y = x^2\sqrt{x^2+1}$

d) $y = \sqrt{x}\,(\ln x)$

e) $y = x^2(2x+5)^7$

f) $y = \sqrt{x}\,(x-3)^2$

g) $y = e^{2x}(4x+1)$

h) $y = x(x+3)^4$

i) $y = x^2 \ln x$

j) $y = x^2(x+3)^3$

k) $y = e^x(2x-1)^4$

l) $y = \dfrac{x}{x+1}$

Hint for part (l): Let $y = x(x+1)^{-1}$ and use the product rule

m) $y = 6e^{\sqrt{x}}(x+4)$

n) $y = 4x^3\,e^{2x}$

o) $y = 2e^x \ln 2x$

p) $y = \dfrac{x-2}{x+2}$

q) $y = \dfrac{2x+3}{1-5x}$

r) $y = (x^2+1)\ln(x^2+1)$.

2. Find the gradient of the tangent to the curve $y = x^2 e^{-x}$ at the point $\left(1, \frac{1}{e}\right)$.

3. Find the equation of the tangent to the curve $y = x(1 + x^2)^3$ when $x = 2$.

4. Show that $\dfrac{dy}{dx} = \dfrac{2x^2 + 4x - 1}{\sqrt{(x^2 - 1)}}$ when $y = (x + 4)\sqrt{(x^2 - 1)}$.

5. Find the exact coordinates of the turning point on the curve $y = e^x(1 + x)$ and determine whether this point is a maximum or a minimum point.

4.4 Differentiating quotients

Consider $y = \dfrac{u}{v}$ where u and v are functions of x.

Let δx be a small increase in x. Let δu, δv, and δy be the corresponding increases in u, v, and y.

If
$$y = \frac{u}{v}$$

$$y + \delta y = \frac{u + \delta u}{v + \delta v}$$

$$\delta y = \frac{u + \delta u}{v + \delta v} - y$$

$$= \frac{u + \delta u}{v + \delta v} - \frac{u}{v}$$

$$= \frac{v(u + \delta u) - u(v + \delta v)}{v(v + \delta v)}$$

$$= \frac{uv + v\delta u - uv - u\delta v}{v^2 + v\delta v}$$

$$\frac{\delta y}{\delta x} = \frac{v\dfrac{\delta u}{\delta x} - u\dfrac{\delta v}{\delta x}}{v^2 + v\delta v}$$

As $\delta x \longrightarrow 0$, $v\,\delta v \longrightarrow 0$, $\dfrac{\delta y}{\delta x} \longrightarrow \dfrac{dy}{dx}$, $\dfrac{\delta u}{\delta x} \longrightarrow \dfrac{du}{dx}$, $\dfrac{\delta v}{\delta x} \longrightarrow \dfrac{dv}{dx}$.

Thus, $\dfrac{dy}{dx} = \dfrac{v\dfrac{du}{dx} - u\dfrac{dv}{dx}}{v^2}$

$$\boxed{\text{If } y = \frac{u}{v}, \text{ then } \frac{dy}{dx} = \frac{v\dfrac{du}{dx} - u\dfrac{dv}{dx}}{v^2}}$$

This is known as **The Quotient Rule**.

Example 8

Find $\dfrac{dy}{dx}$ when $y = \dfrac{x^3}{x^2 + 1}$

Let $y = \dfrac{u}{v}$ where $u = x^3$ and $v = x^2 + 1$ ← y is a quotient of two functions.

$\dfrac{du}{dx} = 3x^2$ and $\dfrac{dv}{dx} = 2x$ ← Differentiate u and differentiate v.

$\dfrac{dy}{dx} = \dfrac{v\dfrac{du}{dx} - u\dfrac{dv}{dx}}{v^2}$ ← Use the quotient rule.

$\dfrac{dy}{dx} = \dfrac{(x^2 + 1)(3x^2) - (x^3)(2x)}{(x^2 + 1)^2}$ ← Substitute for each of the 4 unknowns.

$\quad = \dfrac{3x^4 + 3x^2 - 2x^4}{(x^2 + 1)^2}$

$\quad = \dfrac{x^2(x^2 + 3)}{(x^2 + 1)^2}$ ← Factorise by taking out the common x^2.

Example 9

Find $\dfrac{dy}{dx}$ when $y = \dfrac{4x + 3}{\sqrt{(2x - 1)}}$.

Let $y = \dfrac{u}{v}$ where $u = 4x + 3$ and $v = \sqrt{(2x - 1)}$

$\dfrac{du}{dx} = 4$ ← Differentiate u and differentiate v.

$\dfrac{dv}{dx} = \dfrac{1}{2}(2x - 1)^{-\frac{1}{2}}(2) = (2x - 1)^{-\frac{1}{2}}$

$\dfrac{dy}{dx} = \dfrac{v\dfrac{du}{dx} - u\dfrac{dv}{dx}}{v^2}$ ← Use the quotient rule.

$\dfrac{dy}{dx} = \dfrac{(2x-1)^{\frac{1}{2}}(4) - (4x + 3)(2x - 1)^{-\frac{1}{2}}}{2x - 1}$ ← Substitute for each of the 4 unknowns.

$\quad = \dfrac{(2x-1)^{-\frac{1}{2}}[4(2x-1) - (4x + 3)]}{2x - 1}$ **Note:** $(2x-1)^{-\frac{1}{2}} \times (2x-1)^1 = (2x-1)^{\frac{1}{2}}.$

$\quad = \dfrac{4x - 7}{(2x-1)^{\frac{3}{2}}}$ ← $\dfrac{(2x-1)^{-\frac{1}{2}}}{(2x-1)^1} = \dfrac{1}{(2x-1)^{\frac{3}{2}}}$

Exercise 4.4

1. Use the quotient rule to find $\dfrac{dy}{dx}$ when

a) $y = \dfrac{2x}{x+3}$

b) $y = \dfrac{e^{2x}}{x}$

c) $y = \dfrac{x^2}{x+4}$

d) $y = \dfrac{x}{1+x^2}$

e) $y = \dfrac{3x}{2-x}$

f) $y = \dfrac{3x^2}{2e^x - x}$

g) $y = \dfrac{3x-4}{2x+1}$

h) $y = \dfrac{3x+5}{(x+1)^2}$

i) $y = \dfrac{6x^2}{2-x}$

j) $y = \dfrac{4x}{(1-x)^3}$

k) $y = \dfrac{x^2+1}{x-2}$

l) $y = \dfrac{5x^2}{\ln x}$

m) $y = \dfrac{x^3}{\sqrt{(1-2x^2)}}$

n) $y = \dfrac{3x+x^4}{2x^2+1}$

o) $y = \dfrac{x}{1+\sqrt{x}}$

p) $y = \dfrac{x^2+6}{2x-7}$

q) $y = \dfrac{\sqrt{(x+1)^5}}{x}$

r) $y = \dfrac{1-2x^2}{\sqrt{(1+2x^2)}}$.

2. If $y = \dfrac{x^2}{2x-3}$ find the values of x when $\dfrac{dy}{dx} = 0$.

3. Find the equation of the normal to the curve $y = \dfrac{3x^2}{2x-1}$ at the point $(2, 4)$.

4. If $y = \dfrac{x}{x-1}$, find the value of $\dfrac{d^2y}{dx^2}$.

5. Show that the coordinates of the turning point on the curve $y = \dfrac{x+3}{\sqrt{(1+x^2)}}$ is $\left(\dfrac{1}{3}, \sqrt{10}\right)$.

6. Show that if $y = \dfrac{7\ln x - x^3}{e^{3x}}$ then $\dfrac{dy}{dx} = \dfrac{7}{e^3}$ when $x = 1$.

4.5 Differentiating $\sin x$, $\cos x$, and $\tan x$

When we differentiate trigonometric functions, we assume the angle is measured in radians.

Consider $y = \sin x$ where x is measured in radians.

Let δx be a small increase in x. Let δy be the corresponding small increase in y.

If $y = \sin x$

$$y + \delta y = \sin(x + \delta x)$$
$$= \sin x \cos \delta x + \cos x \sin \delta x$$

$\longleftarrow$ $\boxed{\sin(A + B) = \sin A \cos B + \cos A \sin B}$

As $\delta x \to 0$, $\cos \delta x \to 1$ and $\sin \delta x \to \delta x$

Thus, $y + \delta y \approx \sin x + \delta x \cos x$

$$\sin x + \delta y \approx \sin x + \delta x \cos x$$

$$\delta y \approx \delta x \cos x$$

$$\frac{\delta y}{\delta x} \approx \cos x$$

$$\lim_{\delta x \to 0} \left(\frac{\delta y}{\delta x} \right) = \frac{dy}{dx} = \cos x$$

> If $y = \sin x$, then $\dfrac{dy}{dx} = \cos x$

We can use a similar proof to differentiate $\cos x$. The result is stated below.

> If $y = \cos x$, then $\dfrac{dy}{dx} = -\sin x$

Consider $y = \tan x$.   Therefore $y = \dfrac{\sin x}{\cos x}$

Using the quotient rule: $\dfrac{dy}{dx} = \dfrac{\cos x(\cos x) - \sin x(-\sin x)}{\cos^2 x} = \dfrac{\cos^2 x + \sin^2 x}{\cos^2 x} = \dfrac{1}{\cos^2 x} = \sec^2 x$

> If $y = \tan x$, then $\dfrac{dy}{dx} = \sec^2 x$

Example 10

Find $\dfrac{dy}{dx}$ when

a) $y = \cos 2x$ **b)** $y = 5 \sin 3x^2$ **c)** $y = 7 \tan\left(4x + \dfrac{\pi}{2}\right)$.

- -

a) Let $u = 2x \Rightarrow \dfrac{du}{dx} = 2$

$$y = \cos u \Rightarrow \dfrac{dy}{du} = -\sin u = -\sin 2x \quad \longleftarrow \quad \text{Substitute } u = 2x.$$

$$\dfrac{dy}{dx} = \dfrac{dy}{du} \times \dfrac{du}{dx} \quad \longleftarrow \quad \text{Use the chain rule to help you.}$$

$$= (-\sin 2x)(2) = -2 \sin 2x \quad \longleftarrow \quad \text{Substitute for each of the 2 unknowns.}$$

b) Let $u = 3x^2 \Rightarrow \dfrac{du}{dx} = 6x \quad \longleftarrow \quad$ Substitute $u = 3x^2$ and use the chain rule to help you differentiate.

$$y = 5 \sin u \Rightarrow \dfrac{dy}{du} = 5\cos u$$

$$\dfrac{dy}{dx} = (5 \cos u)(6x) \quad \longleftarrow \quad \text{Using the chain rule } \dfrac{dy}{dx} = \dfrac{dy}{du} \times \dfrac{du}{dx}.$$

$$= 30x \cos 3x^2$$

c) Let $u = 4x + \dfrac{\pi}{2} \Rightarrow \dfrac{du}{dx} = 4 \quad \longleftarrow \quad$ Substitute $u = 4x + \dfrac{\pi}{2}$.

$$y = 7 \tan u \Rightarrow \dfrac{dy}{du} = 7 \sec^2 u$$

$$\dfrac{dy}{dx} = (7 \sec^2 u) \times 4 \quad \longleftarrow \quad \text{Using the chain rule } \dfrac{dy}{dx} = \dfrac{dy}{du} \times \dfrac{du}{dx}.$$

$$= 28 \sec^2\left(4x + \dfrac{\pi}{2}\right)$$

Exercise 4.5

1. Find $\dfrac{dy}{dx}$ when

   a) $y = \sin 5x$ **b)** $y = 6 \tan \dfrac{x}{2}$ **c)** $y = \cos(x^2 + 2x)$

   d) $y = 4 \tan^2 x$ **e)** $y = \sin(x^3 - 7)$ **f)** $y = 5 \tan 4x$

   g) $y = \cos^3 x$ **h)** $y = \sin 2x \cos x$ **i)** $y = x^2 \tan 3x$

   j) $y = \dfrac{\cos 2x}{x^2}$ **k)** $y = \sin^3 3x \cos x$ **l)** $y = 4 \tan\sqrt{x}$.

2. **a)** If $y = (\cos 2x - \sin x)^2$, show that $\dfrac{dy}{dx} = -2(\cos 2x - \sin x)(\sin 2x + \cos x)$.

    b) If $y = e^x \tan x$, show that $\dfrac{dy}{dx} = e^x(\sec^2 x + \tan x)$.

    c) If $y = \ln(\cos x)$, show that $\dfrac{dy}{dx} = -\tan x$.

    d) If $y = x\, e^{\sin x}$, show that $\dfrac{dy}{dx} = e^{\sin x}(x \cos x + 1)$.

3. Find the exact gradient of the tangent to the curve $y = 3 \sin 2x$ when $x = \dfrac{\pi}{3}$.

4. Find the equation of the normal to the curve $y = x - \sin x$ at the point $\left(\dfrac{\pi}{2}, \dfrac{\pi}{2} - 1\right)$.

5. Show that $\dfrac{d^2 y}{dx^2} = -\dfrac{1}{2} \sec^{2x}$ when $y = \ln \sqrt{\cos x}$.

6. Find the coordinates of the turning points on the curve $y = 2 \sin x + \cos 2x$ for $0 < x < \pi$ and determine whether these points are maximum or minimum points.

7. Show that the maximum value of the curve $y = x - 2 \sin x$ for $0 \leq x \leq 2\pi$ is $\dfrac{5\pi}{3} + \sqrt{3}$, and find the minimum value of the curve.

8. If $y = \ln \sqrt{1 - \cos x}$, show that $\dfrac{d^2 y}{dx^2} = \dfrac{1}{2(\cos x - 1)}$.

4.6 Implicit differentiation

Functions of the form $y = f(x)$ are **explicit** functions as y is given explicitly in terms of x.

However, some functions that have two variables are not of this form – that is, sometimes we are not given one variable explicitly in terms of the other. We call these functions **implicit functions**.

Examples of implicit functions include:

(1) $x^2 + 3xy^2 - 3y = 0$

(2) $x \ln(y + 2) - x^2 = 2y$

(3) $\dfrac{4y^2}{y^2 + x} = 5$

How can we differentiate an implicit function?

We use implicit differentiation when one of the variables is not given explicitly as a function of the other variable.

To illustrate the process of implicit differentiation, we will consider the example $x^2 + 3y = 6xy$.

Differentiating each term with respect to x,

$$\frac{d}{dx}x^2 + \frac{d}{dx}3y = \frac{d}{dx}6xy$$

Thus $2x + \frac{d}{dx}3y = 6x\frac{d}{dx}y + 6y\frac{d}{dx}x$

$\longleftarrow$ Using the product rule on the RHS.

Thus $2x + \left(\frac{d}{dy}3y\right)\left(\frac{dy}{dx}\right) = (6x)\left(\frac{d}{dy}y\right)\left(\frac{dy}{dx}\right) + 6y$

$\longleftarrow$ We can say $\frac{d}{dx} = \frac{d}{dy} \times \frac{dy}{dx}$.

Thus $2x + 3\frac{dy}{dx} = 6x\frac{dy}{dx} + 6y$

Rearranging gives:

$\longleftarrow$ Collect terms in $\frac{dy}{dx}$ on LHS.

$$(3 - 6x)\frac{dy}{dx} = 6y - 2x$$

Thus $\frac{dy}{dx} = \frac{6y - 2x}{3 - 6x}$

It is important to note that when differentiating an implicit equation to find $\frac{dy}{dx}$, we must use the chain rule to differentiate any expression involving y. For example, $\frac{d}{dx}(3y^2) = 6y\frac{dy}{dx}$.

Example 11

Find $\frac{dy}{dx}$ in terms of x and y when $x^2 - 3y^3 = 4y^2$.

$$2x - 9y^2\frac{dy}{dx} = 8y\frac{dy}{dx}$$

$\longleftarrow$ Differentiate each term with respect to x.

$$2x = 8y\frac{dy}{dx} + 9y^2\frac{dy}{dx}$$

$\longleftarrow$ Rearrange terms.

$$\frac{dy}{dx}(8y + 9y^2) = 2x$$

$\longleftarrow$ Factorise.

$$\frac{dy}{dx} = \frac{2x}{8y + 9y^2}$$

$\longleftarrow$ Rearrange for $\frac{dy}{dx}$.

Example 12

Find the equation of the normal to the curve $5x^2 + 6xy - y^2 = 10$ at the point $(1, 5)$.

$10x + 6x\dfrac{dy}{dx} + 6y - 2y\dfrac{dy}{dx} = 0$ ← Use the product rule to differentiate $6xy$.

$(6x - 2y)\dfrac{dy}{dx} = -10x - 6y$ ← Rearrange terms.

At $(1, 5) \Rightarrow (6 - 10)\dfrac{dy}{dx} = -10 - 30$ ← Substitute for $x = 1$ and $y = 5$.

So $\dfrac{dy}{dx} = \dfrac{-40}{-4} = 10$

Thus gradient of tangent to curve at $(1, 5) = 10$

Gradient of normal to curve at $(1, 5) = -\dfrac{1}{10}$ ← $10m_1 \times -\dfrac{1}{m_2} = -1$ for perpendicular lines.

Equation of normal at $(1, 5)$ is

$y - 5 = -\dfrac{1}{10}(x - 1)$ ← $(y - y_1) = m(x - x_1)$ is the equation of a straight line.

or $-10y + 50 = x - 1$ ← Multiply both sides by -10.

or $10y = 51 - x$

Exercise 4.6

1. Find $\dfrac{dy}{dx}$ in terms of x and y when

   a) $x^2 + 2y - y^2 = 5$ b) $x\cos y = y^2 + x$

   c) $x^2 + 4y^2 = 4$ d) $x^3 + xy^2 = 5x$

   e) $x^2 + y^2 = y$ f) $2xy - 3y = y^2 - 7x$

   g) $x\ln y = 1 + x$ h) $x^3 + y^3 = 3xy$

   i) $x^2 + y^2 + 2x - 4y + 4 = 0$ j) $x^4 + x^2y^2 - y^4 = 5$

   k) $x\sin y - y\sin x = 8$ l) $e^{x+y} = 2x$

   m) $x^2y - xy^2 = 4$ n) $y + y^3 = x - x^2$

   o) $xe^y + 2y = 1$ p) $y = x\ln y$

   q) $x^3 - y^3 + y^2 - 3y = 10$ r) $\sin y + x^2y^3 = 2y + \cos x$.

2. Find the gradients of the tangents to the curve $x^2 + 6y^2 = 10$ when $x = 2$.

3. Find the equation of the tangent to the curve $x^3 + y^3 = 2xy$ at the point $(1, 1)$.

4. Show that $\dfrac{dy}{dx} = \dfrac{2y^2 - 2xy^3}{3x^2y^2 - 4xy - 3}$ when $x^2y^3 - 2xy^2 = 3y$.

5. Given that $\ln(xy) = x^2 + y^2$, show that $\dfrac{dy}{dx} = \dfrac{y(2x^2 - 1)}{x(1 - 2y^2)}$.

4.7 Parametric differentiation

When x and y are related via a third variable, t, then t is called **a parameter**.
Equations which state x and y in terms of t are called **parametric equations**.

Suppose $x = t^3$ and $y = t^2$.

We can look at values of x and y for particular values of t.

t	-2	-1	0	1	2
x	-8	-1	0	1	8
y	4	1	0	1	4

If we want to try to relate x and y directly, notice that
$x = t^3 \Rightarrow x^2 = t^6$, and $y = t^2 \Rightarrow y^3 = t^6$.

Therefore $x^2 = y^3$.

Using implicit differentiation, $2x = 3y^2 \dfrac{dy}{dx}$, thus $\dfrac{dy}{dx} = \dfrac{2x}{3y^2} = \dfrac{2t^3}{3t^4} = \dfrac{2}{3t}$

We can also use the chain rule to produce the same result, but without
needing to eliminate t. Consider again $x = t^3$ and $y = t^2$.

Differentiating with respect to $t \Rightarrow \dfrac{dx}{dt} = 3t^2$ and $\dfrac{dy}{dt} = 2t$

Using the chain rule, $\dfrac{dy}{dx} = \dfrac{dy}{dt} \times \dfrac{dt}{dx} = 2t \times \dfrac{1}{3t^2} = \dfrac{2}{3t}$.

> When differentiating parametric equations involving a parameter t,
> $$\frac{dy}{dx} = \frac{dy}{dt} \div \frac{dx}{dt}$$

Example 13

a) If $x = 2t^3$ and $y = 5t^2 - 6$, find $\dfrac{dy}{dx}$ in terms of t.

b) Find the value of $\dfrac{dy}{dx}$ when $x = -16$.

....

a) $\dfrac{dx}{dt} = 6t^2$, $\dfrac{dy}{dt} = 10t$ ←   Differentiate x and y with respect to t.

$\dfrac{dy}{dx} = \dfrac{10t}{6t^2} = \dfrac{5}{3t}$ ← $\dfrac{dy}{dx} = \dfrac{dy}{dt} \div \dfrac{dx}{dt}$

b) When $x = -16 \Rightarrow -16 = 2t^3$ ←   Use $x = -16$ to find the value of t.

$\Rightarrow t = -2$

$\dfrac{dy}{dx} = \dfrac{5}{-6} = -\dfrac{5}{6}$ ←   Substitute this value of t into the expression you found in (**a**).

When differentiating parametric equations where the parameter is an angle, θ,
$$\frac{dy}{dx} = \frac{dy}{d\theta} \div \frac{dx}{d\theta}$$ where θ is given in **radians**.

Example 14

The parametric equations of a curve are given by $x = 2\cos\theta$, $y = \sin^2\theta$, $0 \le \theta \le 2\pi$.
Find the values of θ when the tangent to the curve is parallel to the x-axis.

$\frac{dx}{d\theta} = -2\sin\theta$, $\frac{dy}{d\theta} = 2\sin\theta\cos\theta$ — Differentiate x and y with respect to θ.

$\frac{dy}{dx} = \frac{2\sin\theta\cos\theta}{-2\sin\theta} = -\cos\theta$ — $\frac{dy}{dx} = \frac{dy}{d\theta} \div \frac{dx}{d\theta}$

When $-\cos\theta = 0$, then — $\frac{dy}{dx} = 0$ when tangent parallel to x-axis.

$\theta = \frac{\pi}{2}$ and $\frac{3\pi}{2}$ — $0 \le \theta \le 2\pi$

Exercise 4.7

1. Find $\frac{dy}{dx}$ for the following parametric equations. Give your answer in its simplest form.

   a) $x = 2t^3$, $\quad y = 4t^2 + 1$

   b) $x = t^3 - 6t$, $\quad y = t + \frac{2}{t}$

   c) $x = 3\cos t$, $y = 2\sin t$

   d) $x = 4t^2 - 7$, $\quad y = 7 + 8t$

   e) $x = t - \sin t$, $y = 2 - \cos t$

   f) $x = \frac{1}{1-t}$, $\quad y = t^2 - 9$

   g) $x = t^2 - 3$, $\quad y = 2 + \sqrt{t}$

   h) $x = 3\cos^3\theta$, $\quad y = 3\sin^3\theta$

   i) $x = t^2 + 3t - 1$, $\quad y = 3(t - 2)$

   j) $x = \cos 2\theta$, $\quad y = 4\sin\theta$

   k) $x = a\cos\theta$, $y = b\sin\theta$

   l) $x = 2 - t^2$, $\quad y = 5 + 4t - t^3$.

2. The parametric equations of a curve are $x = t + \frac{1}{t}$, $y = t - \frac{1}{t}$. Show that $\frac{dy}{dx} = \frac{t^2 + 1}{t^2 - 1}$.

3. The parametric equations of a curve are $x = 4\sin t$, $y = \cos 2t$.
   Show that $\frac{dy}{dx} = -\sin t$.

4. The parametric equations of a curve are $x = 2 + t$, $y = \dfrac{1}{t^2} - 6$. Show that $\dfrac{dy}{dx} = -2t^{-3}$.

5. The parametric equations of a curve are $x = \cos 2\theta$, $y = 2\theta + \sin 2\theta$. Show that $\dfrac{dy}{dx} = -\cot \theta$.

6. The parametric equations of a curve are $x = 2t + 3$, $y = t^2 - 4$.
   Find the gradient of the curve at the point for which $x = 5$.

7. The parametric equations of a curve are $x = 3t$, $y = \dfrac{3}{t}$.

   Find the equation of the normal to the curve at the point $(-1, -9)$.
   Give your answer in the form $ax + by + c = 0$, where a, b, and c are integers.

8. Show that $\dfrac{dy}{dx} = \dfrac{\sin\theta}{1 + \cos\theta}$ when $x = a(\theta + \sin \theta)$ and $y = a(1 - \cos \theta)$, where a is a constant.

9. The parametric equations of a curve are $x = 2t$, $y = 4t^3 - 8t^2$.

   Find the two values of t for which $\dfrac{dy}{dx} = 0$.

10. The parametric equations of a curve are: $x = 5\cos 3t$, $y = 2\sin 3t$, $0 < t < \dfrac{\pi}{2}$.

    a) Find $\dfrac{dy}{dx}$ in terms of t.

    b) Find the coordinates of the stationary point on the curve.

Summary exercise 4

1. Find $\dfrac{dy}{dx}$ when

   a) $y = e^{-5x}$

   b) $y = \ln(4x^2 + 5)$

   c) $y = x(1 + x)^3$

   d) $y = \dfrac{x - 1}{2x - 3}$

   e) $y = \dfrac{\sin x}{e^x}$

   f) $y = (x + 1)^2(x - 2)^3$

   g) $y = \dfrac{x^2}{\sqrt{(x^2 - 1)}}$

   h) $y = 6\ln\sqrt{(3x - 4)}$

   i) $y = xy^2 - x^3$

   j) $y = 7\ln e^{2x}$

   k) $y = \dfrac{xe^x}{\ln x}$

   l) $y = x^2\sqrt{(1 + x^2)}$

   m) $x = 3t - 5$, $\quad y = \dfrac{1}{t}$

   n) $y = -2e^{\sqrt{x}}$

   o) $2x^3 + 3xy^2 - y^3 = 6$

   p) $y = \dfrac{(2e^x - 3)^2}{e^x}$

   q) $y = e^{\tan^2 x}$

   r) $y = x^2 e^{3x-2}$

   s) $x = 1 + 2\cos\theta$, $\quad y = 3 - 6\sin\theta$

   t) $y = \log_e(3x + 1)$

   u) $x = (t + 2)^4$, $\quad y = 8t^3 - 2t$

   v) $y = x\ln x$

   w) $y = \dfrac{\sqrt{x}}{1 - x^2}$

   x) $x^3 + xy^2 - y^3 = 1$.

EXAM-STYLE QUESTIONS

2. Find the exact coordinates of the point on the curve $y = 3xe^{2x}$ at which $\dfrac{d^2y}{dx^2} = 0$.

3. Show that if $y = \ln(\sin\sqrt{x})$ then $\dfrac{dy}{dx} = \dfrac{\cot\sqrt{x}}{2\sqrt{x}}$.

4. The parametric equations of a curve are
$x = (1 + t)^2, y = (1 - t)^2$.

   a) Express $\dfrac{dy}{dx}$ in terms of t, simplifying your answer.

   b) Find the point on the curve where the gradient is parallel to the x-axis.

5. Find the equation of the tangent to the curve $y = \dfrac{x^2 + 2x}{x^2 - 3}$ at the point $(2, 8)$.

6. The equation of a curve is $x \ln y^3 = 6$. Find the gradient of the curve at the point $(2, e)$.

7. A curve has equation $2x^2 + y^2 + 5x - 3y = 1$. Find the equation of the normal to the curve at the point $(-1, 4)$. Give your answer in the form $ax + by + c = 0$, where a, b, and c are integers.

8. The parametric equations of a curve are $x = t^2, y = t^3$.
Find the equation of the tangent to the curve at the point $(1, 1)$.

9. The curve with equation $y = \dfrac{\cos x}{e^x}$ has one stationary point in the interval $0 \le x \le \pi$. Find the exact x coordinate of this point.

10. If $y = 2\cos(\ln x)$, show that
$$x^2\dfrac{d^2y}{dx^2} + x\dfrac{dy}{dx} + y = 0.$$

11. Given that $y = \dfrac{5x^2 - 10x + 9}{(x - 1)^2}$, $x \ne 1$, show that
$\dfrac{dy}{dx} = -\dfrac{8}{(x - 1)^3}$.

12. The parametric equations of a curve are
$x = t + 1, y = 4 - t^2$.
Find the equation of the normal to the curve at the point where $x = 2$.

13. If $y = \sqrt{5x^2 + 3}$, find the value of $\left(\dfrac{dy}{dx}\right)^2$ when $x = -1$.

14. Find the exact coordinates of the stationary point on the curve with equation $y = x\ln 3x$, giving your answer in terms of e.

15. Prove that there are no stationary points on the curve $y = \dfrac{ax + b}{cx + d}$ where a, b, c, and d are constants and $ad \ne bc$

16. Prove that if $y = e^{-\frac{x}{2}}\sin 2x$, then
$$4\dfrac{d^2y}{dx^2} + 4\dfrac{dy}{dx} + 17y = 0.$$

17. By differentiating $\dfrac{1}{\sin 2\theta}$, show that if
$y = \operatorname{cosec} 2\theta$ then $\dfrac{dy}{d\theta} = -2\operatorname{cosec} 2\theta \cot 2\theta$.

18. Show that the curve with equation $y = \sin x - 8\tan x$ cannot have any stationary values.

19. The equation of a curve is
$x^2 - y^2 - 5x + 3y + 13 = 0$.

   a) Find $\dfrac{dy}{dx}$ in terms of x and y.

   b) Find the maximum and minimum values of y.

20. The parametric equations of a curve are given by
$x = (t^2 - 2)^2, y = t(t - 6)$.
Find the coordinates of the stationary point of the curve.

21. Find the set of values of x satisfying the inequality $|3^x - 8| < 0.5$, giving 3 significant figures in your answer. [4]

Cambridge International AS and A Level Mathematics 9709 Paper 3 Q1 October/November 2006

22. a) Use the substitution $x = \tan \theta$ to show that

$$\int \frac{1 - x^2}{(1 + x^2)} \, dx = \int \cos 2\theta \, d\theta.$$ [4]

b) Hence find the value of

$$\int_0^1 \frac{1 - x^2}{(1 + x^2)^2} \, dx.$$ [3]

Cambridge International AS and A Level Mathematics 9709 Paper 3 Q4 May/June 2005

Chapter summary

Differentiating the exponential function

- If $y = e^x$, then $\dfrac{dy}{dx} = e^x$

- If $y = e^{kx}$, then $\dfrac{dy}{dx} = ke^{kx}$

- If $y = e^{f(x)}$, then $\dfrac{dy}{dx} = f'(x) \, e^{f(x)}$

Differentiating the natural logarithmic function

- If $y = \ln x$, then $\dfrac{dy}{dx} = \dfrac{1}{x}$

- If $y = \ln [f(x)]$, then $\dfrac{dy}{dx} = \dfrac{f'(x)}{f(x)}$

Differentiating products

- If $y = uv$, then $\dfrac{dy}{dx} = u\dfrac{dv}{dx} + v\dfrac{du}{dx}$

Differentiating quotients

- If $y = \dfrac{u}{v}$, then $\dfrac{dy}{dx} = \dfrac{v\dfrac{du}{dx} - u\dfrac{dv}{dx}}{v^2}$

Differentiating sin x, cos x, and tan x

- If $y = \sin x$, then $\dfrac{dy}{dx} = \cos x$

- If $y = \cos x$, then $\dfrac{dy}{dx} = -\sin x$

- If $y = \tan x$, then $\dfrac{dy}{dx} = \sec^2 x$

Implicit differentiation

- We use implicit differentiation when one of the variables is not given explicitly as a function of the other variable.
- When differentiating an implicit equation to find $\dfrac{dy}{dx}$, we must use the chain rule to differentiate any expression involving y. For example, $\dfrac{d}{dx}(3y^2) = 6y\dfrac{dy}{dx}$.

Parametric differentiation

- When differentiating parametric equations involving a parameter t,
  $$\frac{dy}{dx} = \frac{dy}{dt} \div \frac{dx}{dt}.$$

- When differentiating parametric equations where the parameter is an angle, θ,
  $$\frac{dy}{dx} = \frac{dy}{d\theta} \div \frac{dx}{d\theta}$$ where θ is given in **radians**.

Integration is the basic building block for solving any equation which describes how things move or change. Such equations are called differential equations and you will meet them in Chapter 10. Integration is used to design and construct dams like this one, which need to be built to withstand extreme stresses caused by the water it holds back.

Objectives

- Extend the idea of 'reverse differentiation' to include the integration of e^{ax+b}, $\dfrac{1}{ax+b}$, $\sin(ax+b)$, $\cos(ax+b)$, and $\sec^2(ax+b)$ (knowledge of the general method of integration by substitution is not required).

- Use trigonometrical relationships (such as double-angle formulae) to facilitate the integration of functions such as $\cos^2 x$.

- Use the trapezium rule to estimate the value of a definite integral, and use sketch graphs in simple cases to determine whether the trapezium rule gives an over-estimate or an under-estimate.

Before you start

You should know how to:

1. Differentiate standard functions,

   e.g. a) $\dfrac{d}{dx}(e^{3x+2})$

   $= 3e^{3x+2}$

   e.g. b) $\dfrac{d}{dx}(\sin 3x + \tan 2x)$

   $= 3\cos 3x + 2\sec^2 2x.$

2. Use trigonometrical relationships,

   e.g. Show

   $\cos 4x = \cos^4 x + \sin^4 x - 6\sin^2 x \cos^2 x.$

   $\cos^4 x = \cos^2 2x - \sin^2 2x$

   $\quad = (\cos^2 x - \sin^2 x)^2 - (2\sin x \cos x)^2$

   $\quad = \cos^4 x + \sin^4 x - 2\sin^2 x \cos^2 x$

   $\qquad\qquad\qquad - 4\sin^2 x \cos^2 x$

   $\quad = \cos^4 x + \sin^4 x - 6\sin^2 x \cos^2 x$

Skills check:

1. Differentiate

   a) $y = e^{2x+1} + e^{-x}$

   b) $y = \sin 4x - 3\cos 2x$

   c) $y = \ln(3x + 7).$

2. Show that $\sin\theta \tan\theta + \cos\theta = \sec\theta.$

In P1 you saw that $y = \int ax^n \, dx = \dfrac{ax^{n+1}}{n+1} + c$ for $n \neq -1$, and more generally that

$\int (ax+b)^n \, dx = \dfrac{(ax+b)^{n+1}}{a(n+1)} + c \quad (n \neq -1).$

In Chapter 4 we looked at how to differentiate a number of other functions. In this chapter we will extend the reverse process of differentiation – that is, integration – to cover many other functions.

5.1 Integration of e^{ax+b}

In **5.1** you saw that $\dfrac{d}{dx}\left(e^{ax+b}\right) = ae^{ax+b}$. This means that

$$\int \left(e^{ax+b}\right) dx = \frac{1}{a}e^{ax+b} + c$$

> It is extremely important you do not forget to write the $+c$ at the end if you are not integrating between two definite limits.

Example 1

Find $\displaystyle\int \left(e^{2x+3}\right) dx$

$\displaystyle\int \left(e^{2x+3}\right) dx = \frac{1}{2}\left(e^{2x+3}\right) + c$

⟵ The coefficient of x in $2x+3$ will be a multiplying factor when the (e^{2x+3}) term is differentiated, so you need to include the $\dfrac{1}{2}$ to compensate for this.

When the integral has no limits (called an *indefinite integral*) as in Example 1, then the integral is determined only up to a constant (the '$+c$' term). However, all the functions you will learn to integrate in this chapter can appear either as definite integrals (where you integrate between two limits), or else as improper integrals (where the function is not defined at one end of the interval). Example 2 looks at an example of a definite integral, and Example 3 at an improper integral.

Example 2

Calculate $\displaystyle\int_0^1 \left(e^{2x+3}\right) dx$

$\displaystyle\int_0^1 \left(e^{2x+3}\right) dx = \left[\frac{1}{2}\left(e^{2x+3}\right)\right]_0^1$

⟵ The function is as in Example 1, but the $+c$ is not needed because we are integrating between two definite limits.

$= \dfrac{1}{2}e^5 - \dfrac{1}{2}e^3$

⟵ Evaluating the expression at the top and bottom limits.

$= \dfrac{1}{2}\left(e^5 - e^3\right)$

> If we were to put the constant term $+c$ in the integral above, it would just cancel out when we evaluate the expression at the top and bottom limits. As such, we do not need to include the $+c$ when integrating any expression between two definite limits.

Example 3

Calculate $\int_{-\infty}^{1} (e^{2x+3}) \, dx$

$\int_{-\infty}^{1} (e^{2x+3}) \, dx = \left[\frac{1}{2}(e^{2x+3}) \right]_{-\infty}^{1}$

> The integral is as in Example 2, but the lower limit is different.

$= \frac{1}{2}e^{5} - \lim_{a \to -\infty} \left(\frac{1}{2}e^{2a+3} \right)$

> Remember that (formally) this is the justification for saying the lower limit is 0.

$= \frac{1}{2}e^{5} - 0$

$= \frac{1}{2}e^{5}$

Example 4

Calculate $\int \sqrt{(e^{x+1})} \, dx$

$\int \sqrt{(e^{x+1})} \, dx = \int (e^{x+1})^{\frac{1}{2}} \, dx$

> The function needs to be expressed with a linear power.

$= \int \left(e^{\frac{1}{2}(x+1)} \right) dx$

> The derivative of the power is $\frac{1}{2}$ so you need to multiply by 2 to compensate.

$= 2e^{\frac{1}{2}(x+1)} + c$

Example 5

Find the area bounded by the graph of $y = e^{2x}$, the x-axis, the y-axes, and the line $x = 1$.

$Area = \int_{0}^{1} e^{2x} \, dx$

> Sketching the graphs helps identify the required integral.

$= \left[\frac{1}{2}e^{2x} \right]_{0}^{1}$

> Integrating and substituting the limits gives the area we're asked for.

$= \frac{1}{2}e^{2} - \frac{1}{2}$

Exercise 5.1

1. Find the following indefinite integrals.

   a) $\displaystyle\int e^{3x-1}\,dx$

   b) $\displaystyle\int e^{4x+1}\,dx$

   c) $\displaystyle\int e^{0.5x+1}\,dx$

   d) $\displaystyle\int e^{\pi x+3}\,dx$

   e) $\displaystyle\int e^{4(x+1)}\,dx$

   f) $\displaystyle\int \sqrt{(e^{x+1})}\,dx$

   g) $\displaystyle\int e^{2-x}\,dx$

   h) $\displaystyle\int e^{3-2x}\,dx$

   i) $\displaystyle\int \sqrt{(e^{1-x})}\,dx$

2. Calculate the following definite integrals, leaving your answers in terms of e.

   a) $\displaystyle\int_0^1 e^{3x}\,dx$

   b) $\displaystyle\int_0^{0.25} e^{4x}\,dx$

   c) $\displaystyle\int_1^2 e^{x-1}\,dx$

   d) $\displaystyle\int_0^1 4e^{4x}\,dx$

   e) $\displaystyle\int_0^1 12e^{3x}\,dx$

   f) $\displaystyle\int_0^2 2e^{4x}\,dx$

   g) $\displaystyle\int_0^1 e^{-x}\,dx$

   h) $\displaystyle\int_0^2 2e^{-2x}\,dx$

   i) $\displaystyle\int_{-1}^1 e^{0.5x}\,dx$

3. Calculate the following integrals, giving your answers in terms of e, or as exact values.

   a) $\displaystyle\int_0^{\ln 2} e^x\,dx$

   b) $\displaystyle\int_0^{2\ln 3} e^x\,dx$

   c) $\displaystyle\int_{\ln 2}^{\ln 3} e^x\,dx$

   d) $\displaystyle\int_0^{\ln 3} 4e^{4x}\,dx$

   e) $\displaystyle\int_0^{\ln 4} 2e^{2x}\,dx$

   f) $\displaystyle\int_0^{\ln 5} e^{-x}\,dx$

   g) $\displaystyle\int_0^1 e^{x\ln 3}\,dx$

   h) $\displaystyle\int_{-2}^1 e^{x\ln 2}\,dx$

   i) $\displaystyle\int_{-1}^4 e^{-x\ln 3}\,dx$

4. Find the area bounded by the graph of $y = e^x$, the x- and y-axes, and the line $x = 3$.

5. Find the area bounded by the graph of $y = e^{-2x}$, the x- and y-axes, and the line $x = 2$.

6. Find the area bounded by the graphs of $y = e^x$, $y = e^{-x}$, the y-axis, and the line $x = 2$.

7. Find $\displaystyle\int_0^k e^{-x}\,dx$ and hence find $\displaystyle\int_0^\infty e^{-x}\,dx$.

8. Calculate the following improper integrals, leaving your answers in terms of e where appropriate.

   a) $\displaystyle\int_0^\infty e^{-3x}\,dx$

   b) $\displaystyle\int_1^\infty e^{-4x}\,dx$

   c) $\displaystyle\int_1^\infty e^{1-x}\,dx$

   d) $\displaystyle\int_{-\infty}^0 4e^{4x}\,dx$

   e) $\displaystyle\int_{-\infty}^1 6e^{3x}\,dx$

   f) $\displaystyle\int_{-\infty}^2 \sqrt{e^x}\,dx$

9. The line $y = e^{-x}$ between $x = 0$ and $x = 3$ is rotated through $360°$ about the x-axis. Find the volume of the solid obtained.

5.2 Integration of $\dfrac{1}{ax+b}$

In **5.1** you saw that $\dfrac{d}{dx}(\ln x) = \dfrac{1}{x}$. This means that $\displaystyle\int\left(\dfrac{1}{x}\right)dx = \ln x + c$.

However, the function $\ln x$ is only defined on the interval $(0, \infty)$. Does this mean that the

integral $\displaystyle\int\left(\dfrac{1}{x}\right)dx$ is only defined on the interval $(0, \infty)$ too?

In actual fact, the integral $\displaystyle\int\left(\dfrac{1}{x}\right)dx$ is defined for all real x, and

we will explain how this is the case by using the graph on the right.

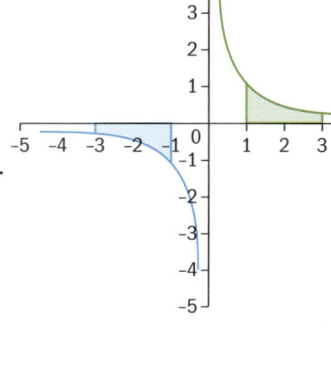

If you look at the graph of $y = \dfrac{1}{x}$ and consider the rotational

symmetry about the origin, it is clear that the shaded areas are

the same size. Therefore the integral $\displaystyle\int\left(\dfrac{1}{x}\right)dx$ from -3 to -1

should just be the negative of the integral from 1 to 3

(it's negative because it lies below the x axis). The resolution of this dilemma

is surprisingly simple – but the first step is to do something trivial:

write $\dfrac{1}{x}$ as $\dfrac{-1}{-x}$, which is obviously true provided $x \neq 0$.

Then $\displaystyle\int\left(\dfrac{-1}{-x}\right)dx = \ln(-x) + c$.

So $\displaystyle\int\left(\dfrac{1}{x}\right)dx = \ln x + c$ or, $\displaystyle\int\ln\left(\dfrac{1}{x}\right)dx = \ln(-x) + c$ which can be summarised as

$$\int\left(\dfrac{1}{x}\right)dx = \ln|x| + c$$

More generally you saw that $\dfrac{d}{dx}(\ln(ax+b)) = \dfrac{a}{ax+b}$, giving the general result

$$\int\left(\dfrac{1}{ax+b}\right)dx = \dfrac{1}{a}\ln|ax+b| + c$$

> It is normal practice only to write the modulus sign in the case of definite integrals.
> For indefinite integrals it is normal to restrict the function by the condition $ax + b > 0$.

Evaluating the two integrals above,

$$\int_1^3 \dfrac{1}{x}dx = \left[\ln|x|\right]_1^3 = \ln 3 - \ln 1 = \ln 3 \quad\text{and}\quad \int_{-3}^{-1} \dfrac{1}{x}dx = \left[\ln|x|\right]_{-3}^{-1} = \ln 1 - \ln 3 = -\ln 3$$

so this not only gives the correct absolute value, but gives the correct sign for the function being below the axis.

Note that

- ln$x$ gets large without bound as x tends to infinity and so there will not be any improper integrals involving logarithms.
- Notice also that ln0 does not exist, which means that definite integrals in this form need to be defined on an interval $[\alpha, \beta]$ such that $ax + b \neq 0$ anywhere in the interval.

Example 6

Find $\displaystyle\int \frac{2}{2x - 3}\,dx$ and state the values of x for which the answer is valid.

$\displaystyle\int \frac{12}{2x - 3}\,dx = \ln(2x - 3) + c$;   valid for $x > \dfrac{3}{2}$ $\longleftarrow$ The derivative of the denominator is 2, so it is an exact integral, and $2x - 3 > 0$ when $x > \dfrac{3}{2}$.

Example 7

a) Sketch the graph of $y = \dfrac{1}{x - 2}$

b) Mark on your sketch the two areas represented by $\displaystyle\int_{-2}^{-1} \frac{1}{x - 2}\,dx$ and $\displaystyle\int_{5}^{6} \frac{1}{x - 2}\,dx$

c) Explain, using the symmetry of the reciprocal function why $\displaystyle\int_{-2}^{-1} \frac{1}{x - 2}\,dx = -\int_{5}^{6} \frac{1}{x - 2}\,dx$

d) Calculate the two integrals in part (c) and verify the result.

a) and b)

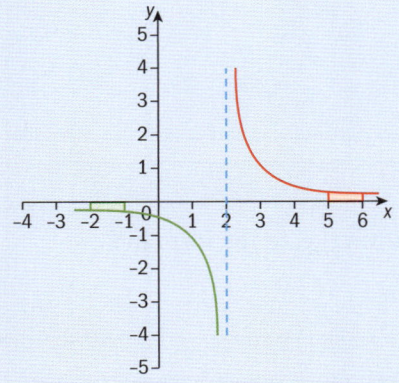

c) The vertical asymptote is at $x = 2$, so the two intervals $(-2, -1)$ and $(5, 6)$ are symmetrically placed on either side of the asymptote. Therefore the two shaded areas are equal.

d) $\displaystyle\int_{-2}^{-1} \frac{1}{x - 2}\,dx = \Big[\ln|x - 2|\Big]_{-2}^{-1} = \ln 3 - \ln 4 = \ln\left(\frac{3}{4}\right)$

$\displaystyle\int_{5}^{6} \frac{1}{x - 2}\,dx = \Big[\ln|x - 2|\Big]_{5}^{6} = \ln 4 - \ln 3 = \ln\left(\frac{4}{3}\right) = -\ln\left(\frac{3}{4}\right)$ as required.

Example 8

Find $\displaystyle\int \frac{2}{5-x}\,dx$ and state the values of x for which the answer is valid.

$$\int \frac{2}{5-x}\,dx = -2\ln(5-x) + c; \quad x < 5$$

The derivative of the denominator is -1, so the integral requires a multiplicative constant of -2 and the answer is valid when $5 - x > 0$, i.e. when $x < 5$.

Example 9

a) Sketch the graph of $y = \dfrac{1}{x+2}$

b) Mark on your sketch the area represented by $\displaystyle\int_{-3}^{1} \frac{1}{x+2}\,dx$

c) Explain why this integral cannot be calculated.

a) and b)

c) The vertical asymptote is at $x = -2$, so the interval $[-3, 1]$ includes the asymptote, and you cannot integrate across an asymptote as the area underneath is of infinite size.

Both the area above the x-axis and the area below the x-axis are infinite.

Example 10

Calculate $\displaystyle\int_{3}^{5} \left(\frac{1}{3x-2}\right)dx$ giving your answer as a single logarithm (i.e. in the form $\ln k$, where k is real).

$$\int_{3}^{5} \left(\frac{1}{3x-2}\right)dx = \left[\frac{1}{3}\ln|3x-2|\right]_{3}^{5}$$

Use the standard form of the logarithmic integral.

$$= \frac{1}{3}\ln 13 - \frac{1}{3}\ln 7$$

Evaluate the expression at the top and bottom limits.

$$= \frac{1}{3}\ln\left(\frac{13}{7}\right) = \ln\left(\sqrt[3]{\frac{13}{7}}\right)$$

First combine the two logarithms, then use the laws of logarithms to take the $\frac{1}{3}$ into the logarithm as the cube root.

Integration of $\dfrac{1}{ax+b}$

Example 11

Calculate $\displaystyle\int_{-2}^{1} \left(\frac{1}{7-2x}\right) dx$ giving your answer as a single logarithm (i.e. in the form $\ln k$, where k is real).

$$\int_{-2}^{1} \left(\frac{1}{7-2x}\right) dx = \left[-\frac{1}{2}\ln|7-2x|\right]_{-2}^{1}$$

Use the standard from of the logarithmic integral, noting a is negative.

$$= -\frac{1}{2}\ln 5 - \left(-\frac{1}{2}\ln 11\right)$$

Evaluate the expresssion at the top and bottom limits.

$$= \frac{1}{2}\ln\left(\frac{11}{5}\right) = \ln\left(\sqrt{2.2}\right)$$

First combine the two logarithms, then use the laws of logarithms to take the $\frac{1}{2}$ into the logarithm as the square root.

Example 12

Calculate $\displaystyle\int_{4}^{6} \left(\frac{1}{7-2x}\right) dx$ giving your answer as a single logarithm (i.e. in the form $\ln k$, where k is real).

$$\int_{4}^{6} \left(\frac{1}{7-2x}\right) dx = \left[-\frac{1}{2}\ln|7-2x|\right]_{4}^{6}$$

The integral is the same as in Example 11, but with different limits.

$$= -\frac{1}{2}\ln 5 - \left(-\frac{1}{2}\ln 1\right)$$

Evaluating the expression at the top and bottom limits now requires taking the absolute values of −5 and −1.

$$= \frac{1}{2}\ln\left(\frac{1}{5}\right) = \ln\left(\sqrt{0.2}\right)$$

As before, first combine the two logarithms, then use the laws of logarithms to take the $\frac{1}{2}$ into the logarithm as the square root.

When ln1 occurs in calculations, it can either be replaced by 0, or you can combine it into a single logarithm using the standard laws.

Exercise 5.2

1. Find the following indefinite integrals and state the values of x for which the answer is valid.

a) $\displaystyle\int \frac{1}{2x-1} dx$

b) $\displaystyle\int \frac{2}{2x-1} dx$

c) $\displaystyle\int \frac{7}{7x+3} dx$

d) $\displaystyle\int \frac{10}{5-2x} dx$

e) $\displaystyle\int \frac{-6}{1-3x} dx$

f) $\displaystyle\int \frac{-2}{6x+5} dx$

g) $\displaystyle\int \frac{-3}{x+1} dx$

h) $\displaystyle\int \frac{1}{7-x} dx$

i) $\displaystyle\int \frac{2}{x+e} dx$

j) $\displaystyle\int \frac{-0.5}{2x+1} dx$

k) $\displaystyle\int \frac{-12}{9x+16} dx$

l) $\displaystyle\int \frac{-2}{10x-3} dx$

2. Calculate the following definite integrals. Note that parts **(a)** to **(c)** use the same indefinite integrals as you calculated in parts **(a)** to **(c)** of question 1.

a) $\displaystyle\int_2^3 \frac{1}{2x-1}\,dx$

b) $\displaystyle\int_5^7 \frac{2}{2x-1}\,dx$

c) $\displaystyle\int_0^1 \frac{7}{7x+3}\,dx$

d) $\displaystyle\int_0^1 \frac{4}{5-2x}\,dx$

e) $\displaystyle\int_0^1 \frac{9}{4-2x}\,dx$

f) $\displaystyle\int_0^2 \frac{-7}{5-x}\,dx$

g) $\displaystyle\int_2^4 \frac{2}{x-7}\,dx$

h) $\displaystyle\int_{-6}^{-4} \frac{-4}{7-2x}\,dx$

i) $\displaystyle\int_{-1}^1 \frac{6}{5-3x}\,dx$.

3. Find the area bounded by the graph of $y = \dfrac{1}{3+x}$, the x- and y-axes, and the line $x = 3$.

4. Find the area bounded by the graph of $y = \dfrac{-2}{x+3}$, the x- and y-axes, and the line $x = 2$.

5. Find the area bounded by the graphs of $y = \dfrac{1}{x-3}$, $y = \dfrac{1}{2x+3}$, the y-axis, and the line $x = 2$.

5.3 Integration of sin (ax + b), cos (ax + b), sec² (ax + b)

In **5.5** you saw that

$$\frac{d}{dx}(\sin(ax+b)) = a\cos(ax+b)$$

$$\frac{d}{dx}(\cos(ax+b)) = -a\sin(ax+b)$$

$$\frac{d}{dx}(\tan(ax+b)) = a\sec^2(ax+b).$$

This means that the following are the standard trigonometric integral forms.

$$\int \sin(ax+b)\,dx = -\frac{1}{a}\cos(ax+b) + c;$$

$$\int \cos(ax+b)\,dx = \frac{1}{a}\sin(ax+b) + c;$$

$$\int \sec^2(ax+b)\,dx = \frac{1}{a}\tan(ax+b) + c$$

Remember that if you are using calculus with trigonometric functions then x (or equivalent) is measured in radians and not in degrees.

Since all the trigonometric functions are periodic, there will not be a limit to which an integral approaches as x tends to infinity, so there will be no improper integrals of this sort, but care does need to be taken with the discontinuities that occur with $\tan x$ and also with all the reciprocal trigonometric functions.

Example 13

Find $\displaystyle\int \cos 2x\,dx$.

$$\int \cos 2x\,dx = \frac{1}{2}\sin 2x + c$$

$\cos 2x$ comes from differentiating $\sin 2x$

Divide by the coefficient of x (i.e. 2).

Example 14

Find $\int \sin\left(3x + \frac{\pi}{4}\right) dx$.

· ·

$\int \sin\left(3x + \frac{\pi}{4}\right) dx = -\frac{1}{3}\cos\left(3x + \frac{\pi}{4}\right) + c$ ◀—— $\sin\left(3x + \frac{\pi}{4}\right)$ sin comes from differentiating $-\cos\left(3x + \frac{\pi}{4}\right)$.

Divide by the coefficient of x (i.e. 3).

Example 15

a) Sketch the graph of $y = \sec x$ and use it to sketch the graph of $y = \sec^2 x$.

b) On your sketch, shade the area corresponding to $\int_0^{\frac{\pi}{4}} \sec^2 x\, dx$ and estimate to 1 significant figure the value of the area.

c) Calculate $\int_0^{\frac{\pi}{4}} \sec^2 x\, dx$.

· ·

a) $y = \sec x$ $y = \sec^2 x$

b) The area is approx. 0.8 wide $\left(\text{the exact width is } \frac{\pi}{4}\right)$ and between 1 and 2 high, mostly closer to 1, so estimate is 1 correct to 1 significant figure.

c) $\int_0^{\frac{\pi}{4}} \sec^2 x\, dx = [\tan x]_0^{\frac{\pi}{4}} = 1 - 0 = 1$

Example 16

By first sketching an appropriate graph, find the area bounded by the graph of $y = \sin\left(x - \frac{\pi}{3}\right)$, the x- and y-axes and the line $x = \pi$.

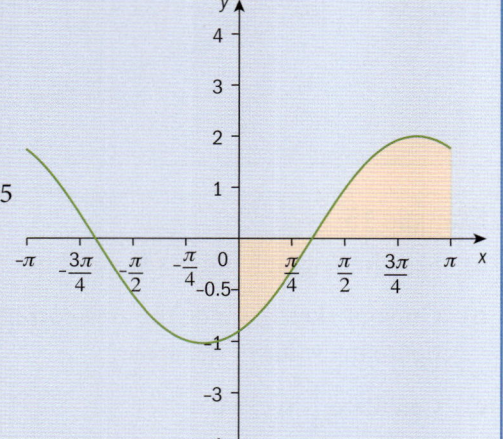

The sketch shows that the area required is partly above and partly below the x-axis, so the integral will need to be done in two sections:

i) from 0 to $\frac{\pi}{3}$,   ii) from $\frac{\pi}{3}$ to π.

$$\int_{0}^{\frac{\pi}{3}} \sin\left(x - \frac{\pi}{3}\right)dx = \left[-\cos\left(x - \frac{\pi}{3}\right)\right]_{0}^{\frac{\pi}{3}} = -1 - (-0.5) = -0.5$$

so the first area is 0.5 (below the axis)

$$\int_{\frac{\pi}{3}}^{\pi} \sin\left(x - \frac{\pi}{3}\right)dx = \left[-\cos\left(x - \frac{\pi}{3}\right)\right]_{\frac{\pi}{3}}^{\pi} = 0.5 - (-1) = 1.5$$

Thus, the total area required is $0.5 + 1.5 = 2$.

Exercise 5.3

1. Find these integrals.

   a) $\displaystyle\int \cos(2x)\,dx$

   b) $\displaystyle\int \sin\left(2x + \frac{\pi}{4}\right)dx$

   c) $\displaystyle\int \sec^2\left(3x + \frac{\pi}{12}\right)dx$

   d) $\displaystyle\int \cos(\pi - x)\,dx$

   e) $\displaystyle\int \cos\left(\frac{\pi}{2} - \frac{x}{4}\right)dx$

   f) $\displaystyle\int -\sin\left(2x - \frac{\pi}{4}\right)dx$

   g) $\displaystyle\int \sec^2\left(\frac{\pi}{12} - 5x\right)dx$

   h) $\displaystyle\int -\frac{1}{2}\sin\left(\frac{1}{2}x\right)dx$

   i) $\displaystyle\int \cos\left(4 - \frac{1}{3}x\right)dx.$

2. Calculate the following definite integrals. Note that parts **(a)** to **(c)** use the same indefinite integral as you calculated in parts **(a)** to **(c)** of question 1.

   a) $\displaystyle\int_{0}^{\frac{\pi}{4}} \cos(2x)\,dx$

   b) $\displaystyle\int_{0}^{\frac{\pi}{2}} \sin\left(2x + \frac{\pi}{4}\right)dx$

   c) $\displaystyle\int_{0}^{\frac{\pi}{12}} \sec^2\left(3x + \frac{\pi}{12}\right)dx$

   d) $\displaystyle\int_{0}^{\frac{\pi}{2}} \cos(2x)\,dx$

   e) $\displaystyle\int_{0}^{\pi} \cos\left(\frac{3\pi}{2} - \frac{x}{3}\right)dx$

   f) $\displaystyle\int_{0}^{\frac{\pi}{2}} \sin\left(3x - \frac{\pi}{4}\right)dx$

   g) $\displaystyle\int_{?}^{\frac{\pi}{12}} \sec^2\left(\frac{\pi}{12} - 3x\right)dx$

   h) $\displaystyle\int_{?}^{\frac{\pi}{4}} -\frac{1}{3}\cos\left(\frac{1}{2}x\right)dx$

   i) $\displaystyle\int_{1}^{3} \cos\left(2 - \frac{1}{2}x\right)dx.$

3. A curve passes through the point $\left(\frac{\pi}{4}, 2\right)$ and its gradient function is $\sec^2 x$. Find the equation of the curve.

4. Find the area bounded by the graph of $y = \cos\left(x + \dfrac{\pi}{4}\right)$, the x- and y-axes, and the line $x = \pi$.

5. Find the area bounded by the graphs of $y = \cos\left(x + \dfrac{\pi}{4}\right)$, $y = \cos\left(x + \dfrac{\pi}{2}\right)$, the y-axis, and the line $x = \dfrac{\pi}{2}$.

6. Find the area bounded by the graph of $y = \sec^2\left(2x + \dfrac{\pi}{2}\right)$, the x-axis, and the lines $x = \dfrac{\pi}{4}$ and $x = \dfrac{\pi}{3}$.

5.4 Extending integration of trigonometric functions

In Chapter 4, you met a number of trigonometric identities which can be used to extend the range of trigonometric functions you can integrate. Those identities were

- $1 + \tan^2 \theta \equiv \sec^2 \theta$

- $1 + \cot^2 \theta \equiv \csc^2 \theta$

- $\sin 2A \equiv 2 \sin A \cos A$

- $\cos 2A \equiv \cos^2 A - \sin^2 A$
  $\equiv 2 \cos^2 A - 1$
  $\equiv 1 - 2 \sin^2 A$

- $\tan 2A \equiv \dfrac{2 \tan A}{1 - \tan^2 A}$

> These are formulae you should know well enough that if you are faced with an integral involving trigonometric functions which you cannot integrate, you can manipulate the expression into one which can be integrated using these formulae.
>
> The double angle formulae for $\cos 2A$ is particularly useful in the forms $\sin^2 A = \dfrac{1}{2}(1 - \cos 2A)$ and $\cos^2 A = \dfrac{1}{2}(1 + \cos 2A)$.
>
> The other formulae developed in Chapter 4, will normally be hinted at, or given, if you need to use them in an integration problem.

Example 17

Find the integrals

a) $\displaystyle\int \cos^2 x \, dx$ **b)** $\displaystyle\int 6 \sin^2 x \, dx.$

a) $\displaystyle\int \cos^2 x \, dx = \int \dfrac{1}{2}(1 + \cos 2x)\,dx$ ← Use the double angle formula to get an integrable form.

$= \dfrac{1}{2}x + \dfrac{1}{4}\sin 2x + c$ ← Remember to integrate the number part as well as the trigonometric function, and include the $+ c$.

b) $\displaystyle\int 12 \sin^2 x \, dx = \int 12 \times \dfrac{1}{2}(1 - \cos 2x)\,dx$ ← Again the double angle formula transforms this into an integrable form.

$= 6x - 3 \sin 2x + c$

Example 18

Calculate the exact value of $\int_0^{\frac{\pi}{6}}(4\tan^2\theta)\,d\theta$.

$\int_0^{\frac{\pi}{6}}(4\tan^2\theta)\,d\theta = \int_0^{\frac{\pi}{6}}(4(\sec^2\theta-1))\,d\theta$ — The use of the identity makes this a function you know how to integrate.

$= [4\tan\theta - 4\theta]_0^{\frac{\pi}{6}}$ — These are standard integrals.

$= \left(\dfrac{4}{\sqrt{3}} - \dfrac{4\pi}{6}\right) - (0 - 0)$ — Evaluate the expression at the top and bottom limits explicitly to show your working.

$= \left(\dfrac{4}{\sqrt{3}} - \dfrac{2\pi}{3}\right)$

Example 19

a) By writing $2x = 3x - x$, and $4x = 3x + x$, show that $\sin 3x \cos x = \dfrac{1}{2}(\sin 2x + \sin 4x)$.

b) Hence calculate $\int_0^{\frac{\pi}{6}}\sin 3x \cos x\,dx$.

a) $\sin 2x = \sin(3x - x) = \sin 3x \cos x - \cos 3x \sin x$ — Using the compound angle formulae for $\sin x$ gives functions that you know how to integrate.

$\sin 4x = \sin(3x + x) = \sin 3x \cos x + \cos 3x \sin x$

Adding the two expressions $\Rightarrow$

$\Rightarrow \sin 2x + \sin 4x = 2\sin 3x \cos x$

$\Rightarrow \sin 3x \cos x = \dfrac{1}{2}(\sin 2x + \sin 4x)$

b) $\int_0^{\frac{\pi}{6}}\sin 3x \cos x\,dx = \int_0^{\frac{\pi}{6}}\dfrac{1}{2}(\sin 2x + \sin 4x)\,dx$ — Using part (a).

$= \left[-\dfrac{1}{4}\cos 2x - \dfrac{1}{8}\cos 4x\right]_0^{\frac{\pi}{6}}$ — Integrate the standard forms.

$= \left(-\dfrac{1}{4}\times\dfrac{1}{2} - \dfrac{1}{8}\times\left(-\dfrac{1}{2}\right)\right) - \left(-\dfrac{1}{4} - \dfrac{1}{8}\right)$

$= \dfrac{5}{16}$

Extending integration of trigonometric functions

Example 20

a) Show that $(3 \sin x - \cos x)^2 = 5 - 3 \sin 2x - 4 \cos 2x$.

b) Hence calculate $\int_0^{\frac{\pi}{2}} (3 \sin x - \cos x)^2 \, dx$.

a) $(3 \sin x - \cos x)^2 = 9 \sin^2 x - 6 \sin x \cos x + \cos^2 x$ ← Expand the square.

$$= 9 \times \frac{1}{2}(1 - \cos 2x) - 3 \sin 2x + \frac{1}{2}(1 + \cos 2x)$$ ← Use the standard formulae.

$$= 5 - 3 \sin 2x - 4 \cos 2x$$ ← Simplify to the required form.

b) $\int_0^{\frac{\pi}{2}} (3 \sin x - \cos x)^2 \, dx = \int_0^{\frac{\pi}{2}} (5 - 3 \sin 2x - 4 \cos 2x) dx$ ← Using part (a).

$$= \left[5x + \frac{3}{2} \cos 2x - 2 \sin 2x \right]_0^{\frac{\pi}{2}}$$ ← Integrate the standard forms.

$$= \left(\frac{5\pi}{2} - \frac{3}{2} + 0 \right) - \left(0 + \frac{3}{2} + 0 \right)$$ Be careful not to assume the limit at 0 is simply '0'.

$$= \frac{5\pi}{2} - 3$$ ← Simplify the answer.

Example 21

a) Find $\frac{d}{dx}(\cos x + x \sin x)$

b) Hence calculate $\int_0^{\frac{\pi}{2}} (x \cos x) dx$

Remember 'hence' means that you are expected to use a previous result.

a) $\frac{d}{dx}(\cos x + x \sin x) = -\sin x + \sin x + x \cos x = x \cos x$ ← Use the product rule.

b) $\int_0^{\frac{\pi}{2}} (x \cos x) dx = \int_0^{\frac{\pi}{2}} \left(\frac{d}{dx}(\cos x + x \sin x) \right) dx$ ← Use part (a).

$$= \left[\cos x + x \sin x \right]_0^{\frac{\pi}{2}}$$ ← Integration is the reverse process to differentiation.

$$= \left(0 + \frac{\pi}{2} \right) - (1 + 0)$$ ← Substitute at upper and lower limits.

$$= \frac{\pi}{2} - 1$$

Example 22

a) Show that $\dfrac{\cot^2 \theta}{1 + \cot^2 \theta} = \cos^2 \theta.$

b) Hence calculate $\displaystyle\int_0^{\frac{\pi}{3}} \left(\dfrac{\cot^2 \theta}{1 + \cot^2 \theta} \right) d\theta.$

a) $\dfrac{\cot^2 \theta}{1 + \cot^2 \theta} = \dfrac{\cot^2 \theta}{\operatorname{cosec}^2 \theta}$ — Use the standard identity.

$= \dfrac{\left(\dfrac{\cos^2 \theta}{\sin^2 \theta} \right)}{\left(\dfrac{1}{\sin^2 \theta} \right)}$ — Expressing in terms of sin and cos only.

$= \cos^2 \theta$

b) $\displaystyle\int_0^{\frac{\pi}{3}} \left(\dfrac{\cot^2 \theta}{1 + \cot^2 \theta} \right) d\theta = \int_0^{\frac{\pi}{3}} (\cos^2 \theta) d\theta$ — Using part (a).

$= \displaystyle\int_0^{\frac{\pi}{3}} \left(\dfrac{1 + \cos 2\theta}{2} \right) d\theta$ — Using the standard form to convert to an integrable function.

$= \left[\dfrac{\theta}{2} + \dfrac{\sin 2\theta}{4} \right]_0^{\frac{\pi}{3}}$ — Integrating.

$= \left(\dfrac{\pi}{6} + \dfrac{\sqrt{3}}{8} \right) - (0 + 0)$ — Substituting at upper and lower limits.

$= \dfrac{\pi}{6} + \dfrac{\sqrt{3}}{8}$

Example 23

a) Show that $4 \sin^4 \theta = \dfrac{3}{2} - 2 \cos 2\theta + \dfrac{\cos 4\theta}{2}.$

b) Hence calculate $\displaystyle\int_0^{\frac{\pi}{2}} (4 \sin^4 \theta) d\theta$

a) $4 \sin^4 \theta = (2 \sin^2 \theta)^2$

$= (1 - \cos 2\theta)^2$

$= 1 - 2 \cos 2\theta + \cos^2 2\theta$ — Using the double angle cosine formula twice to remove powers of $\cos \theta$ and $\sin \theta$ gives the required result. An alternative approach is to start with the right-hand side and work down to single angle expressions.

$= 1 - 2 \cos 2\theta + \left(\dfrac{1 + \cos 4\theta}{2} \right)$

$= \dfrac{3}{2} - 2 \cos 2\theta + \dfrac{\cos 4\theta}{2}$

b) $\displaystyle\int_0^{\frac{\pi}{2}} (4 \sin^4 \theta) d\theta = \int_0^{\frac{\pi}{2}} \left(\dfrac{3}{2} - 2 \cos 2\theta + \dfrac{\cos 4\theta}{2} \right) d\theta$ — Using part (a).

$= \left[\dfrac{3}{2} x - \sin 2\theta + \dfrac{\sin 4\theta}{8} \right]_0^{\frac{\pi}{2}}$ — Integration is the reverse process to differentiation.

$= \left(\dfrac{3\pi}{4} - 0 - 0 \right) - (0 + 0 + 0) = \dfrac{3\pi}{4}$ — Substitute at upper and lower limits.

Exercise 5.4

1. Find these integrals.

    a) $\displaystyle\int (\cos^2 2x)dx$
    b) $\displaystyle\int (\sin^2 3x)dx$
    c) $\displaystyle\int (1 + \tan^2 2x)dx$

    d) $\displaystyle\int (\sin x \cos x)dx$
    e) $\displaystyle\int (\sin^2 x \cos^2 x)dx$
    f) $\displaystyle\int (\sin x + \cos x)^2 dx$

2. Calculate the following definite integrals. Note that parts (a) to (c) use the same indefinite integral as you calculated in parts (a) to (c) of question 1.

    a) $\displaystyle\int_0^{\frac{\pi}{4}} (\cos^2 2x)dx$
    b) $\displaystyle\int_0^{\frac{\pi}{2}} (\sin^2 3x)dx$
    c) $\displaystyle\int_0^{\frac{\pi}{12}} (1 + \tan^2 2x)dx$

    d) $\displaystyle\int_0^{\frac{\pi}{2}} (1 - \cos 2x)^2 dx$
    e) $\displaystyle\int_0^{\pi} (1 + 2\sin x)^2 dx$
    f) $\displaystyle\int_0^{\frac{\pi}{2}} (\sin x - \cos x)^2 dx$

3. A curve passes through the origin and its gradient function is $\cos^2 x$. Find the equation of the curve.

4. Find the area bounded by the graph of $y = \sec^2 x$, the x axis, and the lines $x = \dfrac{\pi}{6}$ and $x = \dfrac{\pi}{3}$.

5. Find the area bounded by the graphs of $y = \cos^2 x$, $y = \sin^2 x$, the y axis, and the line $x = \dfrac{\pi}{4}$.

6. a) By writing $2x = 5x - 3x$, and $8x = 5x + 3x$, show that $\sin 3x \cos 5x = \dfrac{1}{2}(\sin 8x - \sin 2x)$.

    b) Hence calculate $\displaystyle\int_0^{\frac{\pi}{6}} \sin 3x \cos 5x \, dx$.

7. a) Show that $(\sin x + 5 \cos x)^2 = 13 + 5\sin 2x + 12\cos 2x$.

    b) Hence calculate $\displaystyle\int_0^{\frac{\pi}{2}} (\sin x + 5\cos x)^2 dx$.

8. a) Show that $\dfrac{d}{dx}(\cos 2x + 2x \sin 2x - 2x^2 \cos 2x) = 4x^2 \sin 2x$.

    b) Hence calculate $\displaystyle\int_0^{\frac{\pi}{2}} (x^2 \sin 2x)dx$.

9. a) Show that $\dfrac{2\tan\theta}{1 + \tan^2\theta} = \sin 2\theta$.
    b) Hence calculate $\displaystyle\int_0^{\frac{\pi}{6}} \left(\dfrac{2\tan\theta}{1 + \tan^2\theta} \right)^2 d\theta$.

10. a) Show that $\dfrac{1 - \tan^2 x}{1 + \tan^2 x} = \cos 2x$.

    Hence calculate

    b) $\displaystyle\int_0^{\frac{\pi}{6}} \left(\dfrac{1 - \tan^2 x}{1 + \tan^2 x} \right)^2 dx$ and c) $\displaystyle\int_0^{\frac{\pi}{6}} \left(\dfrac{1 + \tan^2 x}{1 - \tan^2 x} \right)^2 dx$.

5.5 Numerical integration using the trapezium rule

So far in this chapter you have extended the number of types of function that you can integrate. If you continue to study mathematics you will continue to extend your range of integrable functions, but there will always be functions for which you need to calculate a definite integral, but you do not have a standard technique to use.

There are a number of numerical methods that can be used to approximate the area under the curve, using only the values of the function at particular points. In this course you only need one such method: **the Trapezium Rule**.

We will first consider using the Trapezium Rule with a simple function ($y = x^2$) which you can also integrate. In doing so, we will get some idea of how good the approximation is, and what factors affect the accuracy.

> Sometimes you will not be able to integrate a function because you have not yet met the technique, but at times you may be faced with a function for which no-one has yet discovered a way to integrate it – and it can be as simple as ($x \tan x$).

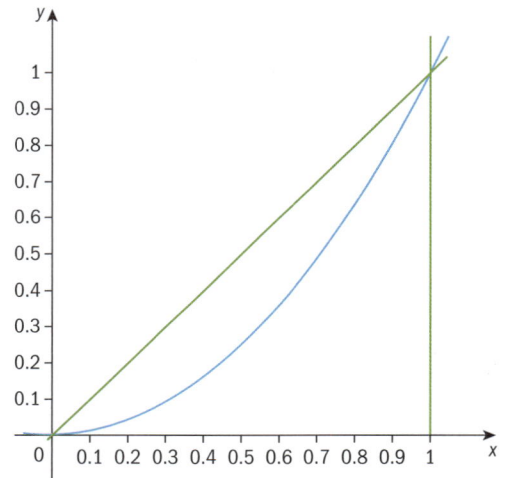

You know the area to the left of the green line under the curve is given by $\int_0^1 x^2 \, dx = \left[\frac{1}{3}x^3 \right]_0^1 = \frac{1}{3} - 0 = \frac{1}{3}$.

We will split the interval [0, 1] up into an increasing number of strips and join the points on the curve by straight lines, creating trapeziums for which we have geometric formulae to calculate areas.

If there was just one strip, the approximation would actually be the area of a triangle because the left hand end starts on the x-axis – at the origin, and it would give an estimate of $\frac{1}{2}$ for the area.

Even with two strips it is clear (visually) that the approximation will be reasonable. The two geometric shapes have areas $\frac{1}{16}$ and $\frac{5}{16}$, which when added together give an estimate of 0.375 for the area under the curve, compared to the true value of 0.333 (a 12.5% overestimate).

> Since the chords joining the points on the curve always lie above the curve itself, it is clear that every estimate with this method will be an overestimate. For most functions this will not be the case.

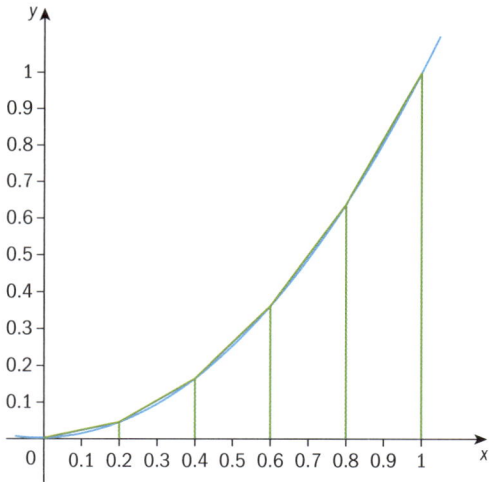

With 5 strips you can see that the chords joining the points on the curve follow the curve very closely so it is no surprise that the area now is 0.34 which is overestimating the true area by only 2%. With 10 strips the overestimate is down to 0.5%.

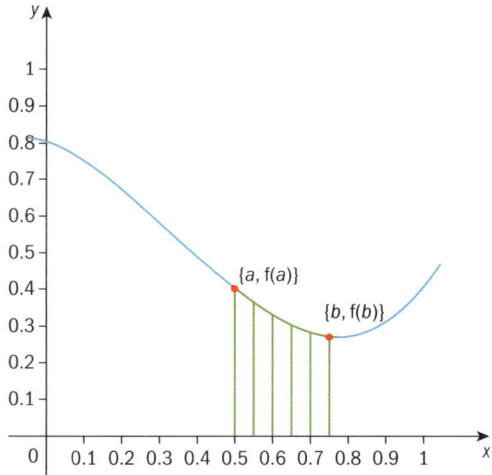

Let us consider the simplest form of the trapezium rule: using a single interval to estimate the integral of a function f(x) over the interval [a, b].

Then $\int_{a}^{b} f(x)\,dx \approx (b-a)\left(\dfrac{f(a)+f(b)}{2}\right)$

(the area of a trapezium is the height multiplied by the average length of the parallel sides).

You have seen how the accuracy can improve if we divide the interval up into a number of strips , however we then have to calculate the area of each strip separately.

A less time-consuming way is the following:

If we choose equal strip widths, that will be a common factor in each area, and then consider the lengths which make the parallel sides of the trapeziums, apart from the first one (at a) and the last one (at b), each occurs as the end of one strip and again as the start of the next one. We can therefore express the estimate obtained by the trapezium rule in a formula rather than actually working out all the areas separately.

Some formal notation is needed to allow us to express this as a general rule. If the interval [a, b] is divided into n intervals, each of width h, then let $y_0 = f(a)$, $y_1 = f(a + h)$, $y_2 = f(a + 2h)$, ... , $y_n = f(a + nh) = f(b)$.

The trapezium rule can then be stated as

$$A = \frac{1}{2}h\{y_0 + y_n + 2(y_1 + y_2 + \ldots\ldots + y_{n-1})\} \approx \int_a^b f(x)\,dx \text{ where } h = \frac{b-a}{n}.$$

Note that the first y value is labeled y_0 because with n strips there are actually $n + 1$ values being used. It is helpful to think about this formula in words relating to way it is derived: the bracket is 'the sum of the end values plus twice the sum of the other values'. The formula usually refers to n intervals, but sometimes it may refer to $n + 1$ ordinates. (the x-coordinate is known as the abscissa and the y-coordinate as the ordinate of a point in the plane).

Example 24

a) Use the Trapezium rule with 5 intervals to estimate $\int_0^{\frac{\pi}{2}}(x \cos x)\,dx$, giving your answer correct to 3 decimal places.

b) In Example 21, you saw that the exact value of this integral is $\frac{\pi}{2} - 1$. Calculate the percentage error in the estimate using the Trapezium rule with 5 intervals.

a) In order to get the answer correct to 3 decimal places we need to use at least 4 decimal places in our working. The example below uses 5 decimal places.

n	0	1	2	3	4	5
x	0	$\frac{\pi}{10}$	$\frac{2\pi}{10} = \frac{\pi}{5}$	$\frac{3\pi}{10}$	$\frac{4\pi}{10} = \frac{2\pi}{5}$	$\frac{5\pi}{10} = \frac{\pi}{2}$
$y = f(x)$	0	0.29878	0.50832	0.55397	0.38832	0

$$A = \frac{1}{2} \times \frac{\pi}{10} \times \{0 + 0 + 2 \times (0.29878 + 0.50832 + 0.55397 + 0.38832)\}$$
$$= 0.54958.... = 0.550 \text{ (3 d.p.)}$$

You may find it helpful to underline the values that are to be added twice.

Show the value obtained in your calculation to more accuracy than required and then give the corrected value.

▶ Continued on the next page

b) $\% \, error = \dfrac{\left(\dfrac{\pi}{2} - 1\right) - 0.550}{\left(\dfrac{\pi}{2} - 1\right)} \times 100\% = 3.6\%$

This is the actual error divided by the true value, expressed as a percentage.

The diagram on the right shows the function $y = x \cos x$ in the required interval, along with the 5 trapeziums whose areas have been used to calculate the estimate. You can see that multiplying $\cos x$ by x means the symmetry of the wave function has been lost.

It is also clear from the sketch that the trapezium rule will give an underestimate of the integral for any number of intervals.

Example 25

a)   Use the Trapezium rule with 5 ordinates to estimate $\displaystyle\int_{1}^{3} \frac{1}{x}\,dx$, giving your answer correct to 3 decimal places.

b)   Show that the exact value of this integral is $\ln 3$.

c)   Calculate the percentage error in the estimate found in part (**a**).

a)

n	0	1	2	3	4
x	1	1.5	2	2.5	3
$y = f(x)$	1	$\frac{2}{3}$	$\frac{1}{2}$	$\frac{2}{5}$	$\frac{1}{3}$

$A = \dfrac{1}{2} \times \dfrac{1}{2} \times \left\{ 1 + \dfrac{1}{3} + 2 \times \left(\dfrac{2}{3} + \dfrac{1}{2} + \dfrac{2}{5} \right) \right\}$

$\quad = \dfrac{67}{60} = 1.11666.... = 1.117 \,(3 \, d.p.)$

In this instance, the values of $f(x)$ can be left as fractions and the exact value of A can be calculated before rounding to the required accuracy.

b) $\displaystyle\int_{1}^{3} \frac{1}{x}\,dx = \left[\ln x \right]_{1}^{3} = \ln 3 - \ln 1 = \ln 3$

c) $\% \, error = \dfrac{\left(\ln 3 - \dfrac{67}{60} \right)}{\ln 3} \times 100\% = -1.6\%$

so the percentage error is 1.6%.

This is the actual error divided by the true value, expressed as a percentage. The negative sign only shows that the estimate is larger than the true value, so the absolute value is given.

You may well be given a sketch of the function being integrated and asked whether you can tell if the trapezium rule gives an overestimate or underestimate. However, you will not normally be asked about the percentage error in the estimate.

Example 26

a) Use the Trapezium rule with 2 intervals to estimate
$\int_0^{\frac{1}{2}\pi} \sqrt{1-\sin x}\ dx$ giving your answer correct to 2 decimal places.

b) The diagram shows the graph of $y = \sqrt{1-\sin x}$. State, with a reason, whether your answer to part **(a)** is an overestimate or an underestimate of the integral.

- -

a)

n	0	1	2
x	0	$\frac{\pi}{4}$	$\frac{\pi}{2}$
y	1	0.5412	0

$A = \frac{1}{2} \times \frac{\pi}{4} \times \{1 + 0 + 2 \times 0.5412\}$
$= 0.8177.... = 0.82\ (2\ d.p.)$

Two intervals only gives one value to be multiplied by 2 in the formula. It may feel different, but the process is exactly the same. Two intervals is used quite often in examinations to cut down on 'busy work'.

b) The graph is curving in such a way that no chord lies above the graph at any point in the interval $\left[0, \frac{\pi}{2}\right]$, so the Trapezium rule will give an underestimate for the integral.

Example 27

Use the Trapezium rule with 2 intervals to estimate $\int_0^1 e^{-x^2}\ dx$ giving your answer correct to 2 decimal places.

- -

n	0	1	2
x	0	0.5	1
y	1	0.7788	0.3679

$A = \frac{1}{2} \times \frac{1}{2} \times \{1 + 0.3679 + 2 \times 0.7788\}$

$= 0.7313.... = 0.73\ (2\ d.p.)$

If you study statistics you may recognise the shape of this curve as that of the Normal distribution.

Exercise 5.5

1. For each of the following, use the Trapezium rule with two intervals to estimate the value of the definite integral.

   a) $\displaystyle\int_0^{\frac{\pi}{4}} (\cos^3 2x)\,dx$

   b) $\displaystyle\int_0^{\frac{\pi}{2}} (\sqrt{1+\sin 3x})\,dx$

   c) $\displaystyle\int_0^{\frac{\pi}{6}} \left(\sqrt{1+\tan^2 2x}\right)dx$

   d) $\displaystyle\int_0^1 \left(\sqrt{1+\sqrt{4x}}\right)dx$

   e) $\displaystyle\int_0^1 \left(\frac{e^{2x}}{1+e^x}\right)dx$

   f) $\displaystyle\int_0^1 \left(\frac{1}{1+9x^2}\right)dx$

2. For each of the following, use the Trapezium rule with the specified number of intervals to estimate the value of the definite integral.

   a) $\displaystyle\int_0^{\frac{\pi}{4}} (\cos^3 2x)\,dx$ [4 intervals]

   b) $\displaystyle\int_0^{\frac{\pi}{2}} \left(\sqrt{1-\sin 2x}\right)dx$ [3 intervals]

   c) $\displaystyle\int_0^{\pi} \left(\sqrt{1+\cos 2x}\right)dx$ [6 intervals]

   d) $\displaystyle\int_0^1 \left(\sqrt{1+\sqrt{4x}}\right)dx$ [5 intervals]

   e) $\displaystyle\int_0^1 \left(\frac{e^{2x}}{1+e^{x^2}}\right)dx$ [4 intervals]

   f) $\displaystyle\int_0^1 \left(\frac{1}{1+9x^3}\right)dx$ [5 intervals]

3. The diagram shows the graph of $y = \sqrt{1+x^2}$. The region R is bounded by the curve, the x- and y-axes and the line $x = 2$.

   a) Use the Trapezium rule with 5 ordinates to obtain an approximation for the area of R, giving your answer to a suitable degree of accuracy.

   b) Explain, with the aid of a sketch, whether the approximation is an overestimate or underestimate.

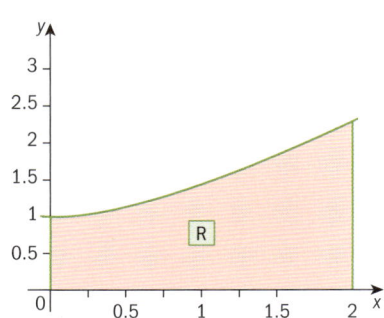

4. The diagram shows the graph of $y = xe^{-x}$.
   The region R is bounded by the curve, the x-axis and the line $x = 2$.

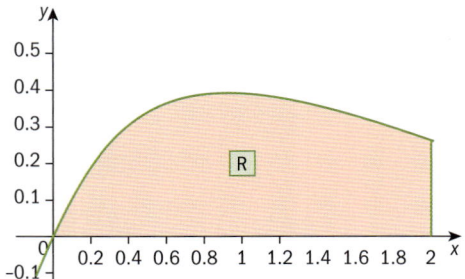

a) Use the Trapezium rule with 11 ordinates to obtain an approximation for the area of R, giving your answer to 3 significant figures.

b) Explain, with the aid of a sketch, whether the approximation is an overestimate or underestimate.

5. The diagram show the graph of $y = \dfrac{1}{1+x}$.
   The region R is bounded by the curve, the x- and y-axes, and the line $x = 2$.

a) Use the trapezium rule with
   i) 2 intervals and
   ii) 10 intervals to obtain approximations for the area of R, giving your answers to 3 significant figures.

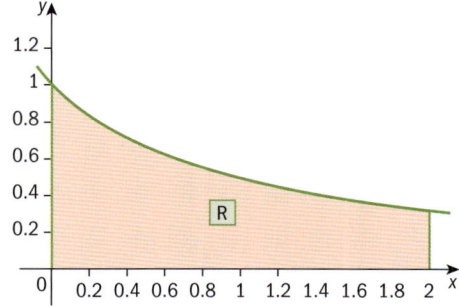

b) Calculate the exact value of $\displaystyle\int_{0}^{2} \dfrac{1}{1+x}\,dx$ and comment on the accuracy of your estimates.

Summary exercise 5

1. Find the following indefinite integrals.

   a) $\int e^{2x-3}\,dx$ **b)** $\int \dfrac{-3}{6x+2}\,dx$

   c) $\int \sin 4x\,dx$ **d)** $\int (\sin^2 6x)\,dx$

   e) $\int \sqrt{e^{2-x}}\,dx$ **f)** $\int \dfrac{12}{3x-4}\,dx$

   g) $\int (\sin x - 2\cos x)^2\,dx$

   h) $\int \dfrac{-6}{3x+5}\,dx$ **i)** $\int -\cos\left(3x+\dfrac{\pi}{3}\right)dx$

2. Calculate the following definite integrals, leaving your answers as exact values.

   a) $\int_0^2 e^{5x}\,dx$ **b)** $\int_0^1 \dfrac{4}{4x-2}\,dx$

   c) $\int_0^{\frac{1}{6}\pi} \sec^2\left(2x-\dfrac{\pi}{6}\right)dx$

   d) $\int_0^{\frac{1}{4}\pi} (1-3\sin 2x)^2\,dx$

   e) $\int_0^{\frac{1}{4}\pi} \sin\left(\dfrac{\pi}{4}-\dfrac{1}{2}x\right)dx$

   f) $\int_{\ln 2}^{\ln 3} e^{2x}\,dx$

   g) $\int_0^{\frac{1}{6}\pi} (\sin^2 x - \cos^2 x)\,dx$

   h) $\int_{-1}^2 e^{-2x}\,dx$ **i)** $\int_{-1}^1 \left(\dfrac{5}{3x-5}\right)dx$

EXAM-STYLE QUESTIONS

3. Find the area bounded by the graph of $y = e^{-2x+1}$, the x- and y-axes, and the line $x = 2$.

4. Find the area bounded by the graph of $y = \dfrac{1}{2+x}$, the x- and y-axes, and the line $x = 1$.

5. A curve passes through the point $\left(\dfrac{\pi}{4},0\right)$ and its gradient function is $\sin^2 x$. Find the equation of the curve.

6. Find the area bounded by the graphs of $y = \dfrac{1}{x-5}$, $y = \dfrac{1}{2x+1}$, the y-axis, and the line $x = 3$.

7. Find the area bounded by the graphs of $y = \dfrac{1}{2x-7}$, $y = 2 + e^{-x}$, the y-axis, and the line $x = 2$.

8. Calculate the following improper integrals, leaving your answers in terms of e where appropriate.

   a) $\int_0^{\infty} e^{-2x}\,dx$

   b) $\int_1^{\infty} e^{-0.5x}\,dx$

   c) $\int_{-\infty}^{-1} \sqrt{e^x}\,dx$

EXAM-STYLE QUESTIONS

9. The line $y = 4e^{-2x}$ between $x = 1$ and $x = 2$ is rotated through 360° about the x-axis. Find the volume of the solid obtained.

10. **a)** By writing $x = 4x - 3x$, $7x = 4x + 3x$, show that $\sin 3x \sin 4x = \dfrac{1}{2}(\cos x - \cos 7x)$.

    b) Hence calculate $\int_0^{\frac{1}{3}\pi} \sin 3x \sin 4x\,dx$.

11. **a)** Show that $(\sin 2x - 3\cos 2x)^2 = 5 - 3\sin 4x + 4\cos 4x$.

    b) Hence calculate $\int_0^{\frac{1}{4}\pi} (\sin 2x - 3\cos 2x)^2\,dx$.

12. **a)** Show that $\dfrac{1+\tan^2 \theta}{1-\tan^2 \theta} = \sec 2\theta$.

    b) Hence calculate $\int_0^{\frac{\pi}{6}} \left(\dfrac{1+\tan^2 \theta}{1-\tan^2 \theta}\right)^2 d\theta$.

13. The diagram shows the graph of $y = \sqrt{1 - \sin x}$. The region R is bounded by the curve, the x- and y-axes and the line $x = \frac{3\pi}{2}$. (R is in two sections).

a) Use the Trapezium rule with 7 ordinates to obtain an approximation for the total area of R, giving your answer correct to 2 decimal places.

b) Explain, with the aid of a sketch, whether the approximation is an overestimate or underestimate.

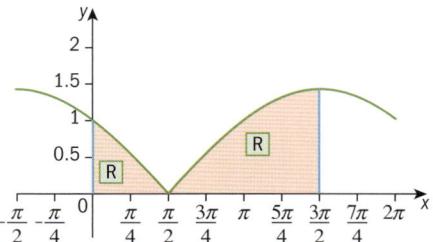

14. The diagram show the graph of $y = x\ln x$. The region R is bounded by the curve, the x-axis, and the line $x = 2$.

a) Explain why $\displaystyle\int_0^2 x\ln x\ dx$ (if you were able to calculate it) would not give the area of R.

b) Calculate an estimate for the area of R using the Trapezium rule and intervals of width 0.5.

15.

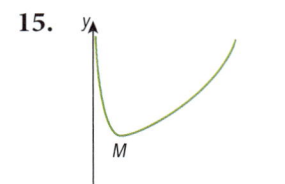

The diagram shows the part of the curve $y = \dfrac{e^{2x}}{x}$ for $x > 0$, and its minimum point M.

a) Find the coordinates of M. [5]

b) Use the trapezium rule with 2 intervals to estimate the value of

$$\int_1^2 \frac{e^{2x}}{x}\ dx$$

giving your answer correct to 1 decimal place. [3]

c) State, with a reason, whether the trapezium rule gives an under-estimate or an over-estimate of the true value of the integral in part (**b**). [1]

Cambridge International AS & A Level Mathematics 9709 Paper 2 Q6 October/November 2006

16. a) Given that $y = \tan 2x$, find $\dfrac{dy}{dx}$. [2]

b) Hence, or otherwise, show that

$$\int_0^{\frac{1}{6}\pi} \sec^2 2x\ dx = \frac{1}{2}\sqrt{3},$$

and, by using an appropriate trigonometrical identity, find the exact value of $\displaystyle\int_0^{\frac{1}{6}\pi} \tan^2 2x\ dx$. [6]

c) Use the identity $\cos 4x = 2\cos^2 2x - 1$ to find the exact value of

$$\int_0^{\frac{1}{6}\pi} \frac{1}{1 + \cos 4x}\ dx. \qquad [2]$$

Cambridge International AS & A Level Mathematics 9709 Paper 2 Q7 October/November 2006

Chapter summary

Integrating exponentials

- $\int (e^{ax+b})\,dx = \dfrac{1}{a}e^{ax+b} + c$

Integrating functions of the form $\dfrac{1}{ax+b}$

- $\int \left(\dfrac{1}{x}\right)dx = \ln|x| + c$

- $\int \left(\dfrac{1}{ax+b}\right)dx = \dfrac{1}{a}\ln|ax+b| + c$

Integrating standard trigonometric functions

- $\int \sin(ax+b)\,dx = -\dfrac{1}{a}\cos(ax+b) + c$

- $\int \cos(ax+b)\,dx = \dfrac{1}{a}\sin(ax+b) + c$

- $\int \sec^2(ax+b)\,dx = \dfrac{1}{a}\tan(ax+b) + c$

Integrating more trigonometric functions

- Standard trigonometrical relationships, such as identities and double angle formulae, can be used to rearrange functions so that they are integrable.

The trapezium rule

The trapezium rule is a method to calculate a numerical approximation to a definite integral:

$A = \dfrac{1}{2}h\{y_0 + y_n + 2(y_1 + y_2 + \ldots\ldots + y_{n-1})\} \approx \int_a^b f(x)dx$ where $h = \dfrac{b-a}{n}$

Computational mathematics is one of the fastest growing employment areas for mathematicians because of the ever-increasing ability of computers to work quickly through large quantities of calculations, and the increasingly sophisticated algorithms which are used. For example, numerical weather prediction uses mathematical models of the atmosphere and oceans to predict the weather based on current climate conditions. This chapter is an introduction to some of the basic techniques which are the foundation of that broad branch of mathematics.

Objectives

- Locate approximately a root of an equation, by means of graphical considerations and/or searching for a sign change.
- Understand the idea of, and use the notation for, a sequence of approximations which converges to a root of an equation.
- Understand how a given simple iterative formula of the form $x_{n+1} = F(x_n)$ relates to the equation being solved, and use a given iteration, or an iteration based on a given rearrangement of an equation, to determine a root to a prescribed degree of accuracy.

Before you start

You should know how to:

1. Substitute values into functions, and evaluate a function at a certain point.

   e.g. If $f(x) = \dfrac{\sqrt{(3x^2 + 4)}}{2}$,

   then $f(1) = \dfrac{\sqrt{(3 + 4)}}{2} = \dfrac{\sqrt{7}}{2}$,

   and $f(2) = \dfrac{\sqrt{(12 + 4)}}{2} = 2$.

Skills check:

Give your answers to **1** and **2** in exact form.

1. If $f(x) = \dfrac{\sqrt{(x^3 + 12)}}{2}$, find

   a) f(1)   b) f(2)   c) f(−2).

2. Substitute values into exponential, logarithmic and trigonometric functions, and evaluate these functions.

e.g. a) If $f(x) = e^{3x} + 2\ln(5x)$,

then $f(1) = e^3 + 2\ln(5)$,

and $f(2) = e^6 + 2\ln(10)$,

Also, $f(-1) = e^{-3} + 2\ln(-5)$ which is undefined because $\ln(-5)$ is undefined.

2. If $f(x) = e^{-x} + \ln(5x + 2)$, find

   a) $f(1)$
   b) $f(2)$
   c) $f(-2)$.

3. Be able to rearrange equations.

e.g. If $y = 3\tan^2 x + 1$, express x in terms of y.

$$y = 3\tan^2 x + 1 \Rightarrow \tan^2 x = \left(\frac{y-1}{3}\right)$$

$$\Rightarrow x = \tan^{-1}\left(\sqrt{\frac{y-1}{3}}\right)$$

3. If $y = e^{2x+1} - 3$, express x in terms of y.

6.1 Finding approximate roots by change of sign or graphical methods

In mathematics, the final step to solving a problem is often to solve an equation. As you learn more mathematics, you extend the range of equations to which you can produce an exact solution using algebra. You have previously learned to solve:

- Linear equations e.g. $3x + 7 = 5 \Rightarrow x = -\frac{2}{3}$
- Quadratic equations of different types e.g.

  ○ $x^2 - 3x + 2 = 0 \Rightarrow (x-1)(x-2) = 0 \Rightarrow x = 1$ or 2

  ○ $3x^2 + 2x - 7 = 0 \Rightarrow x = \frac{-2\pm\sqrt{4+84}}{6} = \frac{-1\pm\sqrt{22}}{3}$

- Trigonometric equations e.g. $2\sin\theta = 1 \Rightarrow \sin\theta = \frac{1}{2} \Rightarrow \theta = 30°, 150°$ $\left(\text{or } \frac{\pi}{6}, \frac{5\pi}{6} \text{ radians}\right)$

In the earlier chapters of this book, you have learnt how to solve equations involving modulus, exponential, and logarithmic functions as well as harder trigonometric equations. If you plan to study P3, you will learn how to solve equations with complex numbers.

However, there are many types of equation for which no-one has found an algebraic method of solving exactly, and many more for which there are methods of solving algebraically, but you have not yet studied them. Being able to find approximate values of solutions to these equations is a valuable skill.

You have probably used the simplest of these methods already (perhaps calling it 'trial and improvement'). We want to treat this method more formally now, and put a condition in.

The Sign-change rule

If $f(x)$ is continuous in an interval $\alpha \le x \le \beta$ and if $f(\alpha)$ and $f(\beta)$ have different signs then $f(x) = 0$ has at least one root between α and β.

Note: The condition that $f(x)$ is continuous is important because it means that the function cannot jump over the x-axis without creating a root. The wording 'at least one' is also important because there can be more than one root. Some illustrations are shown below.

$f(1)$ is negative and $f(5)$ is positive so there must be a root between 1 and 5 – in fact there are two – a (double) root where the graph touches the avis at 2 and another where it cuts at 4 – which creates the change of sign.

f(1) is negative and f(3) is positive so there must be a root between 1 and 3.

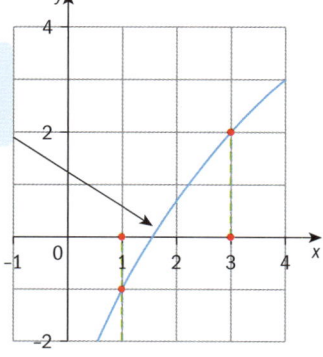

f(1) is negative and f(5) is positive so there must be a root between 1 and 5 – in fact there are three – so the sign actually changes three times in the interval

Note that in a case like this third illustration, the choice of the intermediate values will determine which root we obtain: if we consider f(2) then we will find the root between 1 and 2, whereas if we considered f(3) or f(4) we would find the root between 4 and 5.

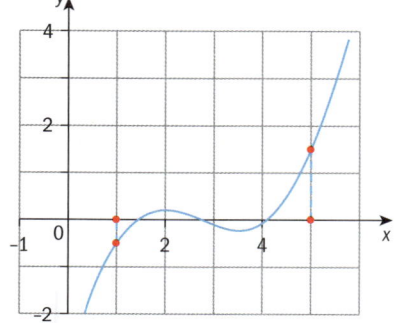

If we have a pair of values giving an interval satisfying the sign-change rule, we can then conduct a systematic process of narrowing the interval until we identify a root correct to any degree of accuracy we want.

The simplest of these processes to narrow the interval is called a **decimal search**, where first we identify a pair of integers between which the root lies, followed by a pair of neighbouring values with in 1 decimal place of each other between which the root lies, followed by a pair of neighbouring values within 2 decimal places of each other between which the root lies, etc. If we keep going with this process we will arrive at the value of the root to the required accuracy.

Example 1

Show that if $f(x) = x^3 - 4x + 5$, then $f(x) = 0$ has a root between $x = -3$ and $x = -2$. Find the root correct to 1 decimal place.

$f(-3) = -27 + 12 + 5 = -10$

$f(-2) = -8 + 8 + 5 = 5$

By the sign-change rule, f has a root between $x = -3$ and $x = -2$. ← State this explicitly each time.

The next step is to trial a value between -3 and -2. If we take -2.5 then we will know whether the root lies in the interval $(-3, -2.5)$ or else in $(-2.5, -2)$.

$f(-2.5) = -0.625 \Rightarrow$ root lies in $(-2.5, -2)$

$f(-2.5) = -0.625$, which is a very small negative number, and tells us that the root is likely to be close to -2.5. The next trial should be -2.4.

$f(-2.4) = 0.776 \Rightarrow$ root lies in $(-2.5, -2.4)$

The sign of the function at the mid-interval point (i.e. at -2.45) will tell us whether -2.5 or -2.4 is closer to the root.

$f(-2.45) = 0.0938...$ So the root is **-2.5 correct to 1 decimal place.**

Note: Just looking at the values of the function at -2.5 and -2.4 is not enough. We need to test the sign of the function at -2.45.

Example 2

Show that if $f(x) = x\cos x - \sin x + 1$ then $f(x) = 0$ has a root between $x = 4$ and $x = 5$, where x is in radians, and find the root correct to 2 decimal places.

$f(4) = -0.85777...$

$f(5) = 3.37723...$ The sign-change rule implies $f(x)$ has a root between $x = 4$ and $x = 5$.

> The next step is to trial a value between 4 and 5. Here the function values suggest the root is closer to 4 than to 5, so 4.2 might be sensible for the next trial value.

$f(4.2) = -0.18751... \Rightarrow$ root lies in $(4.2, 5)$

Try 4.3:

$f(4.3) = 0.19272... \Rightarrow$ root lies in $(4.2, 4.3)$

> The sign-change rule implies f has a root between $x = 4.2$ and $x = 4.3$.

Try 4.25:

$f(4.25) = -0.000882... \Rightarrow$ root lies in $(4.25, 4.3)$

> $f(4.25)$ is extremely small, which suggests 4.25 is close to the root. The next trial should be 4.26.

Try 4.26:

$f(4.26) = 0.03729... \Rightarrow$ root lines in $(4.25, 4.26)$

$f(4.255) = 0.018170...$ so the root is **4.25 correct to 2 decimal places**.

> The actual root is 4.250390... Using a spreadsheet or programmable calculator allows roots of equations like this to be obtained to any degree of accuracy available in the program. There are other roots which are shown in the graph on the left.

Example 3

a) Verify by calculation that $\sqrt{(x^3 + 3)} = 2 - x$ has a root between $x = 0$ and $x = 1$.

b) Find the root correct to 1 decimal place.

a) We first need to write the equation in the form $f(x) = 0$ in order to be able to use the sign-change rule, so let $f(x) = x - 2 + \sqrt{(x^3 + 3)}$

x	$f(x)$
0	−0.26795...
1	1

The sign-change rule implies there is a root between 0 and 1.

> Remember to state this.

b)

x	$f(x)$	
0.3	0.039828...	*The root is between 0 and 0.3*
0.2	−0.06564...	*The root is between 0.2 and 0.3*
0.25	−0.01344...	*The root is between 0.25 and 0.3*

> A table of values like this, showing the search process with the values is often the simplest way of keeping track of where we know the root to be located.

The root is 0.3 correct to 1 decimal place.

Example 4

a) Sketch the graphs of $y = x^2 + 2$ and $y = \dfrac{1}{x}$ on the same set of axes.

b) Use your graphs to explain why there is only one root to the equation $x^3 + 2x - 1 = 0$.

a)

b) The intersection of the graphs $y = x^2 + 2$ and $y = \dfrac{1}{x}$ is the point at which $x^2 + 2 = \dfrac{1}{x}$.

Rearranging gives $x^3 + 2x - 1 = 0$.

Since there is only one point of intersection of the two graphs, there is only one root of $x^3 + 2x - 1 = 0$.

Exercise 6.1

1. Show that if $f(x) = x^3 - 5x^2 + 3x + 2$, then $f(x) = 0$ has a root between $x = 1$ and $x = 2$ and find the root correct to 1 decimal place.

2. Show that if $f(x) = 2^x - 3x - 3$ then, $f(x) = 0$ has a root between $x = 3$ and $x = 4$ and find the root correct to 1 decimal place.

3. Show that if $f(x) = \dfrac{x}{3} - \cos^{-1} x$, where x is in radians, then $f(x) = 0$ has a root between $x = 0.5$ and $x = 1.0$ and find the root correct to 1 decimal place.

4. Show that if $f(x) = \dfrac{\pi}{3} + \dfrac{x}{5} - \tan x$, where x is in radians, then $f(x) = 0$ has a root between $x = 0$ and $x = 1$ and find the root correct to 1 decimal place.

5. **a)** Show that if $f(x) = x^3 - 3x + 1$, then $f(x) = 0$ has a root between $x = 0$ and $x = 1$ and find the root correct to 1 decimal place.

   b) Show that there are two other roots of $f(x) = 0$ which lie in the intervals $(-2, -1)$ and $(1, 2)$, and find those roots correct to 1 decimal place.

6. Sketch the graphs of $y = e^{2x}$ and $y = 7 - 3x$ on the same axes for values of x between -2 and 3. Show that $e^{2x} = 7 - 3x$ has a root between $x = 0$ and $x = 1$ and find the root correct to 2 decimal places.

7. Show that if $f(x) = 5 - xe^{2x}$ then $f(x) = 0$ has a root between $x = 0$ and $x = 1$ and find the root correct to 2 decimal places.

8. Show that if $f(x) = \ln(x^2 + 3) - 2x + 5$ then $f(x) = 0$ has a root between $x = 3$ and $x = 4$ and find the root correct to 2 decimal places.

6.2 Finding roots using iterative relationships

Consider the sequence with nth term given by $x_n = 1-(0.1)^{n-1}$. The sequence is $0,\ 0.9,\ 0.99,\ 0.999,\ 0.9999,\ 0.99999, \ldots$ Every term is getting bigger in the sequence, but we are always subtracting a positive number from 1 so we know there is an upper limit of 1 on the terms in the sequence.

As n increases we can see that the sequence will get as close to 1 as we require it (and will stay close to it), and we say that the sequence **converges to** 1.

> We can put a formal condition on convergence by saying that L is the limit of a sequence $\{x_n\}$, if for any positive value ε, we can find an integer N so that $|x_n - L| < \varepsilon$ for every $n > N$.

> The iterative relation $x_{n+1} = F(x_n)$, together with an initial value x_1, defines a sequence. If that sequence converges to a limit, then the limit will be a solution to the equation $x = F(x)$.

Example 5

a) Show that the equation $\operatorname{cosec} x = 4 - \dfrac{1}{2}x^2$ can be rearranged to $x = \sin^{-1}\left(\dfrac{2}{8-x^2}\right)$.

b) Use the iterative relationship $x_{n+1} = \sin^{-1}\left(\dfrac{2}{8-x_n^2}\right)$, with $x_1 = 0$, to find four further approximations x_2, x_3, x_4, x_5 writing down 6 decimal places from your calculator.

c) State the root of the equation correct to 3 s.f.

a) $\operatorname{cosec} x = 4 - \dfrac{1}{2}x^2$

$\Rightarrow \dfrac{1}{\sin x} = \dfrac{8-x^2}{2}$

$\Rightarrow x = \sin^{-1}\left(\dfrac{2}{8-x^2}\right)$

b) $x_1 = 0$

$x_2 = 0.252680....$

$x_3 = 0.254758.....$

$x_4 = 0.254792......$

$x_5 = 0.254793.......$

c) The root is 0.255 (to 3 s.f.). ⟵ Because of how fast the sequence appears to be converging, the root is probably 0.25479, to 5 s.f., but you will not be asked to make judgements like this.

 A standard scientific calculator allows you to produce each successive iterated value by a single key press once the function has been set up:

- Enter the initial value and press '='.
- Enter the expression $F(x_n)$ using 'Ans' for each occurrence of x_n.
- Press '=' to generate x_2, and again to generate x_3, and so on.

You need to record a number of digits (often specified in the question) to record your working but the calculator is using all the available accuracy when generating the next approximation.

If you have an equation $f(x) = 0$, you can re-arrange into the form $x = F(x)$ for some function $F(x)$. From here, you can define an iterative equation $x_{x+1} = f(x_n)$ which will often converge to the root of $f(x) = 0$.

Example 6

The graph of $y = x^3 - 4x - 1$ is shown.

a) Show that the equation $x^3 - 4x - 1 = 0$ can be rearranged to

$x = \frac{1}{4}(x^3 - 1)$.

b) Use the iterative relationship $x_{n+1} = \frac{1}{4}(x_n^3 - 1)$, with $x_1 = 0$, to find four further approximations x_2, x_3, x_4, x_5 writing down 6 decimal places (where appropriate) from your calculator.

c) State a root correct to 3 s.f.

d) Use the same iterative relationship with $x_1 = -2$ to find values for x_2, x_3, x_4, x_5.

e) Comment on the behaviour of your sequence.

a) $x^3 - 4x - 1 = 0 \Rightarrow 4x = x^3 - 1 \Rightarrow x = \frac{1}{4}(x^3 - 1)$

b) $x_1 = 0$

$x_2 = -0.25$

$x_3 = -0.2539060.....$

$x_4 = -0.254092......$

$x_5 = -0.254101......$

c) The root is -0.254, correct to 3 s.f.

d) $x_1 = -2$

$x_2 = -2.25$

$x_3 = -3.097656.....$

$x_4 = -7.680870......$

$x_5 = -113.5347......$

e) The sequence does not appear to be converging – we see from the graph that there is a root just above -2, but the sequence is taking us further and further away from that root.

Note: You need to know that not all iterative relationships derived by rearranging an equation into the form $x = f(x)$ will converge. In Example 6, the three roots are $-1.861, -0.2541$, and 2.115. Any initial value taken between -1.861 and 2.115 and used in the iterative relationship we derived will converge to the middle root (-0.2541). Any initial value below -1.861 or above 2.115 will diverge.

We can use an iterative method to find a root to a specified degree of accuracy. The number of iterations required for this degree of accuracy will differ from equation to equation, and the value of the first iteration we choose.

Example 7

a) Show that the equation $xe^x = 1$ has a root between 0 and 1.

b) Show that the equation $xe^x = 1$ can be rearranged into the forms i) $x = e^{-x}$ and ii) $x = -\ln x$

c) Use the iterative relationships i) $x_{n+1} = e^{-x_n}$ and ii) $x_{n+1} = -\ln x_n$ with $x_1 = 0.6$ in both cases, to find eight further approximations $x_2, x_3, x_4, \ldots, x_9$, writing down 6 decimal places from your calculator.

a) Writing $xe^x = 1$ as $xe^x - 1 = 0$, let $f(x) = xe^x - 1$.
   Then $f(0) = -1$, $f(1) = e - 1 = 1.718....$, so by the sign-change rule there is a root between 0 and 1.

b) i) $xe^x = 1 \Rightarrow x = \dfrac{1}{e^x} = e^{-x}$ ii) $xe^x = 1 \Rightarrow \ln x + x = 0 \Rightarrow x = -\ln x$

c) i) $x_{n+1} = e^{-x_n}$ ii) $x_{n+1} = -\ln x_n$

   $x_1 = 0.6$ $x_1 = 0.6$

   $x_2 = 0.548811.....$ $x_2 = 0.510825$

   $x_3 = 0.577635....$ $x_3 = 0.671726....$

   $x_4 = 0.561223......$ $x_4 = 0.397903......$

   $x_5 = 0.570510......$ $x_5 = 0.921546......$

   $x_6 = 0.565236....$ $x_6 = 0.081702....$

   $x_7 = 0.568225.....$ $x_7 = 2.504673.....$

   $x_8 = 0.566529......$ $x_8 = -0.918158......$

   $x_9 = 0.567491......$ x_9 gives an error message
                                                  since we cannot take logs of a negative number.

1 In Example 5, the iteration $x_{n+1} = \sin^{-1}\left(\dfrac{2}{8 - x_n^{2}}\right)$ converged very quickly, allowing the root to be confirmed to 3 decimal places after only 4 iterations despite the first iteration ($x_1 = 0$) being some distance away from the root. In Example 7, the iterative relationship $x_{n+1} = e^{-x_n}$ was started much closer to the root than in Example 5, but after 8 iterations we are still not entirely sure of the root to 3 d.p.

2 Consecutive iterations for $x_{n+1} = \sin^{-1}\left(\dfrac{2}{8 - x_n^{2}}\right)$ are alternately above and below the root (which is 0.5671432....). As such, it is easier to conclude what the root will be than where the sequence produced by the iterative relationship approaches the root from one side only (as in Examples 5 and 6). We will look in the next section at how you can tell whether an iterative relationship will alternate or approach from one side.

Example 7 also illustrates that the iterative relationship (b) $x_{n+1} = -\ln x_n$ initially generated a diverging sequence, but after 9 iterations it required an impossible calculation and thus stopped.

In all the examples so far, we have been given the equation to be solved numerically. Questions can be set where you are asked to show that an equation has to be satisfied in a given context.

Example 8

In the diagram on the right, O is the centre of a circle, and AB is a chord with angle AOB = θ radians.

The smaller segment (shaded) created by the chord is one half of the area of the larger segment.

a) Show that θ satisfies the equation $3\theta - 3\sin\theta = 2\pi$.

b) Show that this equation can be rearranged into the

form $\theta = \frac{2}{3}\pi + \sin\theta$.

c) Use the iterative relation $\theta_{n+1} = \frac{2}{3}\pi + \sin\theta_n$, with initial

value $\theta_1 = 2.6$, to determine the angle θ correct to 2 decimal places.

....

a) The minor sector created by radii OA and OB has area $\frac{1}{2}\theta r^2$ and the triangle OAB has

area $\frac{1}{2}r^2\sin\theta$. If the area of the minor segment is one half of the area of the larger segment

then it is one third of the area of the full circle i.e. $\frac{1}{3}\pi r^2$.

$$\frac{1}{2}\theta r^2 - \frac{1}{2}r^2\sin\theta = \frac{1}{3}\pi r^2$$

$$\left(\times \frac{6}{p2}\right) \Rightarrow 3\theta - 3\sin\theta = 2\pi$$

b) $3\theta - 3\sin\theta = 2\pi \Rightarrow \theta = \frac{2}{3}\pi + \sin\theta$

c) $x_1 = 2.6$ $x_2 = 2.60989....$

$x_3 = 2.60139......$ $x_4 = 2.60870.....$

$x_5 = 2.60241.....$ $x_6 = 2.60782....$

$x_7 = 2.60317....$ $x_8 = 2.60717....$

$x_9 = 2.60373.....$ $x_{10} = 2.60668...$

$x_{11} = 2.60415.....$ $x_{12} = 2.60333...$

$x_{13} = 2.60445.....$ $x_{14} = 2.60607...$

$x_{15} = 2.60468.....$ $x_{16} = 2.60587...$

$x_{17} = 2.60485.....$ $x_{18} = 2.60573...$

$x_{19} = 2.60497.....$ $x_{20} = 2.60562...$

$x_{21} = 2.60506.....$

> Exam questions will not require this number of iterations – the convergence is slow and the root lies close to the border between 2.60 and 2.61 radians.
>
> Because the iterative sequence is alternating as it moves towards the root, as soon as the lower sequence exceeded 2.605, it is safe to conclude that the root will be 2.61 correct to 2 d.p.

The angle is 2.61 radians correct to 2 decimal places.

Example 9

The relationship for the golden ratio, x, can be described by the equation $\frac{1}{x} = \frac{x-1}{1}$.

a) Show that this equation can be rearranged into the form $x = \sqrt[3]{x(x+1)}$.

b) Use the iterative relationship $x_{n+1} = \sqrt[3]{x_n(x_n+1)}$, with initial value $x_1 = 1.6$ to find four further approximations to the golden ratio, showing at least 4 decimal places for each approximation.

c) Give an approximation to the golden ratio correct to 2 decimal places.

$\cdots\cdots\cdots\cdots\cdots\cdots\cdots\cdots\cdots\cdots\cdots\cdots\cdots\cdots\cdots\cdots\cdots\cdots\cdots\cdots$

a) $\frac{1}{x} = \frac{x-1}{1} \times (x^2) \Rightarrow x = x^3 - x^2 \Rightarrow x^3 = x^2 + x$

$\Rightarrow x^3 = x(x+1) \Rightarrow x = \sqrt[3]{x(x+1)}$

b) $x_1 = 1.6$
$x_2 = 1.6082....$
$x_3 = 1.6127....$
$x_4 = 1.6151....$
$x_5 = 1.6165....$

c) The golden ratio is 1.62 correct to 2 decimal places.

Exercise 6.2

1. Find the equation that the following iterative relationships provide roots for, assuming that the iterative sequences converge (you do not need to check whether or not the iterative sequences converge).

a) $x_{n+1} = \dfrac{1}{3 + x_n}$

b) $x_{n+1} = \dfrac{\sqrt{x_n^3 + 2}}{3 + x_n}$

c) $x_{n+1} = e^{-2x_n}$

d) $x_{n+1} = 3 - \ln 2x_n$

e) $x_{n+1} = \cos x_n + \dfrac{\pi}{4}$

f) $x_{n+1} = \sqrt[10]{x_n^6 + 3}$

2. Find at least three possible rearrangements for each of the following equations into a form $x = F(x)$ in each case.

a) $x^4 - 3x + 2 = 0$

b) $x^7 - 4x^3 + 1 = 0$

c) $x^5 - x^2 - 1 = 0$

d) $x^2 - 3x - e^{-x} = 0$

e) $e^{x^2} - 3x = 0$

f) $\cot x + 3 - x^2 = 0$

3. For each iterative formula and given initial value, find values of x_2, x_3, x_4, and x_5 (giving the first 5 decimal places of each approximation).

a) $x_{n+1} = \dfrac{1}{3 + x_n};\quad x_1 = 0.5$

b) $x_{n+1} = \dfrac{\sqrt{x_n^3 + 2}}{3 + x_n};\quad x_1 = 0.5$

c) $x_{n+1} = e^{-2x_n};\quad x_1 = 0.5$

d) $x_{n+1} = 3 - \ln 2x_n;\quad x_1 = 1.75$

e) $x_{n+1} = \cos x_n + \dfrac{\pi}{4};\quad x_1 = 1.1$

f) $x_{n+1} = \sqrt[10]{x_n^6 + 3};\quad x_1 = 1.2$

4. a) Show that the equation $3x^2 - x - 5 = 0$ has a root between $x = 1$ and $x = 2$.

b) Show that the equation $3x^2 - x - 5 = 0$ can be rearranged to $x = \sqrt{\dfrac{x+5}{3}}$.

c) Using an initial value of $x_1 = 1.5$, and the iterative relationship $x_{n+1} = \sqrt{\dfrac{x_n + 5}{3}}$, find the root of $3x^2 - x - 5 = 0$ that lies between 1 and 2 correct to 2 decimal places.

5. a) Show that the equation $e^x - x^2 - 5 = 0$ has a root between $x = 2$ and $x = 3$.

b) Show that the equation $e^x - x^2 - 5 = 0$ can be rearranged to $x = \ln(x^2 + 5)$.

c) Using an initial value of $x_1 = 2.5$, and the iterative relationship $x_{n+1} = \ln(x_n^2 + 5)$, find values of x_2, x_3, x_4, and x_5 (giving the first 5 decimal places of each approximation).

6. a) Show that the equation $\theta = \dfrac{3\pi}{5} + \cos\theta$ has a root between $\theta = 1.7$ and $x = 1.8$.

b) Using an initial value of $\theta_1 = 1.75$, and the iterative relationship $\theta_{n+1} = \dfrac{3\pi}{5} + \cos\theta_n$, find values of θ_2, θ_3, θ_4, and θ_5 (giving the first 5 decimal places of each approximation).

c) Explain why you know that this iterative relationship will (eventually) converge.

7. The shaded segment of the circle has an area of $7\,\text{cm}^2$.

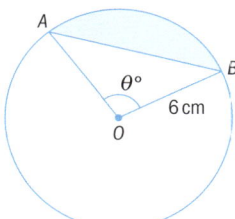

a) Show that angle AOB $= \theta$ (where θ is in radians) satisfies the equation $18\theta - 18\sin\theta = 7$.

b) Use the iterative relationship $\theta_{n+1} = \sin\theta_n + \dfrac{7}{18}$ with initial value $\theta_1 = 1.5$ to find the angle θ in radians correct to 2 decimal places.

8. a) Show that both $x_{n+1} = \cos x_n$ and $x_{n+1} = \cos^{-1} x_n$ are iterative relationships that provide a root to the equation $x - \cos x = 0$ if the sequence converges.

b) Using $x_1 = 0.75$ as an initial value, show that one sequence diverges and the other converges, and find the root correct to 2 decimal places.

9. a) Show that the equation $x^4 - x^2 - 5 = 0$ can be rearranged to each of the following forms.

i) $\quad x = \sqrt{\dfrac{5}{x^2 - 1}}$ **ii)** $\quad x = \dfrac{x^2 + 5}{x^3}$ **iii)** $\quad x = \sqrt[4]{x^2 + 5}$.

b) Using an initial value of $x_1 = 1.7$, show that the iterative relationships based on forms **(i)** and **(ii)** do not converge, and use $x_{n+1} = \sqrt[4]{x_n^2 + 5}$ to find the root correct to 2 decimal places.

10. a) The equation $x^3 + x + 1 = 0$ has one real root. Show by calculation that this root lies between -1 and 0. [2]

b) Show that, if a sequence of values given by the iterative formula

$$x_{n+1} = \frac{2x_n^3 - 1}{3x_n^2 + 1}$$

converges, then it converges to the root of the equation given in part (**a**) [2]

c) Use this iterative formula, with initial value $x_1 = -0.5$, to determine the root correct to 2 decimal places, showing the result of each iteration. [3]

Cambridge International AS and A Level Mathematics 9709, Paper 3, May/June 2004, Question 7

11. a) By sketching a suitable pair of graphs, show that the equation

$$2\cot x = 1 + e^x,$$

where x is in radians, has only one root in the interval $0 < x < \frac{1}{2}\pi$. [2]

b) Verify by calculation that this root lies between 0.5 and 1.0. [2]

c) Show that this root also satisfies the equation

$$x = \tan^{-1}\left(\frac{2}{1 + e^x}\right).$$ [1]

d) Use the iterative formula

$$x_{n+1} = \tan^{-1}\left(\frac{2}{1 + e^{x_n}}\right),$$

with initial value $x_1 = 0.7$, to determine this root correct to 2 decimal places. Give the result of each iteration to 4 decimal places. [3]

Cambridge International AS and A level Mathematics 9709, Paper 3, May/June 2006, Question 6

6.3 Convergence behaviour of iterative functions

Although you are not required to know conditions for convergence of an iterative relationship, there are some simple principles about conditions for convergence which may help you to feel confident that you know what is going on as a sequence progresses; giving you a feeling for how quickly an iterative process may converge, and whether it will approach the root from one side, or the terms will be alternately above and below the root.

> The Cambridge 9709 syllabus does not require you to know conditions for convergence of an iterative relationship, though you do need to know that an iteration may fail to converge.

We know that the equation $x^2 - 3x - 5 = 0$ can be solved using the quadratic

formula, with solutions $x = \frac{3 \pm \sqrt{9 + 20}}{2} = -1.19258\ldots$ or $4.19258\ldots$

The equation $x^2 - 3x - 5 = 0$ can also be rearranged into various forms with $x = F(x)$, that are not normally helpful because to work out x on the left-hand side we need a numerical value for x on the right-hand side. However, rearranging equations into the form $x = F(x)$, and creating an iterative relationship $x_{n+1} = F(x_n)$ will often produce a sequence of approximations which converge to a root of $x = F(x)$.

Example 10

Verify that the equation can be rearranged to $x^2 - 3x - 5 = 0$ in each case.

a) $\ x = \sqrt{(3x+5)}$

b) $\ x = \dfrac{3x+5}{x}$

c) $\ x = \dfrac{5}{(x-3)}$

- -

a) $\ x = \sqrt{(3x+5)}$

$\Rightarrow x^2 = 3x + 5$

$\Rightarrow x^2 - 3x - 5 = 0$

b) $\ x = \dfrac{3x+5}{x}$

$\Rightarrow x^2 = 3x - 5$

$\Rightarrow x^2 - 3x - 5 = 0$

c) $\ x = \dfrac{5}{(x-3)}$

$\Rightarrow x(x-3) = 5$

$\Rightarrow x^2 - 3x - 5 = 0$

Starting close to a particular root does not mean that the iterative relationship will converge to that root (or indeed that it will converge at all). Consider what happens if we form iterative relationships from the three equations in Example 10 and take initial values of -1 and 4:

a) For iterative relationship $x_{n+1} = \sqrt{(3x_n + 5)}$, if we choose $x_1 = 4$ then x_n converges to 4.19. If we choose $x_1 = -1$, we see that x_n still converges to 4.19.

b) Iterative relationship $x_{n+1} = \dfrac{3x_n + 5}{x_n}$ produces the same behaviour as in part **(a)** above.

c) For iterative relationship $x_{n+1} = \dfrac{5}{(x_n - 3)}$, choosing either $x_1 = 4$ or $x_1 = -1$ results in x_n converging to -1.19.

In Example 6, we saw that initial values between the largest and smallest roots gave sequences which converged to the middle root, while any other initial value which was not an exact root itself generated a divergent sequence. So we could start very close to a root and either move towards a different root or not converge to any root at all.

In Example 7, there was an iterative relationship which produced an undefined calculation.

Sketching the graphs of $y = x$ and $y = F(x)$ on the same axes helps us to have some idea of what will happen in any iterative relationship.

- If the gradient of $F(x)$ close to the root is positive, then any convergent sequence will approach the root from one side, and if it is negative the sequence will alternate values above and below the root.

- Generally the smaller the modulus of the gradient, the faster the convergence will be. For example, for small gradients it will take fewer iterations to find the root to a given accuracy.

We will illustrate this by some examples.

Drawing the graphs of $y = x$ and $y = \frac{1}{4}(x^3 - 1)$ (which we studied in Example 6) shows that the gradient close to the middle root is small and positive so we would expect the convergence to be from one side (**a staircase diagram**) and fast:

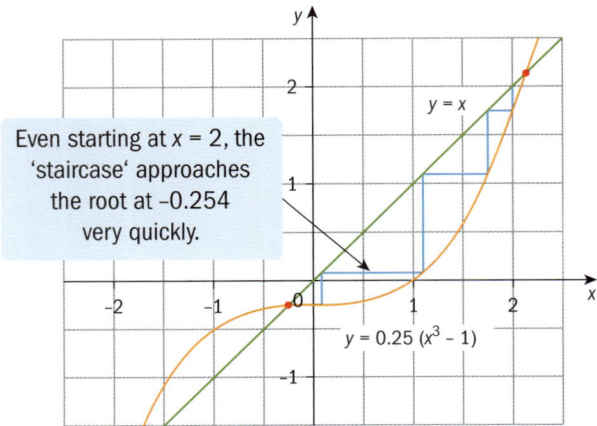

Even starting at $x = 2$, the 'staircase' approaches the root at −0.254 very quickly.

The iterative process in Example 8 was very slow and alternated values above and below the root. The graphs below show the functions $y = x$ and $y = \frac{2}{3}\pi + \sin x$, on a very expanded scale in the neighbourhood of the root ($x = 2.61$). In this neighbourhood, the gradient of $y = \frac{2}{3}\pi + \sin x$ is negative and not numerically small (approximately −0.9). Thus, convergence happens very slowly, and the graph of the iterations produces what is known as **a cobweb diagram**.

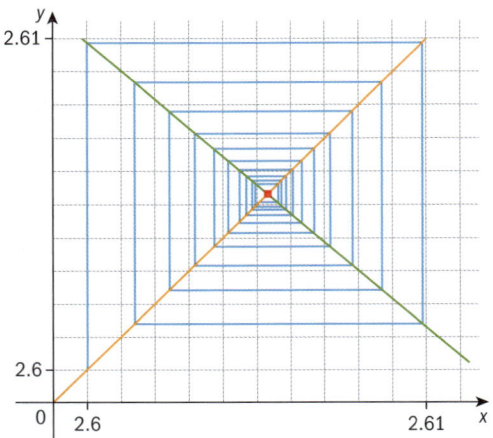

Summary exercise 6

1. Show that if $f(x) = x^3 - 3x^2 + 4x - 1$ then $f(x) = 0$ has a root between $x = 0$ and $x = 1$ and find the root correct to 1 decimal place.

2. Show that if $f(x) = 3^x - 5x - 4$ then $f(x) = 0$ has a root between $x = -1$ and $x = 0$ and another between $x = 2$ and $x = 3$ find the roots correct to 1 decimal place.

3. i) Show that if $f(x) = x^3 + 4x^2 - 2x - 7$ then $f(x) = 0$ has a root between $x = 1$ and $x = 2$ and find the root correct to 1 decimal place.

   ii) Show that there are two other roots for this in the intervals $(-5, -4)$ and $(-2, -1)$ find the roots correct to 1 decimal place.

4. Show that if $f(x) = 6 - x^2 e^x$ then $f(x) = 0$ has a root between $x = 1$ and $x = 2$ and find the root correct to 2 decimal places.

5. Find the equation that the iterative relationship provides a root for, if the sequence converges (you do not need to check whether it does converge):

   i) $x_{n+1} = \dfrac{1}{4 + 2x_n}$

   ii) $x_{n+1} = \sqrt{\dfrac{x_n^3 - 1}{5}}$

6. Find at least three possible rearrangements of the equation into the form $x = F(x)$

   i) $x^4 - 4x^2 + 3 = 0$ ii) $x^5 - 3x^2 + 2 = 0$

7. For each iterative formula, and given initial value, find values of x_2, x_3, x_4 and x_5 (giving the first 5 decimal places of each approximation):

   i) $x_{n+1} = \dfrac{1}{5 - x_n}$; $x_1 = 0.5$

   ii) $x_{n+1} = \dfrac{\sqrt{2x_n^3 + 3}}{5 + x_n}$; $x_1 = 0.5$

EXAM-STYLE QUESTIONS

8. i) Show that the equation $e^x - 2x^2 - 1 = 0$ has a root between $x = 2$ and $x = 3$.

   ii) Show that the equation $e^x - 2x^2 - 1 = 0$ can be rearranged to $x = \ln(2x^2 + 1)$

   iii) Using an initial value of $f(3) = 1.085\ldots$, and the iterative relationship $x_{n+1} = \ln\left(2x_n^2 + 1\right)$ find the root correct to 2 decimal places.

9. i) Show that both $x_{n+1} = \tan x_n - 1$ and $x_{n+1} = \tan^{-1}(x_n + 1)$ are iterative relationships that provide a root to the equation $x - \tan x + 1 = 0$ if the sequence converges.

   ii) Using $x_1 = 0.75$ as an initial value, show that one sequence diverges and the other converges and find the root correct to 2 decimal places.

10.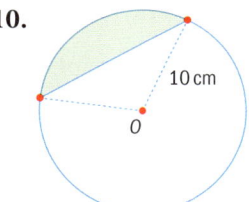
    10 cm
    O

    The shaded segment of the circle has an area of 19 cm².

    i) Show that angle $AOB = \theta$ radians satisfies the equation $50\theta - 50\sin \theta = 19$

    ii) Use the iterative relationship

    $\theta_{n+1} = \sin \theta_n + \dfrac{19}{50}$ with an initial value of 1.5 to find the angle in radians correct to 2 decimal places.

11.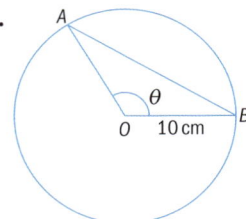

The diagram shows a circle with centre O an radius $10\,\text{cm}$. The chord AB divides the circle into two regions whose areas are in the ratio $1:4$ and it is required to find the length of AB. The angle AOB is θ radians.

i) Show that $\theta = \dfrac{2}{5}\pi + \sin\theta$. [3]

ii) Showing all your working, use an iterative formula, based on the equation in part (**i**), with an initial value of 2.1, to find θ correct to 2 decimal places. Hence find the length of AB in centimetres correct to 1 decimal place. [5]

Cambridge International AS and A Level Mathematics 9709 Paper 31 Q6 May/June 2011

12. i) By sketching a suitable pair of graphs, show that the equation
$$\sec x = 3 - \frac{1}{2}x^2,$$
where x is in radians, has a root in the interval $0 < x < \frac{1}{2}\pi$. [2]

ii) Verify by calculation that this root lies between 1 and 1.4. [2]

iii) Show that this root also satifies the equation
$$x = \cos^{-1}\left(\frac{2}{6-x^2}\right)$$ [1]

iv) Use an iterative formula based on the equation in part (**iii**) to determine the root correct to 2 decimal places. Give the result of each iteration to 4 decimal places. [3]

Cambridge International AS and A Level Mathematics 9709 Paper 31 Q5 October/November 2011

13. The sequence of values given by the iterative formula
$$x_{n+1} = \frac{3x_n}{4} + \frac{15}{x_n^3},$$
with initial value $x_1 = 3$, converges to α.

i) Use this iterative formula to find α correct to 2 decimal places, giving the result of each iteration to 4 decimal places. [3]

ii) State an equation satisfied by α and hence find the exact value of α. [2]

Cambridge International AS and A Level Mathematics 9709 Paper 31 Q3 October/November 2009

14. i) By sketching a suitable pair of graphs, show that the equation
$$\operatorname{cosec} x = \frac{1}{2}x + 1,$$
where x is in radians, has a root in the interval $0 < x < \frac{1}{2}\pi$. [2]

ii) Verify, by calculation, that this root lies between 0.5 and 1. [2]

iii) Show that this root also satisfies the equation
$$x = \sin^{-1}\left(\frac{2}{x+2}\right).$$ [1]

iv) Use the iterative formula
$$x_{n+1} = \sin^{-1}\left(\frac{2}{x_n+2}\right),$$
With initial value $x_1 = 0.75$, to determine this root correct to 2 decimal places. Give the result of each iteration to 4 decimal places. [3]

Cambridge International AS and A Level Mathematics 9709 Paper 3 Q7 May/June 2005

15.

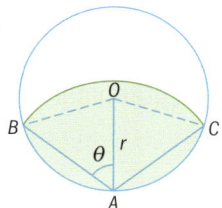

In the diagram, A is a point on the circumference of a circle with centre O and radius r. A circular arc with centre A meets the circumference at B and C. The angle OAB is θ radians. The shaded region is bounded by the circumference of the circle and the arc with centre A joining B and C. The area of the shaded region is equal to half the area of the circle.

i) Show that $\cos 2\theta = \dfrac{2\sin 2\theta - \pi}{4\theta}$. [5]

ii) Use the iterative formula

$$\theta_{n+1} = \frac{1}{2}\cos^{-1}\left(\frac{2\sin 2\theta_n - \pi}{4\theta_n}\right),$$

with initial value $\theta_1 = 1$, to determine θ correct to 2 decimal places, showing the result of each iteration to 4 decimal places. [3]

Cambridge International AS and
A Level Mathematics 9709
Paper 31 Q6 May/June 2013

16. i) By sketching suitable graphs, show that the equation

$$\sec x = 3 - x^2$$

Has exactly one root in the interval $0 < x < \frac{1}{2}\pi$. [2]

ii) Show that, if a sequence of values given by the iterative formula

$$x_{n+1} = \cos^{-1}\left(\frac{1}{3 - x_n^2}\right)$$

converges, then it converges to a root of the equation given in part **(i)**. [2]

iii) Use this iterative formula, with initial value $x_1 = 1$, to determine the root in the interval $0 < x < \frac{1}{2}\pi$ correct to

2 decimal places, showing the result of each iteration. [3]

Cambridge International AS and
A Level Mathematics 9709
Paper 31 Q5 October/November 2003

17.

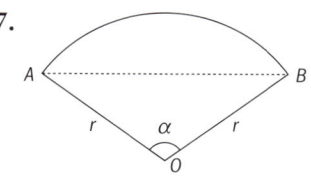

The diagram shows a sector AOB of a circle with centre O and radius r. The angle AOB is α radians, where $0 < \alpha < \pi$. The area of triangle AOB is half the area of the sector.

i) Show that a satisfies the equation

$$x = 2\sin x.$$ [2]

ii) Verify by calculation that α lies between $\frac{1}{2}\pi$ and $\frac{2}{3}\pi$. [2]

iii) Show that, if a sequence of values given by the iterative formula

$$x_{n+1} = \frac{1}{3}\left(x_n + 4\sin x_n\right).$$

converges, then it converges to a root of the equation in part **(i)**. [2]

iv) Use this iterative formula, with initial value $x_1 = 1.8$, to find α correct to 2 decimal places. Give the result of each iteration to 4 decimal places. [3]

Cambridge International AS and
A Level Mathematics 9709 Paper 3
Q6 May/June 2007

18. i) Use the substitution $x = \sin^2\theta$ to show that

$$\int \sqrt{\left(\frac{x}{1-x}\right)}\,dx = \int 2\sin^2\theta\,d\theta.$$ [4]

ii) Hence find the exact value of

$$\int_0^{\frac{1}{4}} \sqrt{\left(\frac{x}{1-x}\right)}\,dx.$$ [4]

Cambridge International AS and
A Level Mathematics 9709
Paper 3 Q6 October/November 2005

Chapter summary

Locating a root of an equation approximately using the sign-change rule or a graphical method

- The Sign-change rule:

  If f(x) is continuous in an interval $\alpha \leq x \leq \beta$ and if f(α) and f(β) have different signs then f(x) = 0 has at least one root between α and β.

- The simplest process to narrow an interval in which the root of an equation lies is called a **decimal search**. First we identify a pair of integers between which the root lies, followed by a pair of neighbouring values within 2 decimal places, etc. If we keep going with this process we will arrive at the value of the root to the required accuracy.

Finding roots using iterative relationships

- The iterative relation $x_{n+1} = F(x_n)$, together with an initial value x_1, defines a sequence. If that sequence converges to a limit, then the limit will be a solution to the equation $x = F(x)$.

- If you have an equation f(x) = 0, you can re-arrange into the form $x = F(x)$ for some function F(x). From here, you can define an iterative equation $x_{n+1} = F(x_n)$ which will often converge to a root of f(x) = 0. Use an iterative relationship and an initial value to produce further approximations to a root.

- We can use an iterative method to find a root to a specified degree of accuracy. The number of iterations required for this degree of accuracy will differ from equation to equation, and the value of the first iteration we choose.

Convergence behaviour of iterative functions

- If the gradient of F(x) close to the root is positive, then any convergent sequence will approach the root from one side, and if it is negative the sequence will alternate values above and below the root.

- Generally the smaller the modulus of the gradient, the faster the convergence will be. For example, for small gradients it will take fewer iterations to find the root to a given accuracy.

Review exercise B – Pure 2

1. Find the value of $\dfrac{dy}{dx}$ when $x = 4$ in each of the following cases:

   i) $y = x \ln(x - 3)$, [4]

   ii) $y = \dfrac{x-1}{x+1}$. [3]

Cambridge International AS & A Level Mathematics 9709 Paper 21 Q5 May/June 2011

2. Show that $\displaystyle\int_{-\frac{1}{4}\pi}^{\frac{1}{4}\pi}(3 + 2\tan^2\theta)\,d\theta = \dfrac{1}{2}(8 + \pi)$. [4]

Cambridge International AS & A Level Mathematics 9709 Paper 21 Q6b May/June 2011

3. i) By sketching a suitable pair of graphs, show that the equation
   $$e^{2x} = 14 - x^2$$
   has exactly two real roots. [3]

   ii) Show by calculation that the positive root lies between 1.2 and 1.3. [2]

   iii) Show that this root also satisfies the equation
   $$x = \dfrac{1}{2}\ln(14 - x^2).$$ [1]

   iv) Use an iteration process based on the equation in part c), with a suitable starting value, to find the root correct to 2 decimal places. Give the result of each step of the process to 4 decimal places. [3]

Cambridge International AS & A Level Mathematics 9709 Paper 21 Q7 May/June 2011

4. The parametric equations of a curve are
   $$x = 1 + \ln(t - 2), \quad y = t + \dfrac{9}{t}, \quad \text{for } t > 2.$$

   i) Show that $\dfrac{dy}{dx} = \dfrac{(t^2 - 9)(t - 2)}{t^2}$. [3]

   ii) Find the coordinates of the only point on the curve at which the gradient is equal to 0. [3]

Cambridge International AS & A Level Mathematics 9709 Paper 21 Q4 October/November 2010

5.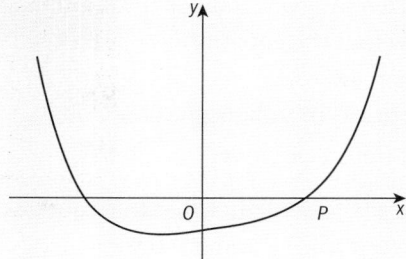

The diagram shows the curve $y = x^4 + 2x - 9$. The curve cuts the positive x-axis at the point P.

   i) Verify by calculation that the x-coordinate of P lies between 1.5 and 1.6. [2]

   ii) Show that the x-coordinate of P satisfies the equation
   $$x = \sqrt[3]{\left(\dfrac{9}{x} - 2\right)}.$$ [1]

   iii) Use the iterative formula
   $$x_{n+1} = \sqrt[3]{\left(\dfrac{9}{x_n} - 2\right)}$$
   to determine the x-coordinate of P correct to 2 decimal places. Give the result of each iteration to 4 decimal places. [3]

Cambridge International AS & A Level Mathematics 9709 Paper 21 Q2 October/November 2013

6. The equation of a curve is
   $y = \dfrac{1}{2}e^{2x} - 5e^x + 4x$. Find the exact x-coordinate of each of the stationary points of the curve and determine the nature of each stationary point. [6]

Cambridge International AS & A Level Mathematics 9709 Paper 21 Q3 October/November 2013

7.

i) By sketching a suitable pair of graphs, show that the equation
$$\cot x = 4x - 2,$$
where x is in radians, has only one root for $0 \le x \le \frac{1}{2}\pi$. [2]

ii) Verify by calculation that this root lies between $x = 0.7$ and $x = 0.9$. [2]

iii) Show that this root also satisfies the equation
$$x = \frac{1+2\tan x}{4\tan x}.$$
[1]

iv) Use the iterative formula $x_{n+1} = \dfrac{1+2\tan x_n}{4\tan x_n}$ to determine this root correct to 2 decimal places.

Give the result of each iteration to 4 decimal places. [3]

Cambridge International AS &
A Level Mathematics 9709 Paper 21
Q6 May/June 2013

8. The parametric equations of a curve are
$$x = \ln(1-2t), \quad y = \frac{2}{t}, \quad \text{for } t < 0.$$

i) Show that $\dfrac{dy}{dx} = \dfrac{1-2t}{t^2}$. [3]

ii) Find the exact coordinates of the only point on the curve at which the gradient is 3. [3]

Cambridge International AS &
A Level Mathematics 9709 Paper 21
Q4 October/November 2012

9. A curve has parametric equations
$$x = \frac{1}{(2t+1)^2}, \quad y = \sqrt{(t+2)}.$$

The point P on the curve has parameter p and it is given that the gradient of the curve at P is -1.

i) Show that $p = (p+2)^{\frac{1}{6}} - \dfrac{1}{2}$. [6]

ii) Use an iterative process based on the equation in part **a)** to find the value of p correct to 3 decimal places. Use a starting value of 0.7 and show the result of each iteration to 5 decimal places. [3]

Cambridge International AS &
A Level Mathematics 9709 Paper 21
Q6 May/June 2012

10. Show that $\displaystyle\int_2^6 \frac{2}{4x+1}\,dx = \ln\frac{5}{3}$. [5]

Cambridge International AS &
A Level Mathematics 9709 Paper 21
Q2 October/November 2011

11. The parametric equations of a curve are
$$x = e^{3t}, \quad y = t^2 e^t + 3.$$

i) Show that $\dfrac{dy}{dx} = \dfrac{t(t+2)}{3e^{2t}}$. [4]

ii) Show that the tangent to the curve at the point $(1, 3)$ is parallel to the x-axis. [2]

iii) Find the exact coordinates of the other point on the curve at which the tangent is parallel to the x-axis. [2]

Cambridge International AS &
A Level Mathematics 9709 Paper 21
Q7 October/November 2011

12.

i) By first expanding $\cos(2x + x)$, show that
$$\cos 3x \equiv 4\cos^3 x - 3\cos x.$$
[5]

ii) Hence show that
$$\int_0^{\frac{1}{6}\pi} (2\cos^3 x - \cos x)\,dx = \frac{5}{12}.$$
[5]

Cambridge International AS &
A Level Mathematics 9709 Paper 21
Q8 October/November 2011

13.

The diagram shows the curve $y = x\sin x$, for $0 \le x \le \pi$. The point $Q\left(\frac{1}{2}\pi, \frac{1}{2}\pi\right)$ lies on the curve.

i) Show that the normal to the curve at Q passes through the point $(\pi, 0)$. [5]

ii) Find $\dfrac{\mathrm{d}}{\mathrm{d}x}(\sin x - x\cos x)$. [2]

iii) Hence evaluate $\displaystyle\int_0^{\frac{1}{2}\pi} x \sin x\, \mathrm{d}x$. [3]

**Cambridge International AS &
A Level Mathematics 9709 Paper 21
Q8 October/November 2010**

14. The equation of a curve is
$$x^2 y + y^2 = 6x.$$

i) Show that $\dfrac{\mathrm{d}y}{\mathrm{d}x} = \dfrac{6 - 2xy}{x^2 + 2y}$. [4]

ii) Find the equation of the tangent to the curve at the point with coordinates $(1, 2)$, giving your answer in the form $ax + by + c = 0$. [3]

**Cambridge International AS &
A Level Mathematics 9709 Paper 21
Q6 May/June 2010**

15. i) By sketching a suitable pair of graphs, show that the equation
$$e^{2x} = 2 - x$$
has only one root. [2]

ii) Verify by calculation that this root lies between $x = 0$ and $x = 0.5$. [2]

iii) Show that, if a sequence of values given by the iterative formula
$$x_{n+1} = \frac{1}{2}\ln(2 - x_n)$$
converges, then it converges to the root of the equation in part **a)**. [1]

iv) Use this iterative formula, with initial value $x_1 = 0.25$, to determine the root correct to 2 decimal places. Give the result of each iteration to 4 decimal places. [3]

**Cambridge International AS &
A Level Mathematics 9709 Paper 21
Q7 May/June 2010**

16. i) By differentiating $\dfrac{\cos x}{\sin x}$, show that if
$$y = \cot x \text{ then } \frac{\mathrm{d}y}{\mathrm{d}x} = -\operatorname{cosec}^2 x.$$ [3]

ii) By expressing $\cot^2 x$ in terms of $\operatorname{cosec}^2 x$ and using the result of part **a)**, show that
$$\int_{\frac{1}{4}\pi}^{\frac{1}{2}\pi} \cot^2 x\, \mathrm{d}x = 1 - \frac{1}{4}\pi.$$ [4]

iii) Express $\cos 2x$ in terms of $\sin^2 x$ and hence show that $\dfrac{1}{1 - \cos 2x}$ can be expressed as $\dfrac{1}{2}\operatorname{cosec}^2 x$.

Hence, using the result of part **a)**, find
$$\int \frac{1}{1 - \cos 2x}\,\mathrm{d}x.$$ [3]

**Cambridge International AS &
A Level Mathematics 9709 Paper 21
Q8 May/June 2010**

17. The curve with equation $y = x \ln x$ has one stationary point.

i) Find the exact coordinates of this point, giving your answers in terms of e. [5]

ii) Determine whether this point is a maximum or a minimum point. [2]

**Cambridge International AS &
A Level Mathematics 9709 Paper 21
Q6 October/November 2009**

18.

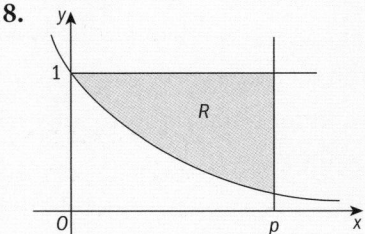

The diagram shows the curve $y = e^{-x}$. The shaded region R is bounded by the curve and the lines $y = 1$ and $x = p$, where p is a constant.

i) Find the area of R in terms of p. [4]

ii) Show that if the area of R is equal to 1 then
$$p = 2 - e^{-p}.$$ [1]

iii) Use the iterative formula

$$p_{n+1} = 2 - e^{-p_n},$$

with initial value $p_1 = 2$, to calculate the value of p correct to 2 decimal places. Give the result of each iteration to 4 decimal places. [3]

Cambridge International AS & A Level Mathematics 9709 Paper 21 Q7 October/November 2009

19. The equation of a curve is $y^2 + 2xy - x^2 = 2$.

i) Find the coordinates of the two points on the curve where $x = 1$. [2]

ii) Show by differentiation that at one of these points the tangent to the curve is parallel to the x-axis. Find the equation of the tangent to the curve at the other point, giving your answer in the form $ax + by + c = 0$. [7]

Cambridge International AS & A Level Mathematics 9709 Paper 21 Q8 October/November 2009

20.

The diagram shows the curve $y = \dfrac{1}{1+\sqrt{x}}$ for values of x from 0 to 2.

i) Use the trapezium rule with two intervals to estimate the value of

$$\int_0^2 \frac{1}{1+\sqrt{x}}\,dx,$$

giving your answer correct to 2 decimal places. [3]

ii) State, with a reason, whether the trapezium rule gives an under-estimate or an over-estimate of the true value of the integral in part **a)**. [1]

Cambridge International AS & A Level Mathematics 9709 Paper 2 Q3 May/June 2009

21.

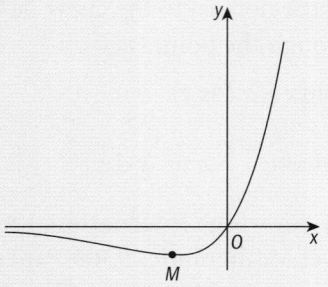

The diagram shows the curve $y = xe^{2x}$ and its minimum point M.

i) Find the exact coordinates of M. [5]

ii) Show that the curve intersects the line $y = 20$ at the point whose x-coordinate is the root of the equation

$$x = \frac{1}{2}\ln\!\left(\frac{20}{x}\right).$$ [1]

iii) Use the iterative formula

$$x_{n+1} = \frac{1}{2}\ln\!\left(\frac{20}{x_n}\right),$$

with initial value $x_1 = 1.3$, to calculate the root correct to 2 decimal places, giving the result of each iteration to 4 decimal places. [3]

Cambridge International AS & A Level Mathematics 9709 Paper 2 Q7 May/June 2009

22. i) Find the equation of the tangent to the curve $y = \ln(3x - 2)$ at the point where $x = 1$. [4]

ii) a) Find the value of the constant A such that

$$\frac{6x}{3x-2} \equiv 2 + \frac{A}{3x-2}.$$ [2]

b) Hence show that

$$\int_2^6 \frac{6x}{3x-2}\,dx = 8 + \frac{8}{3}\ln 2.$$ [5]

Cambridge International AS & A Level Mathematics 9709 Paper 21 Q8 May/June 2009

23. Find the exact coordinates of the point on the curve $y = xe^{-\frac{1}{2}x}$ at which $\dfrac{d^2y}{dx^2} = 0$. [7]

Cambridge International AS & A Level Mathematics 9709 Paper 2 Q6 October/November 2008

24. i) By sketching a suitable pair of graphs, show that the equation
$$\cos x = 2 - 2x,$$
where x is in radians, has only one root for $0 \leq x \leq \frac{1}{2}\pi$. [2]

ii) Verify by calculation that this root lies between 0.5 and 1. [2]

iii) Show that, if a sequence of values given by the iterative formula
$$x_{n+1} = 1 - \frac{1}{2}\cos x_n$$
converges, then it converges to the root of the equation in part **a)**. [1]

iv) Use this iterative formula, with initial value $x_1 = 0.6$, to determine this root correct to 2 decimal places. Give the result of each iteration to 4 decimal places. [3]

Cambridge International AS & A Level Mathematics 9709 Paper 2 Q7 October/November 2008

25. i) a) Prove the identity
$$\sec^2 x + \sec x \tan x \equiv \frac{1 + \sin x}{\cos^2 x}.$$

b) Hence prove that
$$\sec^2 x + \sec x \tan x \equiv \frac{1}{1 - \sin x}. \quad [3]$$

ii) By differentiating $\dfrac{1}{\cos x}$, show that if $y = \sec x$ then $\dfrac{dy}{dx} = \sec x \tan x$. [3]

iii) Using the results of parts **a)** and **b)**, find the exact value of
$$\int_0^{\frac{1}{4}\pi} \frac{1}{1 - \sin x}\, dx. \quad [3]$$

Cambridge International AS & A Level Mathematics 9709 Paper 2 Q8 October/November 2008

26. The equation of a curve is
$$x^2 + y^2 - 4xy + 3 = 0.$$

i) Show that $\dfrac{dy}{dx} = \dfrac{2y - x}{y - 2x}$. [4]

ii) Find the coordinates of each of the points on the curve where the tangent is parallel to the x-axis. [5]

Cambridge International AS & A Level Mathematics 9709 Paper 2 Q7 May/June 2008

Review exercise B – Pure 3

1. i) Prove that $\cot\theta + \tan\theta \equiv 2\operatorname{cosec}2\theta$. [3]

    ii) Hence show that

    $$\int_{\frac{1}{6}\pi}^{\frac{1}{3}\pi} \operatorname{cosec}2\theta\,\mathrm{d}\theta = \frac{1}{2}\ln 3.$$ [4]

    Cambridge International AS and
    A Level Mathematics 9709
    Paper 31 Q5 October/November 2013

2. For each of the following curves, find the gradient at the point where the curve crosses the y-axis:

    i) $y = \dfrac{1+x^2}{1+e^{2x}}$ [3]

    ii) $2x^3 + 5xy + y^3 = 8$. [4]

    Cambridge International AS and
    A Level Mathematics 9709
    Paper 31 Q5 May/June 2013

3.

    In the diagram, $ABCD$ is a rectangle with $AB = 3a$ and $AD = a$. A circular arc, with centre A and radius r, joins points M and N on AB and CD respectively. The angle MAN is x radians. The perimeter of the sector AMN is equal to half the perimeter of the rectangle.

    i) Show that x satisfies the equation

    $$\sin x = \frac{1}{2}(2+x).$$ [3]

    ii) This equation has only one root in the interval $0 < x < \frac{1}{2}\pi$. Use the iterative formula

    $$x_{n+1} = \sin^{-1}\left(\frac{2+x_n}{4}\right),$$

with initial value $x_1 = 0.8$, to determine the root correct to 2 decimal places. Give the result of each iteration to 4 decimal places. [3]

Cambridge International AS and
A Level Mathematics 9709
Paper 3 Q3 May/June 2008

4. Find the exact value of the constant k for which $\displaystyle\int_1^k \frac{1}{2x-1}\,\mathrm{d}x = 1$. [4]

Cambridge International AS and
A Level Mathematics 9709
Paper 3 Q1 October/November 2007

5. The equation of a curves is $y = x\sin 2x$, where x is in radians. Find the equation of the tangent to the curve at the point where $x = \frac{1}{4}\pi$. [4]

Cambridge International AS and
A Level Mathematics 9709
Paper 3 Q3 May/June 2007

6. i) It is given that $2\tan 2x + 5\tan^2 x = 0$. Denoting $\tan x$ by t, form an equation in t and hence show that either $t = 0$ or $t = \sqrt[3]{(t+0.8)}$. [4]

    ii) It is given that there is exactly one real value of t satisfying the equation $t = \sqrt[3]{(t+0.8)}$. Verify by calculation that this value lies between 1.2 and 1.3. [2]

    iii) Use the iterative formula $t_{n+1} = \sqrt[3]{(t_n + 0.8)}$ to find the value of t correct to 3 decimal places. Give the result of each iteration to 5 decimal places. [3]

    iv) Using the values of t found in previous parts of the question, solve the equation $2\tan 2x + 5\tan^2 x = 0$ for $-\pi \le x \le \pi$. [3]

Cambridge International AS and
A Level Mathematics 9709
Paper 31 Q10 May/June 2012

Pure 3

7. **i)** By sketching a suitable pair of graphs, show that the equation

$$\sec x = 3 - \frac{1}{2}x^2,$$

where x is in radians, has a root in the interval $0 < x < \frac{1}{2}\pi$. [2]

ii) Verify by calculation that this root lies between 1 and 1.4. [2]

iii) Show that this root also satisfies the equation

$$x = \cos^{-1}\left(\frac{2}{6-x^2}\right).$$ [1]

iv) Use an iterative formula based on the equation in part (**iii**) to determine the root correct to 2 decimal places. Give the result of each iteration to 4 decimal places. [3]

Cambridge International AS and A Level Mathematics 9709 Paper 31 Q5 October/November 2011

8. Find $\dfrac{dy}{dx}$ in each of the following cases:

i) $y = \ln(1 + \sin 2x)$, [2]

ii) $y = \dfrac{\tan x}{x}$. [2]

Cambridge International AS and A Level Mathematics 9709 Paper 31 Q2 May/June 2011

9. The curve with equation

$$6e^{2x} + ke^y + e^{2y} = c,$$

Where k and c are constants, passes through the point P with coordinates $(\ln 3, \ln 2)$.

i) Show that $58 + 2k = c$. [2]

ii) Given also that the gradient of the curve at P is -6, find the values of k and c. [5]

Cambridge International AS and A Level Mathematics 9709 Paper 31 Q5 May/June 2011

10.

The diagram shows a semicircle ACB with centre O and radius r. The angle BOC is x radians. The area of the shaded segment is a quarter of the area of the semicircle.

i) Show that x satisfies the equation

$$x = \frac{3}{4}\pi - \sin x.$$ [3]

ii) This equation has one root. Verify by calculation that the root lies between 1.3 and 1.5. [2]

iii) Use the iterative formula

$$x_{n+1} = \frac{3}{4}\pi - \sin x_n$$

to determine the root correct to 2 decimal places. Give the result of each iteration to 4 decimal places. [3]

Cambridge International AS and A Level Mathematics 9709 Paper 31 Q6 May/June 2010

11. A curve has equation $y = e^{-3x}\tan x$. Find the x-coordinates of the stationary points on the curve in the interval $-\frac{1}{2}\pi < x < \frac{1}{2}\pi$. Give your answers correct to 3 decimal places. [6]

Cambridge International AS and A Level Mathematics 9709 Paper 31 Q4 October/November 2009

12. The parametric equations of a curve are

$$x = a(2\theta - \sin 2\theta), \ y = a(1 - \cos 2\theta).$$

Show that $\dfrac{dy}{dx} = \cot\theta$. [5]

Cambridge International AS and A Level Mathematics 9709 Paper 3 Q4 October/November 2008

Maths in real-life

Nature of Mathematics

The Fibonacci sequence is extremely simple to define: it starts 1, 1 and then the next term is generated by the sum of the previous two terms – so the third is 2 = 1 + 1 and the fourth is 3 = 1 + 2. It continues infinitely:

1, 1, 2, 3, 5, 8, 13, 21, 34, 55, 89, 144, …, …

Perhaps surprisingly, these numbers occur frequently in nature. Exploring some geometry will help see why this happens, but here are some examples.

Flower heads with seeds often look like this. This sort of arrangement seems to optimize the packing of seeds which are the same size – and it looks as though there are spirals both curving right and left from the centre. The numbers of these spirals in each direction are usually neighbouring Fibonacci numbers.

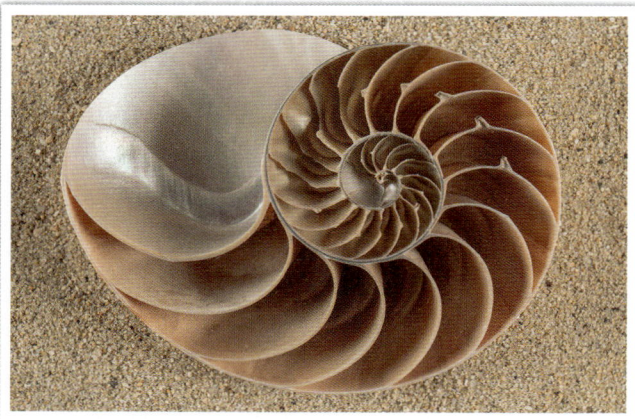

As the Nautilus shell grows the size of the chambers follows the pattern of the Fibonacci numbers. The shape of the Nautilus shell suggests that this is also an extremely efficient way for growth to occur.

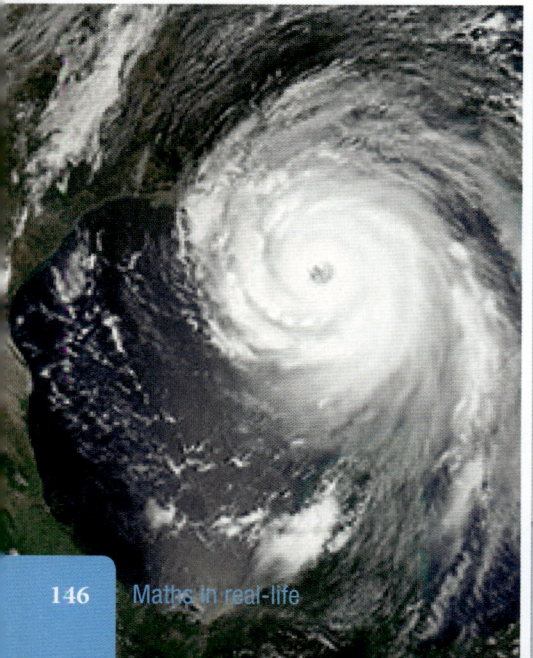

This picture of Hurricane Irene looks very like the shape of the Nautilus shell. The shape is built up by adding squares on the new longest side, generating the Fibonacci numbers.

This movement seems to be the way that hurricanes limit the energy losses from the system.

Similar patterns appear on the grandest scale imaginable – in the way galaxies are clustered in our universe.

Look at how the geometry grows: starting with a square side 1, add another above it of side 1. Then to the right add a square of side 2, below a square of side 3, left a square of side 5 and so on. Moving round, getting larger each time, creates the spiral shape seen in the Nautilus shell, the hurricane and the galaxies.

As the Fibonacci series continues, the ratio of successive terms approaches the 'golden ratio' – which was felt by the Ancient Greeks to be the most pleasing on the eye and they used it a lot in architecture – for example in the Parthenon (shown here).

The golden ratio occurs when the short side (AB) of a rectangle (ABCD) has unit length, and the long side (AD) has length x if, when you remove a square (ABEF) whose sides have length 1, you are left with a rectangle (CDEF) whose sides (lengths 1 and $x - 1$) are in the same ratio as the sides in the original rectangle (ABCD) were. The quadratic equation satisfied by the golden ratio, comes from solving

$AB = AF = 1$

$AD = x$

$(FD = x - 1)$

$$\frac{1}{x} = \frac{x-1}{1}$$

$$\Rightarrow 1 = x(x - 1)$$

$$\Rightarrow x^2 - x - 1 = 0$$

using the quadratic formula gives:

$$x = \frac{1 \pm \sqrt{1+4}}{2} = \frac{1 + \sqrt{5}}{2} = 1.618 \text{ (3 d.p.)}$$

Note that the second solution is discarded because it is negative. Also, some sources reference the golden ratio as 0.61803... = 1 ÷ 1.618.....

7 Further algebra

Real-world applications of algebra include the use of binomial expansions in the area of computing, for example, the automatic distribution of internet provider addresses.

Economists use binomial expansions to help them make realistic predictions about the way economies will behave in the next few years. Architects and engineers also use them in designing infrastructure and in calculating the costs and time associated in bringing their plans to realisation while at the same time ensuring that projects are profitable.

Objectives

- Recall an appropriate form for expressing rational functions in partial fractions, and carry out the decomposition, in cases where the denominator is no more complicated than $(ax + b)(cx + d)(ex + f)$ or $(ax + b)(cx + d)^2$ or $(ax + b)(x^2 + c^2)$, and where the degree of the numerator does not exceed that of the denominator.

- Use the expansion of $(1 + x)^n$, where n is a rational number and $|x| < 1$ (finding a general term is not included, but adapting the standard series to expand e.g. $\left(2 - \frac{1}{2}x\right)^{-1}$ is included).

Before you start

You should know how to:

1. Equate coefficients,

    e.g. Given $3x^2 - Bx + 5 = Ax^2 + 2x - C$, find the values of A, B, and C.

    Coefficient of x^2: $A = 3$

    Coefficient of x : $B = -2$

    Constants: $C = -5$

    e.g. Given $2Ax^2 - 9x + C - 8 = 6x^2 + 3Bx - 7$ find the values of A, B, and C.

    Coefficient of x^2: $2A = 6 \Rightarrow A = 3$

    Coefficient of x : $-9 = 3B \Rightarrow B = -3$

    Constants: $C - 8 = -7 \Rightarrow C = 1$

Skills check:

For each of the following equations, find the values of A, B and C.

1. $Ax^2 - 4x + 9 = x^2 + Bx - C$

2. $3Ax^2 - 2Bx - 2 = 6x^2 + 8x - 2 + C$

3. $(1 - A)x^2 - 2x + 5C = 8x^2 - 3Bx - 10$

4. $12x^2 + 2Bx = 4Ax^2 + x - 7C$

5. $2Ax^2 - 3Bx + 6 = 6x^2 + 4 - 2C$

7.1 Partial fractions

We use the method of **partial fractions** to split a rational function into simpler component parts. We choose which method to use by looking at the denominator of the function.

> **Partial fractions type 1:** Algebraic fractions with linear factors in the denominator can be split into partial fractions in the following way:
> $$\frac{1}{(x + p)(x + q)} \equiv \frac{A}{(x + p)} + \frac{B}{(x + q)}.$$

Example 1

Express $\dfrac{(x + 3)}{(x - 2)(x + 1)}$ in partial fractions.

$\dfrac{x + 3}{(x - 2)(x + 1)} = \dfrac{A}{x - 2} + \dfrac{B}{x + 1}$ ⟵ The denominator has 2 linear factors so we use two fractions.

$x + 3 \equiv A(x + 1) + B(x - 2)$ ⟵ Multiply each term by $(x - 2)(x + 1)$.

Method 1: Choose appropriate values for x.

We use the symbol $\equiv$ to mean it is true for all values of x.

Let $x = 2$ ⟵ Choose a value of x that makes one bracket $= 0$.

$5 = A(3) + 0 \implies A = \dfrac{5}{3}$

Let $x = -1$ ⟵ Choose a value of x that makes the other bracket $= 0$.

$2 = 0 + B(-3) \implies B = -\dfrac{2}{3}$

Hence $\dfrac{x + 3}{(x - 2)(x + 1)} = \dfrac{5}{3(x - 2)} - \dfrac{2}{3(x + 1)}$ ⟵ Substitute the values of A and B into the fractions.

Method 2: Expand the brackets and equate coefficients.

$x + 3 \equiv A(x + 1) + B(x - 2)$

$x + 3 \equiv Ax + A + Bx - 2B$ ⟵ Expand the brackets.

$(1)x + (3) \equiv (A + B)x + (A - 2B)$

Hence $A + B = 1$ and $A - 2B = 3$ ⟵ Equate the coefficients of x.

Solving simultaneously gives $A = \dfrac{5}{3}$ and $B = -\dfrac{2}{3}$

Hence $\dfrac{x + 3}{(x - 2)(x + 1)} = \dfrac{5}{3(x - 2)} - \dfrac{2}{3(x + 1)}$ ⟵ Put the minus sign (–) before the second fraction.

Example 2

Express $\dfrac{x}{(1 - x - 2x^2)}$ in partial fractions.

$\dfrac{x}{(1 + x)(1 - 2x)} = \dfrac{A}{1 + x} + \dfrac{B}{1 - 2x}$ ⟵   Factorise the denominator into two linear factors.

$x \equiv A(1 - 2x) + B(1 + x)$ ⟵   Multiply each term by $(1 + x)(1 - 2x)$.

Let $x = -1$

$\quad -1 = A(3) + 0 \;\Rightarrow\; A = -\dfrac{1}{3}$

Let $x = \dfrac{1}{2}$ ⟵   When $1 - 2x = 0$, $x = \dfrac{1}{2}$

$\quad \dfrac{1}{2} = 0 + B\left(\dfrac{3}{2}\right) \;\Rightarrow\; B = \dfrac{1}{3}$

Hence $\dfrac{x}{(1 + x)(1 - 2x)} = -\dfrac{1}{3(1 + x)} + \dfrac{1}{3(1 - 2x)}$ ⟵   Substitute the values of A and B into the fractions.

Note: You could also use the method of equating coefficients.

Exercise 7.1A

Express the following in partial fractions.

1. $\dfrac{x - 4}{(x - 1)(x - 2)}$

2. $\dfrac{x}{(x + 1)(x - 4)}$

3. $\dfrac{3x + 5}{x^2 + 2x - 3}$

4. $\dfrac{2x - 1}{x(2x + 1)}$

5. $\dfrac{2x^2 + 17x + 21}{(x + 2)(x + 3)(x - 3)}$

Partial fractions type 2: Algebraic fractions with a **repeated factor in the denominator** can be split into partial fractions in the following way:

$$\frac{1}{(x+p)^2} = \frac{A}{(x+p)} + \frac{B}{(x+p)^2}$$

Example 3

Express $\dfrac{3x^2 + 2}{x(x-1)^2}$ in partial fractions.

· ·

$\dfrac{3x^2 + 2}{x(x-1)^2} \equiv \dfrac{A}{x} + \dfrac{B}{x-1} + \dfrac{C}{(x-1)^2}$ Separate the repeated factor $(x-1)^2$ into two denominators.

$3x^2 + 2 \equiv A(x-1)^2 + Bx(x-1) + Cx$

Let $x = 0$ This means B and C are eliminated.

$\quad 2 = A(-1)^2 + 0 + 0 \Rightarrow A = 2$

Let $x = 1$

$\quad 5 = 0 + 0 + C \quad \Rightarrow \quad C = 5$ This means A and B are eliminated.

Equating coefficients of x^2:

$3 = A + B \Rightarrow 3 = 2 + B \Rightarrow B = 1$ Substitute for $A = 2$ in $3 = A + B$.

$\dfrac{3x^2 + 2}{x(x-1)^2} \equiv \dfrac{2}{x} + \dfrac{1}{x-1} + \dfrac{5}{(x-1)^2}$ **Note:** You could also equate coefficients of x or equate the constant terms.

Example 4

Express $\dfrac{2x}{(x+3)^2}$ in partial fractions.

· ·

$\dfrac{2x}{(x+3)^2} \equiv \dfrac{A}{x+3} + \dfrac{B}{(x+3)^2}$ Separate the repeated factor $(x+3)^2$ into two denominators.

$2x \equiv A(x+3) + B$

Let $x = -3$ This means A is eliminated.

$\quad -6 = B \Rightarrow B = -6$

Equating coefficients of x: Instead, you could equate the constants, i.e. $0 = 3A + B$.

$2 = A \quad \Rightarrow \quad A = 2$

$\dfrac{2x}{(x+3)^2} \equiv \dfrac{2}{x+3} - \dfrac{6}{(x+3)^2}$

Exercise 7.1B

Express the following in partial fractions.

1. $\dfrac{x}{(x-2)^2}$

2. $\dfrac{3}{x(3x-1)^2}$

3. $\dfrac{x}{(x-1)(x-2)^2}$

4. $\dfrac{1}{(x+1)(x-1)^2}$

5. $\dfrac{1}{x(x^2-2x+1)}$

> **Partial fractions type 3:** Algebraic fractions with a **quadratic factor in the denominator** can be split into partial fractions in the following way:
> $$\frac{1}{(x+p)(x^2+q)} \equiv \frac{A}{x+p} + \frac{Bx+C}{x^2+q}.$$

Example 5

Express $\dfrac{2x^2 - x + 6}{(x+1)(x^2+2)}$ in partial fractions.

· ·

$\dfrac{2x^2 - x + 6}{(x+1)(x^2+2)} \equiv \dfrac{A}{x+1} + \dfrac{Bx+C}{x^2+2}$ ← Put $Bx + C$ as the numerator of $x^2 + 2$.

$2x^2 - x + 6 \equiv A(x^2+2) + (Bx+C)(x+1)$

Let $x = -1$ ← This means B and C are eliminated.

$\quad 9 = A(3) + 0 \quad \Rightarrow \quad A = 3$

Equating coefficients of x^2:

$2 = A + B \quad \Rightarrow \quad 2 = 3 + B$ ← Substitute for $A = 3$ in $2 = A + B$.

$\qquad\qquad \Rightarrow \quad B = -1$

Equating coefficients of the constant terms:

$6 = 2A + C \quad \Rightarrow \quad 6 = 6 + C$ ← Substitute for $A = 3$ in $6 = 2A + C$.

$\qquad\qquad \Rightarrow \quad C = 0$

$\dfrac{2x^2 - x + 6}{(x+1)(x^2+2)} \equiv \dfrac{3}{x+1} - \dfrac{x}{x^2+2}$

> **Note:** You could also equate coefficients of x, but it is simpler in this case to equate coefficients of x^2 and equate the constant terms.

Example 6

Express $\dfrac{5x^2 - 5x + 4}{(2x - 1)(3 - x^2)}$ in partial fractions.

$\dfrac{5x^2 - 5x + 4}{(2x - 1)(3 - x^2)} \equiv \dfrac{A}{2x - 1} + \dfrac{Bx + C}{3 - x^2}$ ⟵ Put $Bx + C$ as the numerator of $3 - x^2$

$5x^2 - 5x + 4 \equiv A(3 - x^2) + (Bx + C)(2x - 1)$

Let $x = \dfrac{1}{2}$ ⟵ This means B and C are eliminated.

$\dfrac{11}{4} = A\left(3 - \dfrac{1}{4}\right) + 0 \quad \Rightarrow \quad A = 1$

Equating coefficients of x^2:

$5 = -A + 2B \Rightarrow 5 = -1 + 2B$ ⟵ Substitute for $A = 1$ in $5 = -A + 2B$.

$\Rightarrow B = 3$

Equating coefficients of the constant terms:

$4 = 3A - C \quad \Rightarrow \quad 4 = 3 - C$ ⟵ Substitute for $A = 3$ in $4 = 3A - C$.

$\Rightarrow C = -1$

$\dfrac{5x^2 - 5x + 4}{(2x - 1)(3 - x^2)} \equiv \dfrac{1}{2x - 1} + \dfrac{3x - 1}{3 - x^2}$

Note: You could also equate coefficients of x.

Exercise 7.1C

Express the following in partial fractions

1. $\dfrac{1 - 5x - 2x^2}{(1 - x)(x^2 + 2)}$

2. $\dfrac{2(1 - x^2)}{x(2 + x^2)}$

3. $\dfrac{5}{(x^2 + 4)(x + 1)}$

4. $\dfrac{3 - x}{(1 - 2x)(1 + x^2)}$

5. $\dfrac{7x - 4x^2}{(3 - x^2)(x - 2)}$

Partial fractions type 4: Improper algebraic fractions occur when a polynomial of degree n is divided by another polynomial of degree n.

They can be split into partial fractions by first doing long division, and then splitting the remainder into partial fractions using one of the techniques discussed in types 1, 2, or 3:

$$\frac{x^2}{(x + p)(x + q)} \equiv A + \frac{Bx + C}{(x + p)(x + q)}.$$

Example 7

Express $\dfrac{x^2 - x + 5}{(x - 1)(x + 2)}$ in partial fractions.

We have a polynomial of degree 2 divided by another polynomial of degree 2, so the leading term in the quotient will be a constant. Let this constant be A.

$$\frac{x^2 - x + 5}{(x - 1)(x + 2)} \equiv A + \frac{B}{x - 1} + \frac{C}{x + 2}$$

> There are now no improper fractions.

$$x^2 - x + 5 \equiv A(x - 1)(x + 2) + B(x + 2) + C(x - 1)$$

> Multiply each term by $(x - 1)(x + 2)$

Method 1: Assigning values to x and equating coefficients.

Let $x = 1$: $5 = 0 + B(3) + 0 \Rightarrow B = \dfrac{5}{3}$

Let $x = -2$. $11 = 0 + 0 + C(-3) \Rightarrow C = -\dfrac{11}{3}$

Equating coefficients of x^2: $1 = A$

> It is easiest to equate coefficients of x^2 first.

$$\frac{x^2 - x + 5}{(x - 1)(x + 2)} \equiv 1 + \frac{5}{3(x - 1)} - \frac{11}{3(x + 2)}$$

Method 2: Using long division, $(x^2 - x + 5) \div (x^2 + x - 2) = 1$ remainder $-2x + 7$

Thus $\dfrac{x^2 - x + 5}{(x - 1)(x + 2)} = 1 + \dfrac{-2x + 7}{(x - 1)(x + 2)}$

> This is a **type 1** partial fraction.

Consider $\dfrac{-2x + 7}{(x - 1)(x + 2)} \equiv \dfrac{A}{x - 1} + \dfrac{B}{x + 2}$

$$-2x + 7 \equiv A(x + 2) + B(x - 1)$$

Let $x = 1$: $5 = A(3) + 0 \Rightarrow A = \dfrac{5}{3}$

> Do not forget to put the constant term in the answer!

Let $x = -2$: $11 = 0 + B(-3) \Rightarrow B = -\dfrac{11}{3}$

$$\frac{x^2 - x + 5}{(x - 1)(x + 2)} \equiv 1 + \frac{5}{3(x - 1)} - \frac{11}{3(x + 2)}$$

Example 8

Find the values of the constants A, B, C, and D such that

$$\frac{3x^3 + 2x^2 + 6x + 4}{x^2(x+1)} = A + \frac{B}{x} + \frac{C}{x^2} + \frac{D}{x+1}$$

We have a polynomial of degree 3 which is divided by another polynomial of degree 3.

Using long division, $(3x^3 + 2x^2 + 6x + 4) \div (x^3 + x^2) = 3$ remainder $- x^2 + 6x + 4$

Hence $\dfrac{3x^3 + 2x^2 + 6x + 4}{x^2(x+1)} = 3 + \dfrac{-x^2 + 6x + 4}{x^2(x+1)}$

Thus $A = 3$.

Consider $\dfrac{-x^2 + 6x + 4}{x^2(x+1)} \equiv \dfrac{B}{x} + \dfrac{C}{x^2} + \dfrac{D}{x+1}$ ← Repeated factor and linear factor.

$-x^2 + 6x + 4 \equiv Bx(x+1) + C(x+1) + Dx^2$

Let $x = -1$: $-3 = D(1) \implies D = -3$

Let $x = 0$: $4 = C(1) \implies C = 4$

Coeff x^2: $-1 = B + D \implies B = 2$

$A = 3$, $B = 2$, $C = 4$, $D = -3$ ← Remember to state the values of A, B, C, and D at the end of your working.

Exercise 7.1D

1. Express $\dfrac{x^2 + 5}{(x-2)(x-3)}$ in partial fractions.

2. Express $\dfrac{x^2}{x^2 - 1}$ in partial fractions.

3. Express $\dfrac{3x^3 - 5x^2 - 3x - 40}{(x^2 + 4)(x-3)}$ in partial fractions.

4. Express $\dfrac{-x^2 - 8x - 11}{(x+3)^2}$ in partial fractions.

5. Find the values of the constants A, B, and C such that $\dfrac{2x - 2x^2}{x^2 - 4} = A + \dfrac{B}{x+2} + \dfrac{C}{x-2}$.

Exercise 7.1E

This exercise covers all four types of partial fractions that you have studied in this chapter so far.
Express the following in partial fractions.

1. $\dfrac{x-1}{(x+1)(x+3)}$ 2. $\dfrac{x^2-4x+3}{(x+1)(x+2)}$ 3. $\dfrac{2x-7}{(x-5)^2}$ 4. $\dfrac{11x+3}{(1-5x)(1+x^2)}$

5. $\dfrac{x-8}{(x+1)(x-2)}$ 6. $\dfrac{3x^2+5x+3}{x(x-3)}$ 7. $\dfrac{1+3x^2}{x(x-1)^2}$ 8. $\dfrac{2x^2-3}{x(x^2+2)}$

9. $\dfrac{x-1}{x^2+x}$ 10. $\dfrac{x(x+2)}{(x+3)(x-3)}$ 11. $\dfrac{8x^2+5x+3}{(x-2)(x+1)^2}$ 12. $\dfrac{x-3}{(x-1)(x^2+1)}$

13. $\dfrac{x}{(x-1)(x-2)^2}$ 14. $\dfrac{x^2+1}{x^2-1}$ 15. $\dfrac{2x^2+2x+3}{(x+2)(x^2+3)}.$

7.2 Binomial expansions of the form $(1 + x)^n$ when n is not a positive integer

In unit P1, you learned that $(1+x)^n = 1 + nx + \dfrac{n(n-1)}{2!}x^2 + \dfrac{n(n-1)(n-2)}{3!}x^3 + \ldots + x^n$,
where n was a positive integer.

We now consider the expansion when n is **not** a positive integer.

$$(1+x)^n = 1 + nx + \frac{n(n-1)}{2!}x^2 + \frac{n(n-1)(n-2)}{3!}x^3 + \ldots$$

where n is any integer or fraction and $|x| < 1$.

In the case where n is **not** a positive integer, the Binomial expansion of $(1 + x)^n$
does not terminate (it forms an infinite series). In this case, we must restrict the
domain of the expansion to $|x| < 1$ in order for the infinite series to converge. Most
often we use this expansion to find an **approximation** to $(1 + x)^n$ up to a certain order of n.

Example 9

Expand $(1 - 2x)^{-2}$ in ascending powers of x, up to and including the term in x^3,
simplifying the coefficients.

$(1-2x)^{-2} = 1 + (-2)(-2x) + \dfrac{(-2)(-3)}{2\times 1}(-2x)^2 + \dfrac{(-2)(-3)(-4)}{3\times 2\times 1}(-2x)^3 + \ldots$ ← $n = -2$, and we use $-2x$ instead of x.

$= 1 + 4x + 12x^2 + 32x^3$

valid for $|-2x| < 1$ ← Instead of $|x| < 1$ we write $|-2x| < 1$.

i.e. valid for $-\dfrac{1}{2} < x < \dfrac{1}{2}$ ← Write this condition in the form shown.

Example 10

a) Expand $\left(1 - \frac{x}{2}\right)^{\frac{1}{3}}$ in ascending powers of x, up to and including the term in x^3.

b) Use your expansion to evaluate $\sqrt[3]{0.999}$ giving your answer to 6 decimal places.

a) $\left(1 - \frac{x}{2}\right)^{\frac{1}{3}} = 1 + \left(\frac{1}{3}\right)\left(-\frac{x}{2}\right) + \frac{\left(\frac{1}{3}\right)\left(-\frac{2}{3}\right)}{2 \times 1}\left(-\frac{x}{2}\right)^2 + \frac{\left(\frac{1}{3}\right)\left(-\frac{2}{3}\right)\left(-\frac{5}{3}\right)}{3 \times 2 \times 1}\left(-\frac{x}{2}\right)^3 + \ldots$

$= 1 - \frac{1}{6}x - \frac{1}{36}x^2 - \frac{5}{648}x^3 + \ldots$ provided $\left|-\frac{x}{2}\right| < 1$, i.e. $-2 < x < 2$

> Do not forget to write this condition.

b) Let $\left(1 - \frac{x}{2}\right)^{\frac{1}{3}} = \sqrt[3]{0.999} = (0.999)^{\frac{1}{3}}$

$1 - \frac{x}{2} = 0.999$

$1 - 0.999 = \frac{x}{2}$

$x = 0.002$

> We use this value of x to find $\sqrt[3]{0.999}$.

$\sqrt[3]{0.999} = 1 - \frac{1}{6}(0.002) - \frac{1}{36}(0.002)^2 - \frac{5}{648}(0.002)^3 + \ldots$

> Substitute $x = 0.002$ into your expansion in **(a)**.

$= 1 - 0.000\,333\,333 - 0.000\,000\,111 - 0.000\,000\,02666$

$= 0.999\,666\,5293 = 0.999\,667$ to 6 d.p.

Example 11

Find the coefficient of x^3 in the expansion of $\dfrac{3 + x}{1 + 3x}$.

$\dfrac{3 + x}{1 + 3x} = (3 + x)(1 + 3x)^{-1}$

> The denominator is the same as $(1 + 3x)^{-1}$

$(1 + 3x)^{-1} = 1 + (-1)(3x) + \frac{(-1)(-2)}{2 \times 1}(3x)^2 + \frac{(-1)(-2)(-3)}{3 \times 2 \times 1}(3x)^3 + \ldots$

$= 1 - 3x + 9x^2 - 27x^3 + \ldots$

> Expand the denominator.

$(3 + x)(1 + 3x)^{-1} = (3 + x)(1 - 3x + 9x^2 - 27x^3 + \ldots)$

Term in x^3 is $(3)(-27x^3) + (x)(9x^2) = -72x^3$

> You do not need to fully expand the brackets.

Coefficient of x^3 is -72

> Do not put x^3 in your answer.

> **Example 12**
>
> Find the first 4 terms in the expansion of $(1 + x^2)^{-3}$.
>
> ....
>
> $(1 + x^2)^{-3} = 1 + (-3)\, x^2 + \dfrac{(-3)(-4)}{2 \times 1}\,(x^2)^2 + \dfrac{(-3)(-4)(-5)}{3 \times 2 \times 1}\,(x^2)^3 + \ldots$ ⟵ Use x^2 instead of x in the expansion.
>
> $\qquad\qquad = 1 - 3x^2 + 6x^4 - 10x^6 + \ldots$ for $-1 < x < 1$

Exercise 7.2

1. Expand the following in ascending powers of x, up to and including the term in x^3, and simplify the coefficients. State the range of values of x for which each expansion is valid.

   a) $(1 - x)^{-4}$ 　　　　b) $(1 + 3x)^{-5}$ 　　　　c) $\sqrt[3]{(1 + x)}$

   d) $\left(1 + \dfrac{x}{2}\right)^{-3}$ 　　　　e) $\dfrac{x}{1 - x}$ 　　　　f) $\dfrac{1}{(1 - 5x)^2}$

2. Find the first 3 terms in the expansion of $\dfrac{1}{\sqrt{1 + x^2}}$ in ascending powers of x.

3. a) Find the first 4 terms in the expansion of $\dfrac{1}{(1 - x)^2}$ in ascending powers of x.

   b) Hence find the 50th term of this expansion.

4. a) Expand $\sqrt{(1 - 2x)}$ in ascending powers of x, up to and including the term in x^4.

   b) Use your expansion to evaluate $\sqrt{(0.8)}$, giving your answer to 4 decimal places.

5. Find the first 3 terms in the expansion of $\dfrac{1 - x}{\sqrt{1 + x}}$ in ascending powers of x.

6. Find the first 4 terms in the expansion of $(1 + 4x^2)^{-2}$.

7. Expand $\dfrac{1 + 2x}{(1 - x)^2}$ in ascending powers of x, up to and including the term in x^3.

8. a) Expand $(1 + x)^{\frac{1}{2}}$ in ascending powers of x, up to and including the term in x^3.

   b) Use your expansion to evaluate $\sqrt{(1.08)}$ giving your answer to 4 decimal places.

9. The first 3 terms in the expansion of $(1 + ax)^n$ are $1 + 4x + 10x^2$.

   a) Find the value of a and the value of n.

   b) Hence find the coefficient of the term in x^3.

10. Expand $\dfrac{1 - 3x}{1 + x^2}$ in ascending powers of x, up to and including the term in x^4.
    State the values of x for which the expression is valid.

11. The first 3 terms in the expansion of $(1 + kx)^n$ are $1 + 12x + 81x^2$.

   a) Find the value of k and the value of n. **b)** Hence find the term in x^3.

12. Find the first 5 terms in the expansion of $\dfrac{1 - 2x}{1 + 2x^2}$ in ascending powers of x.

7.3 Binomial expansions of the form $(a + x)^n$ where n is not a positive integer

We now consider the expansion when it is of the form $(a + x)^n$ where $a \neq 1$.

We want to use the binomial expansion:

$$(1 + x)^n = 1 + nx + \frac{n(n - 1)}{2!} x^2 + \frac{n(n-1)(n-2)}{3!} x^3 + \ldots$$

where n is any integer or fraction and $|x| < 1$.

> However, when the first term in the bracket is not 1, we must first factorise the expression into the form $a^n \left(1 + \dfrac{x}{a}\right)^n$ and then use the binomial expansion of $(1 + x)^n$, so
>
> $$(ax + x)^n = a^n \left(1 + \frac{x}{a}\right)^n = a^n \left(1 + n\left(\frac{x}{a}\right) + \frac{n(n-1)}{2!}\left(\frac{x}{a}\right)^2 + \ldots\right).$$

Example 13

Express the following in the form $a^n(1 + x)^n$.

a) $(5 - x)^{-3}$ **b)** $\sqrt{(2 - 6x)}$ **c)** $\dfrac{32}{(4 + 2x)^2}$

- -

a) $(5 - x)^{-3} = \left[5\left(1 - \dfrac{x}{5}\right)\right]^{-3}$ ← Take out 5 and divide each term in the bracket by 5.

$\qquad = 5^{-3}\left(1 - \dfrac{x}{5}\right)^{-3}$ ← Remember 5 is also to the power of -3.

$\qquad = \dfrac{1}{125}\left(1 - \dfrac{x}{5}\right)^{-3}$

b) $\sqrt{(2 - 6x)} = [2(1 - 3x)]^{\frac{1}{2}}$ ← It is helpful to use square brackets and round brackets.

$\qquad = 2^{\frac{1}{2}} (1 - 3x)^{\frac{1}{2}}$

c) $\dfrac{32}{(4 + 2x)^2} = 32(4 + 2x)^{-2}$ ← First get rid of the fraction.

$\qquad = 32\left[4\left(1 + \dfrac{1}{2}x\right)\right]^{-2}$

$\qquad = 32\left[4^{-2}\left(1 + \dfrac{1}{2}x\right)^{-2}\right]$

$\qquad = 2\left(1 + \dfrac{1}{2}x\right)^{-2}$ ← $4^{-2} = \dfrac{1}{16}$

Example 14

Expand $\dfrac{1}{\sqrt{(4+x)}}$ in ascending powers of x, up to and including the term in x^2, simplifying the coefficients.

$\dfrac{1}{\sqrt{(4+x)}} = (4+x)^{-\frac{1}{2}}$ ← Rearrange to the form $(a+b)^n$

$= \left[4\left(1+\dfrac{x}{4}\right)\right]^{-\frac{1}{2}} = 4^{-\frac{1}{2}}\left(1+\dfrac{x}{4}\right)^{-\frac{1}{2}}$ ← We want it to be in the form $p(1+q)^n$

$= \dfrac{1}{2}\left(1+\dfrac{x}{4}\right)^{-\frac{1}{2}}$ ← $4^{-\frac{1}{2}} = \dfrac{1}{\sqrt{4}} = \dfrac{1}{2}$

Now $\left(1+\dfrac{x}{4}\right)^{-\frac{1}{2}} = 1 + \left(-\dfrac{1}{2}\right)\left(\dfrac{x}{4}\right) + \dfrac{\left(-\dfrac{1}{2}\right)\left(-\dfrac{3}{2}\right)}{2\times 1}\left(\dfrac{x}{4}\right)^2$ ← $n = -\dfrac{1}{2}$

$= 1 - \dfrac{1}{8}x + \dfrac{3}{144}x^2 + \dots$

Thus $\dfrac{1}{\sqrt{(4+x)}} = \dfrac{1}{2}\left(1 - \dfrac{1}{8}x + \dfrac{3}{144}x^2 + \dots\right)$

$= \dfrac{1}{2} - \dfrac{1}{16}x + \dfrac{3}{288}x^2 + \dots$

for $-4 < x < 4$ ← Do not forget to write this condition.

Exercise 7.3

1. Express the following in the form $a(1+x)^n$.

   a) $(25 - 50x)^{-\frac{1}{2}}$ b) $\dfrac{9}{(3-x)^2}$ c) $\sqrt[3]{(27+4x)}$.

2. Expand $(2-x)^{-5}$ in ascending powers of x, up to and including the term in x^3, simplifying the coefficients.

3. Find the first 3 terms in the expansion of $\sqrt{(4-x)}$ in ascending powers of x.

4. Find the coefficient of the term in x^2 in the expansion of $\dfrac{1}{3+x}$.

5. Find the first 4 terms in the expansion of $\dfrac{1}{(2+x^2)}$ in ascending powers of x.

6. a) Expand $\dfrac{1}{(4-x)^2}$ in ascending powers of x, up to and including the term in x^3, simplifying the coefficients.

   b) Hence find the coefficient of x^3 in the expansion of $\dfrac{2-x}{(4-x)^2}$.

7. Expand $(1+3x)\sqrt{(9-x)}$ in ascending powers of x, up to and including the term in x^3, simplifying the coefficients.

8. Expand $\dfrac{1}{\sqrt{(4 + 2x^2)}}$ in ascending powers of x, up to and including the term in x^6, simplifying the coefficients.

9. The first 3 terms in the expansion of $(2 + px)^{-2}$ are $\dfrac{1}{4} + 6x + qx^2$.

    a) Find the value of p and the value of q.   b) Hence find the term in x^3.

10. In the expansion of $(4 - kx)^{\frac{1}{2}}$ where $k \neq 0$, the coefficient of the term in x is 8 times the coefficient of the term in x^2. Find the value of k.

7.4 Binomial expansions and partial fractions

> We can use partial fractions to help us simplify the expansions of more complex expressions.

Example 15

a) Express $\dfrac{1}{(x + 1)(x - 1)^2}$ in partial fractions.

b) Hence obtain the expansion of $\dfrac{1}{(x + 1)(x - 1)^2}$ in ascending powers of x, up to and including the term in x^2.

....

a) In Exercise 2.1B question 4, you found that

$$\dfrac{1}{(x + 1)(x - 1)^2} = \dfrac{1}{4(x + 1)} - \dfrac{1}{4(x - 1)} + \dfrac{1}{2(x - 1)^2}$$

b) $\dfrac{1}{4(x + 1)} = \dfrac{1}{4}(1 + x)^{-1} = \dfrac{1}{4}[1 + (-1)\,x + \dfrac{(-1)(-2)}{2 \times 1}\,x^2 + \ldots]$

Expand each of the three terms separately.

$\qquad = \dfrac{1}{4}(1 - x + x^2 + \ldots) = \dfrac{1}{4} - \dfrac{1}{4}x + \dfrac{1}{4}x^2 + \ldots$

for $|x| < 1$, i.e. for $-1 < x < 1$.

$\dfrac{1}{4(x - 1)} = \dfrac{1}{4}(-1)^{-1}(1 - x)^{-1} = -\dfrac{1}{4}[1 + (-1)(-x) + \dfrac{(-1)(-2)}{2 \times 1}(-x)^2 + \ldots]$

We want to expand $(1 - x)^{-1}$ not $(x - 1)^{-1}$

$\qquad = -\dfrac{1}{4}(1 + x + x^2 + \ldots) = -\dfrac{1}{4} - \dfrac{1}{4}x - \dfrac{1}{4}x^2 + \ldots$

for $|-x| < 1$, i.e. for $-1 < x < 1$.

$\dfrac{1}{2(x - 1)^2} = \dfrac{1}{2}(-1)^{-2}(1 - x)^{-2} = \dfrac{1}{2}[1 + (-2)(-x) + \dfrac{(-2)(-3)}{2 \times 1}(-x)^2 + \ldots]$

$\qquad = \dfrac{1}{2}[1 + 2x + 3x^2 + \ldots] = \dfrac{1}{2} + x + \dfrac{3}{2}x^2 + \ldots$

for $|-x| < 1$, i.e. for $-1 < x < 1$.

Add the first three terms of each expansion.

Hence,

$\dfrac{1}{(x + 1)(x - 1)^2} = \dfrac{1}{4} - \dfrac{1}{4}x + \dfrac{1}{4}x^2 - \dfrac{1}{4} - \dfrac{1}{4}x + \dfrac{1}{4}x^2 + \dfrac{1}{2} + x + \dfrac{3}{2}x^2 = \dfrac{1}{2} + \dfrac{1}{2}x + \dfrac{3}{2}x^2$

for $-1 < x < 1$.

This is true for all the separate expansions.

Example 16

a) Express $\dfrac{2x^2 - x + 6}{(x + 1)(x^2 + 2)}$ in partial fractions.

b) Hence obtain the expansion of $\dfrac{2x^2 - x + 6}{(x + 1)(x^2 + 2)}$ in ascending powers of x, up to and including the term in x^3.

. . . .

a) See Example 5, where we showed that

$$\frac{2x^2 - x + 6}{(x + 1)(x^2 + 2)} = \frac{3}{x + 1} - \frac{x}{x^2 + 2}$$

b) $\dfrac{3}{x + 1} = 3(1 + x)^{-1} = 3\left[1 + (-1)x + \dfrac{(-1)(-2)}{2 \times 1}x^2 + \dfrac{(-1)(-2)(-3)}{3 \times 2 \times 1}x^3 + \ldots\right]$

$= 3(1 - x + x^2 - x^3) = 3 - 3x + 3x^2 - 3x^3$

for $|x| < 1$, i.e. for $-1 < x < 1$.

$\dfrac{x}{x^2 + 2} = x(2 + x^2)^{-1} = x(2)^{-1}\left(1 + \dfrac{x^2}{2}\right)^{-1} = \dfrac{x}{2}\left(1 + \dfrac{x^2}{2}\right)^{-1}$

> We want to expand a bracket in the form $(1 + x)^n$

$= \dfrac{x}{2}\left[1 + (-1)\left(\dfrac{x^2}{2}\right) + \dfrac{(-1)(-2)}{2 \times 1}\left(\dfrac{x^2}{2}\right)^2 + \dfrac{(-1)(-2)(-3)}{3 \times 2 \times 1}\left(\dfrac{x^2}{2}\right)^3 + \ldots\right]$

> Use $\dfrac{x^2}{2}$ instead of x in the expansion.

$= \dfrac{x}{2}\left[1 - \dfrac{x^2}{2} + \dfrac{x^4}{4} + \ldots\right] = \dfrac{x}{2} - \dfrac{x^3}{4} + \ldots$

for $\left|\dfrac{x^2}{2}\right| < 1$, i.e. for $0 < x^2 < 2$, or $-\sqrt{2} < x < \sqrt{2}$

Hence $\dfrac{2x^2 - x + 6}{(x + 1)(x^2 + 2)} = (3 - 3x + 3x^2 - 3x^3) - \left(\dfrac{x}{2} - \dfrac{x^3}{4}\right)$

> The overlap of $-1 < x < 1$ and $-\sqrt{2} < x < \sqrt{2}$ is $-1 < x < 1$.

$= 3 - \dfrac{7}{2}x + 3x^2 - \dfrac{11}{4}x^3$ for $-1 < x < 1$.

Exercise 7.4

1. a) Express $\dfrac{3x + 5}{(1 - x)(1 + 3x)}$ in partial fractions.

  b) Hence obtain the expansion of $\dfrac{3x + 5}{(1 - x)(1 + 3x)}$ in ascending powers of x, up to and including the term in x^2.

2. a) Express $\dfrac{x^2}{(1 + 2x)(1 + x)^2}$ in partial fractions.

  b) Hence find the coefficient of x^3 in the expansion of $\dfrac{x^2}{(1 + 2x)(1 + x)^2}$.

3. Let $f(x) = \dfrac{x+2}{x^2-1}$.

   a) Express $f(x)$ in partial fractions.

   b) Hence obtain the expansion of $f(x)$ in ascending powers of x, up to and including the term in x^3.

4. a) Express $\dfrac{8 - 11x + 4x^2}{(1-x)^2(2-x)}$ in partial fractions.

   b) Hence find the first 3 terms in the expansion of $\dfrac{8 - 11x + 4x^2}{(1-x)^2(2-x)}$ in ascending powers of x, simplifying your answer.

5. a) Express $\dfrac{1}{(1-x)(x-2)}$ in partial fractions.

   b) Hence find the coefficient of x^2 in the expansion of $\dfrac{1}{(1-x)(x-2)}$.

6. a) Express $\dfrac{1}{(1-x)(1+x^2)}$ in partial fractions.

   b) Hence find the first 3 terms in the expansion of $\dfrac{1}{(1-x)(1+x^2)}$ in ascending powers of x.

7. Let $f(x) = \dfrac{3-x}{(1-2x)(1+x^2)}$.

   a) Express $f(x)$ in partial fractions.

   b) Hence obtain the expansion of $f(x)$ in ascending powers of x, up to and including the term in x^3.

Summary exercise 7

1. Express in partial fractions $\dfrac{3x+11}{x^2-x-6}$

2. Express in partial fractions $\dfrac{2x-1}{(x+1)(x^2+1)}$

3. Express in partial fractions $\dfrac{x^2+3x-1}{(x-2)(x+1)}$

4. Express in partial fractions $\dfrac{3x+1}{(x-1)^2(x+2)}$

5. Express in partial fractions

   $\dfrac{32+45x-8x^2}{10(x+3)(x-2)(2x-1)}$

6. Express in partial fractions $\dfrac{96+12x-5x^2}{(x+4)^2(x-4)}$

7. Express in partial fractions $\dfrac{7x^2+8x-11}{(x^2-2)(x+3)}$

8. Find the values of the constants A, B, C, and D such that

    $$\frac{5x^3 - 11x^2 + 7x - 3}{x(x-1)^2} = A + \frac{B}{x} + \frac{C}{x-1} + \frac{D}{(x-1)^2}$$

9. A function $f(x)$ is defined as $f(x) = (2 - 5x)^{-2}$. Obtain the expansion of $f(x)$ in ascending powers of x, as far as the term in x^3, and simplify your answer.

10. a) Expand $(1 - x)^{\frac{1}{2}}$ in ascending powers of x, up to and including the term in x^3.

    b) Use your expansion to evaluate $\sqrt{(0.9)}$ giving your answer to 4 decimal places.

11. Find the binomial expansion of $\dfrac{1}{\left(4 + 3x\right)^3}$ in ascending powers of x, up to and including the term in x^2. Simplify your answer.

12. a) Expand $\dfrac{1}{\sqrt{4 + 2x}}$ in ascending powers of x, up to and including the term in x^3.

    b) Use your expansion to find the value of $4.02^{-\frac{1}{2}}$.

13. Expand $(2 - x)(\sqrt{1 - 2x})$ in ascending powers of x, up to and including the term in x^3, simplifying the coefficients.

14. a) Express $\dfrac{2x + 5}{(1 + x)(2 + x)}$ in partial fractions.

    b) Hence obtain the expansion of $\dfrac{2x + 5}{(1 + x)(2 + x)}$ in ascending powers of x, up to and including the term in x^3, simplifying your answer.

15. $f(x) = \dfrac{3x^2 - x + 4}{(1 + x)(1 - x)^2}$.

    a) Express $f(x)$ in partial fractions.

    b) Hence obtain the expansion of $f(x)$ in ascending powers of x, up to and including the term in x^2.

16. i) Express $\dfrac{2 - x + 8x^2}{(1 - x)(1 + 2x)(2 + x)}$ in partial fractions. [5]

    ii) Hence obtain the expansion of $\dfrac{2 - x + 8x^2}{(1 - x)(1 + 2x)(2 + x)}$ in ascending powers of x, up to and including the term in x^2. [5]

    Cambridge International AS and A Level Mathematics 9709 Paper 3 Q9 October/November 2007

17. i) Simplify $\left(\sqrt{(1+x)} + \sqrt{(1-x)}\right)\left(\sqrt{1+x} - \sqrt{(1-x)}\right)$, showing your working, and deduce that

    $$\frac{1}{\sqrt{(1+x)} + \sqrt{(1-x)}} = \frac{\sqrt{(1+x)} - \sqrt{(1-x)}}{2x} \quad [2]$$

    ii) Using the result, or otherwise, obtain the expansion of $\dfrac{1}{\sqrt{(1+x)} + \sqrt{(1-x)}}$ in ascending powers of x, up to and including the term in x^2. [4]

    Cambridge International AS and A Level Mathematics 9709 Paper 3 Q5 October/November 2006

18. i) Express $\dfrac{3x^2 + x}{(x + 2)(x^2 + 1)}$ in partial fractions. [5]

    ii) Hence obtain the expansion of $\dfrac{3x^2 + x}{(x + 2)(x^2 + 1)}$ in ascending powers of x, up to and including the term in x^3. [5]

    Cambridge International AS and A Level Mathematics 9709 Paper 3 Q9 October/November 2005

Chapter summary

Partial fractions

- **Type 1:** Algebraic fractions with **linear factors in the denominator** can be split into partial fractions in the following way: $\dfrac{1}{(x+p)(x+q)} \equiv \dfrac{A}{x+p} + \dfrac{B}{x+q}$

- **Type 2:** Algebraic fractions with a **repeated factor in the denominator** can be split into partial fractions in the following way: $\dfrac{1}{(x+p)^2} \equiv \dfrac{A}{x+p} + \dfrac{B}{(x+p)^2}$

- **Type 3:** Algebraic fractions with a **quadratic factor in the denominator** can be split in to partial fractions in the following way: $\dfrac{1}{(x+p)(x^2+q)} \equiv \dfrac{A}{x+p} + \dfrac{Bx+C}{x^2+q}$

- **Type 4: Improper algebraic fractions** occur when a polynomial of degree n is divided by another polynomial of degree n. They can be split into partial fractions by first doing long division, and then splitting the remainder into partial fractions using one of the techniques discussed in Types 1, 2, or 3: $\dfrac{x^2}{(x+1)(x+2)} = 1 + \dfrac{-3x-2}{(x+1)(x+2)}$

Binomial expansions of $(1+x)^n$ where n is not a positive integer

- $(1+x)^n = 1 + nx + \dfrac{n(n-1)}{2!}x^2 + \dfrac{n(n-1)(n-2)}{3!}x^3 + \ldots$ where n is any integer or fraction and $|x| < 1$.

Binomial expansions of $(a+x)^n$ where n is not a positive integer

- First put it in the form $a^n\left(1 + \dfrac{x}{a}\right)^n$

Binomial expansion and partial fractions

- We can use partial fractions to help us simplify the expansions of more complex expressions.

In recent years, many parts of the world have been badly hit by severe flooding caused by changing weather patterns. It often happens when more rain falls on already saturated ground, or rain falls more quickly than it can be absorbed into the ground and runs off to lower levels. Differential equations and integration techniques are important mathematical tools employed in the search for ways to predict when and where flooding might occur and ways to lessen the effects of it.

Objectives

- Integrate rational functions by means of decomposition into partial fractions (restricted to the types of partial fractions specified in Chapter 7).

- Recognise an integrand of the form $\dfrac{kf'(x)}{f(x)}$, and integrate, for example, $\dfrac{x}{x^2 + 1}$ or $\tan x$.

- Recognise when an integrand can usefully be regarded as a product, and use integration by parts to integrate, for example, $x \sin 2x$, $x^2 e^x$ or $\ln x$.

- Use a given substitution to simplify and evaluate either a definite or an indefinite integral.

Before you start

You should know how to:

1. Decompose rational functions into partial fractions,

   e.g. **a)** $\dfrac{x+3}{(x-2)(x+1)} = \dfrac{5}{3(x-2)} - \dfrac{2}{3(x+1)}$

   e.g. **b)** $\dfrac{3x^2+2}{x(x-1)^2} \equiv \dfrac{2}{x} + \dfrac{1}{x-1} + \dfrac{5}{(x-1)^2}$

2. Differentiate the logarithm of a function,

   e.g. $\dfrac{d}{dx}[\ln(e^x - 3x)] = \dfrac{e^x - 3}{e^x - 3x}$

Skills check:

1. Express the following as partial fractions.

   a) $\dfrac{x^2 + 3x - 1}{(x-2)(x+1)}$ **b)** $\dfrac{3x+1}{(x-1)^2(x+2)}$

   c) $\dfrac{7x^2 + 8x - 11}{(x^2 - 2)(x+3)}$

2. Find $\dfrac{dy}{dx}$ when $y = \ln\sqrt{(2x+1)}$

In Chapter 5 you extended integration to include a number of different types of function which you had learned how to differentiate in Chapter 4, including logarithms, exponentials, and trigonometric functions. In Chapter 4 you also learned how to differentiate products and quotients, and techniques of implicit and parametric differentiation. This chapter will extend the range of integrals you can tackle.

8.1 Integration using partial fractions

In **7.1** you learned how to express 4 different types of function in partial fractions. One of the applications of this is that almost all the partial fraction forms are the sum of integrable functions where the initial rational function is not directly integrable. The first four examples here use functions which were split into partial fractions in Chapter 7, so we are looking at the extra to be done when an integral is required.

> To integrate difficult algebraic fractions, first look to see if you can split the fraction into partial fractions, where each partial fraction is integrable.

Example 1 deals with fractions which have linear factors in the denominator.

Example 1

Find $\int \dfrac{x+3}{(x-2)(x+1)}\,dx$

$\dfrac{x+3}{(x-2)(x+1)} = \dfrac{5}{3(x-2)} - \dfrac{2}{3(x+1)}$ ← From Example 1 in Chapter 7.

So $\int \dfrac{x+3}{(x-2)(x+1)}\,dx = \int \left(\dfrac{5}{3(x-2)} - \dfrac{2}{3(x+1)}\right)dx$ ← Two terms which are each integrable as logarithmic functions.

$\int \dfrac{x+3}{(x-2)(x+1)}\,dx = \int \left(\dfrac{5}{3(x-2)} - \dfrac{2}{3(x+1)}\right)dx = \dfrac{5}{3}\ln(x-2) - \dfrac{2}{3}\ln(x+1) + c$

Example 2 deals with fractions which have repeated linear factors in the denominator.

Example 2

Calculate $\int \left(\dfrac{3x^2+2}{x(x-1)^2}\right)dx$

$\dfrac{3x^2+2}{x(x-1)^2} = \dfrac{2}{x} + \dfrac{1}{x-1} + \dfrac{5}{(x-1)^2}$ ← From Example 3 in Chapter 7.

$\int \dfrac{3x^2+2}{x(x-1)^2}\,dx = \int \left(\dfrac{2}{x} + \dfrac{1}{x-1} + \dfrac{5}{(x-1)^2}\right)dx$ ← Three terms which are each integrable.

$= 2\ln x + \ln(x-1) - \dfrac{5}{(x-1)} + c$ ← Integrating each term carefully, and including the constant of integration.

Example 3 deals with fractions which have a quadratic factor in the denominator.

Example 3

Calculate $\displaystyle\int \frac{2x^2 - x + 6}{(x + 1)(x^2 + 2)}\,dx$

$$\frac{2x^2 - x + 6}{(x + 1)(x^2 + 2)} = \frac{3}{x + 1} - \frac{x}{x^2 + 2}$$

From Example 5 in Chapter 7.

$$\int \frac{2x^2 - x + 6}{(x + 1)(x^2 + 2)}\,dx = \int \left(\frac{3}{x + 1}\right)dx - \int \left(\frac{x}{x^2 + 2}\right)dx$$

The first of these you already know how to integrate and the second you will meet in the next section of this chapter.

$$= 3\ln(x + 1) - \int \left(\frac{x}{x^2 + 2}\right)dx$$

Example 4 deals with improper algebraic fractions.

Example 4

Improper algebraic fractions. Calculate $\displaystyle\int \left(\frac{x^2 - x + 5}{(x - 1)(x + 2)}\right)dx$

$$\frac{x^2 - x + 5}{(x - 1)(x + 2)} = 1 + \frac{5}{3(x - 1)} - \frac{11}{3(x + 2)}$$

From Example 7 in Chapter 7.

$$\int \frac{x^2 - x + 5}{(x - 1)(x + 2)}\,dx = \int \left(1 + \frac{5}{3(x - 1)} - \frac{11}{3(x + 2)}\right)dx$$

Three terms which are each integrable.

$$= x + \frac{5}{3}\ln(x - 1) - \frac{11}{3}(x + 2) + c$$

Example 5

By first expressing $\dfrac{1}{(x - 2)(x - 3)}$ in partial fractions find $\displaystyle\int \left(\frac{1}{(x - 2)(x - 3)}\right)dx$

$$\frac{1}{(x - 2)(x - 3)} = \frac{A}{x - 2} + \frac{B}{x - 3}$$

Or any standard technique to put into partial fractions.

$$1 = A(x - 3) + B(x - 2); \quad x = 2 \Rightarrow A = -1; \quad x = 3 \Rightarrow B = 1$$

$$\int \left(\frac{1}{(x - 2)(x - 3)}\right)dx = \int \left(\frac{-1}{x - 2} + \frac{1}{x - 3}\right)dx$$

So
$$= -\ln(x - 2) + \ln(x - 3) + c$$

Integrating each term.

$$= \ln\left(k\left(\frac{x - 3}{x - 2}\right)\right)$$

Combining into a single logarithm.

Example 6

By first expressing $\dfrac{4}{x^2(x-2)}$ in partial fractions find $\displaystyle\int\left(\dfrac{4}{x^2(x-2)}\right)dx$

$\dfrac{4}{x^2(x-2)} = \dfrac{A}{x-2} + \dfrac{B}{x} + \dfrac{C}{x^2}$ ← Separate the repeated factor x^2 into two denominators.

$4 = Ax^2 + Bx(x-2) + C(x-2)$

$x = 2 \Rightarrow 4 = A(4) + 0 \Rightarrow A = 1$ ← This means B and C are eliminated.

$4 = x^2 + Bx^2 - 2x + Cx - 2C \Rightarrow B = -1, C = -2$ ← Substituting for A means B and C can be found by equating coefficients.

$\Rightarrow \dfrac{4}{x^2(x-2)} = \dfrac{1}{x-2} - \dfrac{1}{x} - \dfrac{2}{x^2}$

$\displaystyle\int\left(\dfrac{4}{x^2(x-2)}\right)dx = \int\left(\dfrac{1}{x-2} - \dfrac{1}{x} - \dfrac{2}{x^2}\right)dx$

$= \ln(x-2) - \ln x + \dfrac{2}{x} + k$ ← Integrating each term, avoiding using 'c' as constant of integration, having used 'C' before.

$= \ln\left(\dfrac{x-2}{x}\right) + \dfrac{2}{x} + k$ ← Simplifying the logarithmic term.

Example 7

By first expressing $\dfrac{x+2}{x(x^2+4)}$ in partial fractions find $\displaystyle\int\left(\dfrac{x+2}{x(x^2+4)}\right)dx$

$\dfrac{2x-4}{x(x^2+4)} = \dfrac{A}{x} + \dfrac{Bx+C}{(x^2+4)}$ ← Put $Bx+C$ as the numerator of x^2+4

$2x - 4 = A(x^2+4) + (Bx+C)x$

$x = 0 \Rightarrow -4 = A(4) + 0 \Rightarrow A = -1$ ← This means B and C are eliminated.

$2x - 4 = -x^2 - 4 + Bx^2 + Cx \Rightarrow B = 1, C = 2$ ← Substituting for A means B and C can be found by equating coefficients.

$\Rightarrow \dfrac{2x-4}{x(x^2+4)} = \dfrac{-1}{x} + \dfrac{x+2}{(x^2+4)}$

$\displaystyle\int\left(\dfrac{x+2}{x(x^2+4)}\right)dx = \int\left(\dfrac{1}{x} + \dfrac{x+2}{(x^2+4)}\right)dx$ ← You will learn how to integrate the second term later in this chapter.

$= -\ln x + \displaystyle\int\left(\dfrac{x+2}{(x^2+4)}\right)dx$

Example 8

Calculate $\displaystyle\int\left(\frac{3x^3 - 4x^2 - 2}{x^2(x-1)}\right)dx$

$\dfrac{x^3 - 4x^2 - 2}{x^2(x-1)} = A + \dfrac{B}{x} + \dfrac{C}{x^2} + \dfrac{D}{(x-1)}$ ← There are now no improper fractions.

$3x^3 - 4x^2 - 2 = Ax^2(x-1) + Bx(x-1) + C(x-1) + Dx^2$ ← Multiply each term by $x^2(x-1)$

$x = 0 \Rightarrow -2 = -C \Rightarrow C = 2$

$x = 1 \Rightarrow 3 - 4 - 2 = -3 = D \Rightarrow D = -3$

$3x^3 - 4x^2 - 2 = Ax^2(x-1) + Bx(x-1) + 2(x-1) - 3x^2$ ← Substituting values for C and D.

$\qquad = Ax^3 - Ax^2 + Bx^2 - Bx + 2x - 2 - 3x^2$

$\Rightarrow (x^3): 2 = -1 + A \Rightarrow A = 3; \ (x^2): -4 = -3 + B - 3 \Rightarrow B$ ← Equating coefficients to find the last two constants and using the extra term to check.

check: coefficient of $x: 0 = -B + 2 \Rightarrow B = 2$

$\dfrac{3x^3 - 4x^2 - 2}{x^2(x-1)} = 3 + \dfrac{2}{x} + \dfrac{2}{x^2} - \dfrac{3}{(x-1)}$

$\displaystyle\int\left(\frac{3x^3 - 4x^2 - 2}{x^2(x-1)}\right)dx = \int\left(3 + \frac{2}{x} + \frac{2}{x^2} - \frac{3}{(x-1)}\right)dx$

$\qquad = 3x + 2\ln x - \dfrac{2}{x} - 3\ln(x-1) + c$ ← Integrating each term.

$\qquad = 3x - \dfrac{2}{x} + \ln\left(\dfrac{x^2}{(x-1)^3}\right) + c$ ← Combining the logarithmic terms.

Note: The Cambridge syllabus states that the order of the numerator will be no more than the order of the denominator. As such, in the examination you will get no more than a constant along with the rational terms. You can divide out expression where the numerator had order higher than the denominator – question 3, parts b and e, below are simple examples where it will be $ax + b$ plus the rational expressions. You may choose to omit these questions which take you beyond what will be asked in the examination.

Exercise 8.1

1. By first expressing each rational expression using partial fractions, find

a) $\int \left(\dfrac{7}{(x+3)(2x-1)} \right) dx$

b) $\int \left(\dfrac{2}{(x+5)(x+7)} \right) dx$

c) $\int \left(\dfrac{7}{(x+5)(x-2)} \right) dx$

d) $\int \left(\dfrac{2-x}{(2x+3)(x+5)} \right) dx$

e) $\int \left(\dfrac{3-x}{(2+3x)(1-4x)} \right) dx$

f) $\int \left(\dfrac{4-3x}{(x-2)(3-2x)} \right) dx$

g) $\int_{1}^{2} \left(\dfrac{4}{(x+3)(x+5)} \right) dx$

h) $\int_{-0.5}^{0.5} \left(\dfrac{5}{(3+2x)(1-x)} \right) dx$

i) $\int_{3}^{4} \left(\dfrac{5x}{(x-2)(x+3)} \right) dx.$

2. By first expressing each rational expression using partial fractions, find

a) $\int \left(\dfrac{1}{x(x-1)^2} \right) dx$

b) $\int \left(\dfrac{1}{x(x+1)^2} \right) dx$

c) $\int \left(\dfrac{16x}{(x+1)(x-3)^2} \right) dx$

d) $\int_{7}^{12} \left(\dfrac{4-x}{(x-2)^2} \right) dx$

e) $\int_{4}^{5} \left(\dfrac{25}{(2x-3)^2(1+x)} \right) dx$

f) $\int_{3}^{4} \left(\dfrac{4}{(x^3-4x^2+4x)} \right) dx.$

3. By first expressing each improper fraction using partial fractions, find

a) $\int \left(\dfrac{x^2-2x+6}{(x+1)(x-2)} \right) dx$

b) $\int \left(\dfrac{2x^2-x}{(x+1)(x-3)} \right) dx$

c) $\int_{4}^{5} \left(\dfrac{x^2-3x+10}{x^2-x-6} \right) dx$

d) $\int_{3}^{4} \left(\dfrac{10x^2-26x+10}{(2x^2-5x)} \right) dx.$

4. By first expressing $\dfrac{2\sqrt{3}}{x^2-3}$ in partial fractions show that $\displaystyle\int_{2\sqrt{3}}^{3\sqrt{3}} \left(\dfrac{2\sqrt{3}}{x^2-3} \right) dx = \ln\left(\dfrac{3}{2}\right).$

5. By first expressing $\dfrac{4x^2+9x+4}{2x^2+5x+3}$ in partial fractions show that $\displaystyle\int_{0}^{1} \left(\dfrac{4x^2+9x+4}{2x^2+5x+3} \right) dx = 2 + \ln\left(\dfrac{5}{6}\right).$

6. Find the area bounded by the graph of $y = \dfrac{1}{(x-3)(x-2)}$, the x-axis, the y-axis and the line $x = 1$.

7. Find the area bounded by the graph of $y = \dfrac{x^3+3x^2-7x}{(x+2)(x-1)^2}$, the x-axis and the lines $x = 2$ and $x = 3$.

8. Show that the area bounded by the graph of $y = \dfrac{2x^2-4x}{(1-2x)(1-x^2)}$, the x-axis and the lines $x = -3$

   and $x = -2$ is $\ln(2.1)$

8.2 Integration of $\dfrac{kf'(x)}{f(x)}$

In **4.2** you saw that if $y = \ln[f(x)]$ then $\dfrac{dy}{dx} = \dfrac{f'(x)}{f(x)}$. This means that

$$\int\left(\frac{kf'(x)}{f(x)}\right)dx = k\ln(f(x)) + c$$

Example 9

Find $\displaystyle\int\left(\frac{2x}{x^2 + 2}\right)dx$

$\displaystyle\int\left(\frac{2x}{x^2 + 2}\right)dx = \ln(x^2 + 2) + c$ ⟵ Since $\dfrac{d}{dx}(x^2 + 2) = 2x$

Example 10

Calculate $\displaystyle\int\frac{2x^2 - x + 6}{(x + 1)(x^2 + 2)}dx$

$\displaystyle\int\frac{2x^2 - x + 6}{(x + 1)(x^2 + 2)}dx = 3\ln(x + 1) - \int\left(\frac{x}{x^2 + 2}\right)dx$ ⟵ From Example 3.

$\displaystyle\qquad\qquad\qquad = 3\ln(x + 1) - \frac{1}{2}\ln(x^2 + 2) + c$ ⟵ Using the result from Example 9.

Example 11

Find $\displaystyle\int\left(\frac{e^{2x}}{e^{2x} + 3}\right)dx$

$\displaystyle\int\left(\frac{e^{2x}}{e^{2x} + 3}\right)dx = \frac{1}{2}\ln(e^{2x} + 3) + c$ ⟵ Since $\dfrac{d}{dx}(e^{2x} + 3) = 2e^{2x}$

Example 12

Find $\int \left(\dfrac{1 + 2xe^{x^2}}{x + e^{x^2}} \right) dx$

$\int \left(\dfrac{1 + 2xe^{x^2}}{x + e^{x^2}} \right) dx = \ln(x + e^{x^2}) + c$ ← Since $\dfrac{d}{dx}(x + e^{x^2}) = 1 + 2xe^{x^2}$

Example 13

Find $\int \tan 2x \, dx$

$\int \tan 2x \, dx = \int \left(\dfrac{\sin 2x}{\cos 2x} \right) dx$ ← Writing tan in this form allows you to view it as a logarithmic integral, since $\dfrac{d}{dx}(\cos 2x) = -2\sin 2x$

$\qquad\qquad = -\dfrac{1}{2}\ln(\cos 2x) + c$

$\qquad\qquad = \dfrac{1}{2}\ln(\sec 2x) + c$ ← This is an optional simplification.

Example 14

Find $\int_0^1 \left(\dfrac{e^x - e^{-x}}{e^x + e^{-x}} \right) dx$

$\int_0^1 \left(\dfrac{e^x - e^{-x}}{e^x + e^{-x}} \right) dx = \left[\ln(e^x + e^{-x}) \right]_0^1$ ← Since $\dfrac{d}{dx}(e^x + e^{-x}) = e^x - e^{-x}$

$\qquad\qquad = (\ln(e + e^{-1})) - (\ln(1 + 1))$ ← Substituting the upper and lower limits.

$\qquad\qquad = \ln(e + e^{-1}) - \ln 2$

$\qquad\qquad = \ln\left(\dfrac{e^2 + 1}{2e} \right)$ ← Simplifying the logarithmic function.

Exercise 8.2

1. Find the following integrals

a) $\displaystyle\int \frac{6x^2}{x^3+1}\,dx$

b) $\displaystyle\int \frac{\cos 2x}{1+\sin 2x}\,dx$

c) $\displaystyle\int \frac{x-2}{x^2-4x+3}\,dx$

d) $\displaystyle\int \frac{e^{3x}}{4+2e^{3x}}\,dx$

e) $\displaystyle\int \frac{x}{x^2+4}\,dx$

f) $\displaystyle\int (\cot x)\,dx$

g) $\displaystyle\int_0^{\frac{1}{6}\pi} \left(\frac{\cos 2x}{1+\sin 2x}\right)\,dx$

h) $\displaystyle\int_0^1 \left(\frac{e^{2x}}{1+e^{2x}}\right)\,dx$

i) $\displaystyle\left(\frac{2x+5}{(x+3)(x+2)}\right)\,dx.$

2. Find $\displaystyle\int_0^2 \frac{x-1}{x^2-2x+2}\,dx.$

3. a) Show that $\dfrac{d}{dx}(\sin^2 x) = \sin 2x$.

   b) Hence find $\displaystyle\int \frac{1+\sin 2x}{x+\sin^2 x}\,dx.$

4. a) Show that $\dfrac{d}{dx}\left(\tan x + \frac{1}{3}\tan^3 x\right) = \sec^4 x$.

   b) Hence find $\displaystyle\int_{\frac{\pi}{6}}^{\frac{\pi}{3}} \sec^4 x\,dx.$

5. a) Find $\dfrac{d}{dx}(\tan x - x)$.

   b) Hence find $\displaystyle\int_{\frac{\pi}{6}}^{\frac{\pi}{4}} \frac{\tan^2 x}{\tan x - x}\,dx.$

6. Find $\displaystyle\int \frac{e^{2x}-e^{-2x}}{e^{2x}+e^{-2x}}\,dx.$

7. a) Show that $\dfrac{1-\tan x}{1+\tan x} = \dfrac{\cos x - \sin x}{\cos x + \sin x}$.

   b) Hence show that $\displaystyle\int_0^{\frac{\pi}{4}} \frac{1-\tan x}{1+\tan x}\,dx = \frac{1}{2}\ln 2.$

8. a) Show that $\dfrac{1}{1+e^x} = \dfrac{e^{-x}}{e^{-x}+1}$.

   b) Hence find $\displaystyle\int_1^2 \frac{1}{1+e^x}\,dx$ giving your answer correct to 3 decimal places.

8.3 Integration by parts

In **4.3** you met the product rule for differentiation,

$$\frac{d}{dx}(uv) = u\frac{dv}{dx} + v\frac{du}{dx}$$

Since integration is the reverse process to differentiation, it follows that

(integrating both sides) $uv = \int \left(u\frac{dv}{dx} \right) dx + \int \left(v\frac{du}{dx} \right) dx.$

This can be rewritten to give the rule

> **Integration by parts:** $\int \left(u\frac{dv}{dx} \right) dx = uv - \int \left(v\frac{du}{dx} \right) dx$

However, when you are faced with an integral it is not presented in the form of the left-hand side, it is expressed as a product – something like $\int (xe^{2x})\, dx$ and you have to decide which to treat as u and which as $\frac{dv}{dx}$. The key to making this choice lies in the form of the second term on the right: $-\int \left(v\frac{du}{dx} \right) dx$. Choose u which makes $\int \left(v\frac{du}{dx} \right) dx$ integrable directly, or at least simpler. So in the example choose u to be x and then v is whatever satisfies $\frac{dv}{dx} = e^{2x}$ so $v = \frac{1}{2}e^{2x}$

Using integration by parts then gives:

$$\int (xe^{2x})dx = x\left(\frac{1}{2}e^{2x} \right) - \int 1 \times \left(\frac{1}{2}e^{2x} \right)dx$$

$$= \frac{1}{2}xe^{2x} - \frac{1}{4}e^{2x} + c$$

> Differentiating x gives 1 so the second term now is recognisable as a standard integral form.

Generally if the integral is in the form $\int x^n f(x)\, dx$ and $f(x)$ is an exponential or trigonometric function (sin or cos) then take u as x^n because it will become simpler when it is differentiated (you may need to do integration by parts more than once) and the function $f(x)$ does not get any more complicated when it is integrated repeatedly.

Example 15

Find $\int x \cos 3x \, dx$

$\int x \cos 3x \, dx = x \frac{1}{3} \sin 3x - \int \left(\frac{1}{3} \sin 3x \right) dx$

$\qquad = \frac{1}{3} x \sin 3x + \frac{1}{9} \cos 3x + c$

$u = x; \quad \frac{dv}{dx} = \cos 3x \quad \Rightarrow v = \frac{1}{3} \sin 3x$

Don't forget the '+ c' at the end here and be careful with negatives with the $\frac{\sin}{\cos}$ integrals.

Example 16

Find $\int x^2 e^{2x} \, dx$

$\int x^2 e^{2x} \, dx = x^2 \frac{1}{2} e^{2x} - \int \left(2x \frac{1}{2} e^{2x} \right) dx$

$\qquad = \frac{1}{2} x^2 e^{2x} - \int (x e^{2x}) dx$

$\qquad = \frac{1}{2} x^2 e^{2x} - x \frac{1}{2} e^{2x} + \int \left(\frac{1}{2} e^{2x} \right) dx$

$\qquad = \frac{1}{2} x^2 e^{2x} - \frac{1}{2} x e^{2x} + \frac{1}{4} e^{2x} + c$

$u = x^2; \quad \frac{dv}{dx} = e^{2x} \quad \Rightarrow v = \frac{1}{2} e^{2x}$

This needs integration by parts again.

Be careful about the negative signs when you use parts more than once.

Example 17

Find $\int x^2 \cos 2x \, dx$

$\int x^2 \cos 2x \, dx = x^2 \frac{1}{2} \sin 2x - \int \left(2x \frac{1}{2} \sin 2x \right) dx$

$\qquad = \frac{1}{2} x^2 \sin 2x - \int (x \sin 2x) dx$

$\qquad = \frac{1}{2} x^2 \sin 2x + x \frac{1}{2} \cos 2x - \int \left(\frac{1}{2} \cos 2x \right) dx$

$\qquad = \frac{1}{2} x^2 \sin 2x + \frac{1}{2} x \cos 2x - \frac{1}{4} \sin 2x + c$

$u = x^2; \quad \frac{dv}{dx} = \cos 2x \quad \Rightarrow v = \frac{1}{2} \sin 2x$

This needs integration by parts again.

When integrating by parts more than once involving sin and cos you need to be very careful with the negatives.

All the above examples used x or x^2 as u; the function to be differentiated to provide a simpler integral on the right-hand side of the integration by parts rule. There is one very important exception you need to look at now (and remember). This is where $\ln x$ appears in the integral because $\frac{d}{dx}(\ln x) = \frac{1}{x}$ but you do not know any function whose derivative is $\ln x$, so you can not use $\frac{dv}{dx} = \ln x$.

Example 18

Find $\int (x\ln x)\,dx$

$\int (x\ln x)\,dx = \left(\frac{1}{2}x^2\right)\ln x - \int \left(\frac{1}{2}x^2\right)\left(\frac{1}{x}\right)dx$

$\qquad u = \ln x; \quad \frac{dv}{dx} = x \quad \Rightarrow v = \frac{1}{2}x^2$

$= \left(\frac{1}{2}x^2\right)\ln x - \int \left(\frac{1}{2}x\right)dx$

Now the second term on the right is recognisable as a standard integral.

$= \left(\frac{1}{2}x^2\right)\ln x - \frac{1}{4}x^2 + c$

It is not immediately obvious that writing in the form of a product will be helpful, but often in mathematics these little devices prove the key step in a useful result.

Using the same process we can now find $\int (\ln x)\,dx$ by thinking of $\ln x$ as the product $1 \times \ln x$.

Example 19

Find $\int (\ln x)\,dx$

$\int (\ln x)\,dx = (x)\ln x - \int (x)\left(\frac{1}{x}\right)dx$

$\qquad u = \ln x; \quad \frac{dv}{dx} = 1 \quad \Rightarrow v = x$

$= x\ln x - \int (1)dx$

Again the second term on the right is recognisable as a standard integral.

$= x\ln x - x + c$

Definite integration which requires integration by parts requires attention to detail in notation about the use of the limits. You can do the substitution of limits into the integrated expression at the first opportunity but you will then be switching back and forth between evaluation at limits and more integration.

The example below leaves all the evaluation at limits until after all the integration is complete and then it does it all in one go.

Example 20

Find $\displaystyle\int_1^2 (x^3 \ln x)\,dx$

$$\int_1^2 (x^3 \ln x)\,dx = \left[\left(\frac{1}{4}x^4\right)\ln x\right]_1^2 - \int \left(\frac{1}{4}x^4\right)\left(\frac{1}{x}\right)dx$$

$u = \ln x; \quad \dfrac{dv}{dx} = x^3 \quad \Rightarrow v = \dfrac{1}{4}x^4$

$$= \left[\left(\frac{1}{4}x^4\right)\ln x\right]_1^2 - \int_1^2 \left(\frac{1}{4}x^3\right)dx$$

Now the second term on the right is recognisable as a standard integral.

$$= \left[\left(\frac{1}{4}x^4\right)\ln x\right]_1^2 - \left[\frac{1}{16}x^4\right]_1^2$$

Finishing the integration processes.

$$= (4\ln 2 - 0) - \left(1 - \frac{1}{16}\right)$$

Now substitute the limits into both terms at the same time.

$$= 4\ln 2 - \frac{15}{16}$$

When doing integration, it's important to remember results you've learned in other areas of mathematics. These will often help you to integrate functions you might not think are integrable at first glance.

Example 21

Find $\displaystyle\int (\ln x^3)\,dx$

$$\int (\ln x^3)\,dx = \int (3\ln x)\,dx$$

Simplifying the logarithm.

$$= 3(x\ln x - x) + c$$

Using the result in Example 19.

Example 22

Find $\int (2x^3 e^{x^2}) \, dx$

$\int (2x^3 e^{x^2}) \, dx = \int (x^2 \cdot 2x e^{x^2}) \, dx$

$\qquad = (x^2 e^{x^2}) - \int (2x e^{x^2}) \, dx$

$\qquad = x^2 e^{x^2} + e^{x^2} + c$

$u = x^2; \quad \dfrac{dv}{dx} = 2x e^{x^2} \quad \Rightarrow v = e^{x^2}$

$\dfrac{du}{dx} = 2x$

Now the second term is an exact integral.

The next example illustrates a surprising extension to the range of functions which can be integrated by parts. In all the examples above, we ended up with an integrable function on the right-hand side after using integration by parts once or twice. However, if we try to integrate $\sin x e^x$ or similar then we will never find an integrable function, no matter how many times we integrate by parts. However, there is a way around this. Integrating by parts twice (whilst keeping the exponential and trigonometric functions in the same role each time) we will end up with $-k$ times the expression on the left as the integral appearing on the right-hand side, so by collecting like terms we can then evaluate the integral. Example 23 shows how this is done.

> Remember with an indefinite integral it is important to remember to introduce the constant of integration at the end.

Example 23

Find $\int (\sin x \cdot e^{2x}) \, dx$

$\int (\sin x \cdot e^{2x}) \, dx = \sin x \cdot \dfrac{1}{2} e^{2x} - \int \left(\cos x \cdot \dfrac{1}{2} e^{2x} \right) dx$

$\qquad = \sin x \cdot \dfrac{1}{2} e^{2x} - \cos x \cdot \dfrac{1}{4} e^{2x} + \int \left(-\sin x \cdot \dfrac{1}{4} e^{2x} \right) dx$

$\Rightarrow \dfrac{5}{4} \int \left(\sin x \cdot e^{2x} \right) dx = \sin x \cdot \dfrac{1}{2} e^{2x} - \cos x \cdot \dfrac{1}{4} e^{2x}$

$\Rightarrow \int \left(\sin x \cdot e^{2x} \right) dx = \dfrac{4}{5} \left(\sin x \cdot \dfrac{1}{2} e^{2x} - \cos x \cdot \dfrac{1}{4} e^{2x} \right) + c$

$\qquad = \dfrac{1}{5} \left(2 \sin x \cdot e^{2x} - \cos x \cdot e^{2x} \right) + c$

$u = \sin x; \quad \dfrac{dv}{dx} = e^{2x} \quad \Rightarrow v = \dfrac{1}{2} e^{2x}$

For the second integral use

$u = \cos x; \quad \dfrac{dv}{dx} = \dfrac{1}{2} e^{2x}$

After integrating by parts twice, the last term is the same function as the left-hand side.

Collecting like terms.

Remember the constant.

Exercise 8.3

1. Find the following integrals using integration by parts.

   a) $\displaystyle\int 4xe^{2x}\,dx$

   b) $\displaystyle\int 3x\sin 2x\,dx$

   c) $\displaystyle\int (2x+1)e^{x}\,dx$

   d) $\displaystyle\int (x+3)\cos 3x\,dx$

   e) $\displaystyle\int xe^{3x-1}\,dx$

   f) $\displaystyle\int (x\ln 5x)\,dx$

2. Calculate the following definite integrals. Note that the first three are integrating the same function as parts **(a)** to **(c)** of question 1.

   a) $\displaystyle\int_{0}^{1} 4xe^{2x}\,dx$

   b) $\displaystyle\int_{0}^{\frac{\pi}{2}} 3x\sin 2x\,dx$

   c) $\displaystyle\int_{0}^{2} (2x+1)e^{x}\,dx$

   d) $\displaystyle\int_{0}^{\frac{\pi}{2}} x\cos\!\left(\frac{1}{2}x\right)dx$

   e) $\displaystyle\int_{1}^{e} x^{4}\ln x\,dx$

   f) $\displaystyle\int_{0}^{\frac{\pi}{2}} x\sin x\,dx$

3. Find the following integrals using integration by parts twice.

   a) $\displaystyle\int 18x^{2}e^{3x}\,dx$

   b) $\displaystyle\int 3x^{2}\cos 2x\,dx$

   c) $\displaystyle\int_{0}^{2} (2x^{2}+1)e^{x}\,dx$

   d) $\displaystyle\int_{1}^{e} x^{2}e^{x}\,dx$

   e) $\displaystyle\int_{0}^{\frac{\pi}{2}} x^{2}\sin x\,dx$

   f) $\displaystyle\int_{\frac{\pi}{6}}^{\frac{\pi}{3}} x^{2}\sin 2x\,dx$

   g) $\displaystyle\int_{0}^{1} x^{2}e^{3x}\,dx$

   h) $\displaystyle\int_{-\infty}^{0} x^{2}e^{3x}\,dx$

   i) $\displaystyle\int_{0}^{\frac{\pi}{4}} e^{3x}\cos 2x\,dx$

4. Find the area bounded by the graph of $y = x\sin x$, the x-axis and the line $x = \dfrac{\pi}{2}$ (as shown in the diagram).

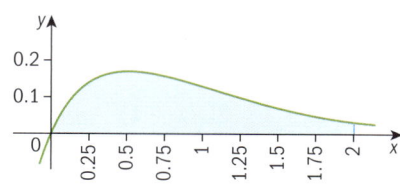

5. Find the area bounded by the graphs of $y = xe^{-2x}$, the x-axis and the line $x = 2$ (as shown in the diagram).

6. **a)** Find the area bounded by the graph of $y = x \cos 2x$, the x-axis and the lines $x = 0$ and $x = \dfrac{\pi}{6}$.

    b) Find the volume of the solid of revolution obtained by rotating this region about the x-axis.

8.4 Integration using substitution

In P1 you met the chain rule for differentiation, $\dfrac{dy}{dx} = \dfrac{dy}{du} \cdot \dfrac{du}{dx}$.

The technique of integration by substitution is essentially the reverse of the chain rule for differentiation.

$\displaystyle\int \dfrac{3x}{\sqrt{x+1}}\,dx$ is not an integral that you can deal with directly at the moment, but using the substitution $u^2 = x + 1$ transforms it into an integral that you can do.

> You could write it as
> $$3\int \dfrac{(x+1)-1}{\sqrt{x+1}}\,dx \text{ and create}$$
> two integrable terms

In order to evaluate this we need to substitute for all the x terms and also for dx (and, for definite integrals, we must also change the limits to limits in the new variable).

> It is sensible, at least at first, to write down everything in the integral in terms of the new variable, u, (including the dx term), and then rewrite the integral completely in one go.

Don't ever write an integral which is a mix of x and u.

$u^2 = x + 1$

$\Rightarrow \sqrt{x+1} = u; \quad 3x = 3(u^2 - 1);$

$2u\dfrac{du}{dx} = 1 \quad \Rightarrow 2u\dfrac{du}{dx}\,dx = dx$

While this looks strange at the moment, when the substitutions are made you will see why the technique works.

Substituting gives:

$\displaystyle\int \dfrac{3x}{\sqrt{x+1}}\,dx = \int \dfrac{3(u^2 - 1)}{u} 2u\dfrac{du}{dx}\,dx$

This is where the chain rule is used to replace $\dfrac{du}{dx}\,dx$ by du.

$\displaystyle\qquad\qquad = \int 6(u^2 - 1)\,du$

$\displaystyle\qquad\qquad = 2u^3 - 6u + c$

Your original integral was in terms of x, so you need to express your answer in terms of x:

$$2u^3 - 6u + c = 2(\sqrt{x+1})^3 - 6\sqrt{x+1} + c$$

So $\displaystyle \int \frac{3x}{\sqrt{x+1}} \, dx = 2(\sqrt{x+1})^3 - 6\sqrt{x+1} + c$ ← There is some simplification which could be done by using $(\sqrt{x+1})^3 = (x+1)\sqrt{x+1}$ but unless you are asked to show a particular form for the answer, this is perfectly acceptable.

In this course, you are not expected to identify appropriate substitutions for yourself, but you will probably be able to spot relationships between the integral form and the substitution you are told to use that would allow you to do without that help later on.

Example 24

Find $\displaystyle \int \sqrt{4-x^2} \, dx$ using the substitution $x = 2\sin\theta$.

$x = 2\sin\theta \implies \sqrt{4-x^2} = \sqrt{4-4\sin^2\theta} = 2\cos\theta$ ← Only one expression here.

$1 = 2\cos\theta\dfrac{d\theta}{dx} \implies dx = 2\cos\theta\dfrac{d\theta}{dx}dx$ ← You need this to be able to complete the transformation of the integral to the new variable.

Substituting gives

$$\int \sqrt{4-x^2} \, dx = \int 2\cos\theta \cdot 2\cos\theta \frac{d\theta}{dx}dx$$

$$= \int 4\cos^2\theta \, d\theta = \int 2(1+\cos 2\theta) \, d\theta$$ ← Using the standard identity.

$$= 2\theta + \sin 2\theta + c = 2\theta + 2\sin\theta\cos\theta + c$$ ← Using the double angle relation.

$$= 2\sin^{-1}\left(\frac{1}{2}x\right) + \frac{1}{2}x\sqrt{4-x^2} + c$$ ← Expressing in terms of the original variable.

When dealing with definite integrals, make sure you express the limits in terms of the new variable, u.

Example 25

Find $\displaystyle\int_1^2 x\sqrt{5x-2}\,\mathrm{d}x$ using the substitution $u = 5x - 2$. Give your answer to 3 significant figures.

$u = 5x - 2 \implies x = \dfrac{u+2}{5}; \ \sqrt{5x-2} = u^{\frac{1}{2}}$

The two terms in the integral.

$x = 1 \implies u = 3; \quad x = 2 \implies u = 8$

Changing the limits.

$5 = \dfrac{\mathrm{d}u}{\mathrm{d}x} \implies \mathrm{d}x = \dfrac{1}{5}\dfrac{\mathrm{d}u}{\mathrm{d}x}\mathrm{d}x$

To be able to complete the transformation of the integral to the new variable.

$\displaystyle\int_1^2 x\sqrt{5x-2}\,\mathrm{d}x = \int_3^8 \left(\dfrac{u+2}{5}\right)u^{\frac{1}{2}}\cdot\dfrac{1}{5}\dfrac{\mathrm{d}u}{\mathrm{d}x}\mathrm{d}x$

Substituting for the limits at the same time.

$= \dfrac{1}{25}\displaystyle\int_3^8 \left(u^{\frac{3}{2}} + 2u^{\frac{1}{2}}\right)\mathrm{d}u$

Taking the common factor outside the integral simplifies things.

$= \dfrac{1}{25}\left[\dfrac{2}{5}u^{\frac{5}{2}} + \dfrac{4}{3}u^{\frac{3}{2}}\right]_3^8$

These are standard integral forms now.

$= \dfrac{1}{25}\left(\dfrac{256}{5}\sqrt{2} + \dfrac{64}{3}\sqrt{2}\right) - \dfrac{1}{25}\left(\dfrac{18}{5}\sqrt{3} + 4\sqrt{3}\right)$

Because the limits are changed at the start, the definite integral does not need to be expressed in terms of x again.

$= \dfrac{1}{25}\left(\dfrac{1088}{15}\sqrt{2} - \dfrac{38}{5}\sqrt{3}\right) = 3.5765.. = 3.58\,(3s.f.)$

Example 26

Find $\displaystyle\int_1^e \left(\dfrac{1}{x}\ln x\right)\mathrm{d}x$ using the substitution $x = e^u$.

$x = e^u \implies \dfrac{1}{x} = e^{-u}; \quad \ln x = u$

Expressing both parts in terms of u.

$x = 1, u = 0; \quad x = e, u = 1$

$1 = e^u\dfrac{\mathrm{d}u}{\mathrm{d}x} \implies \mathrm{d}x = e^u\dfrac{\mathrm{d}u}{\mathrm{d}x}\mathrm{d}x$

To be able to complete the transformation of the integral to the new variable.

Substituting gives

$\displaystyle\int_1^e \left(\dfrac{1}{x}\ln x\right)\mathrm{d}x = \int_0^1 e^{-u}\,ue^u\,\dfrac{\mathrm{d}u}{\mathrm{d}x}\mathrm{d}x$

Substituting for all terms and the limits.

$= \displaystyle\int_0^1 u\,\mathrm{d}u$

Simplifying the function and using the chain rule.

$= \left[\dfrac{1}{2}u^2\right]_0^1$

Integrating as a function of u.

$= \dfrac{1}{2} - 0 = \dfrac{1}{2}$

Evaluating the function at the limits.

Example 27

Find $\displaystyle\int \frac{1}{9+x^2}\,dx$ using the substitution $x = 3\tan\theta$.

$x = 3\tan\theta \;\Rightarrow\; 9 + x^2 = 9 + 9\tan^2\theta = 9\sec^2\theta$ Expressing in its simplest form.

$1 = 3\sec^2\theta\dfrac{d\theta}{dx} \;\Rightarrow\; dx = 3\sec^2\theta\dfrac{d\theta}{dx}dx$ To be able to complete the transformation of the integral to the new variable.

$\displaystyle\int \frac{1}{9+x^2}\,dx = \int\left(\frac{1}{9\sec^2\theta}\right)3\sec^2\theta\,\frac{d\theta}{dx}\,dx$ Substituting for all terms.

$\displaystyle = \int\frac{1}{3}\,d\theta$ Simplifying the function and using the chain rule.

$\displaystyle = \frac{1}{3}\theta + c$ Integrating as a function of θ.

$\displaystyle = \frac{1}{3}\tan^{-1}\left(\frac{x}{3}\right) + c$ Expressing this in terms of the original variable.

Example 28

Find $\displaystyle\frac{1}{\sqrt{5-4x^2}}\,dx$ using the substitution $\int x = \frac{\sqrt{5}}{2}\sin\theta$.

$x = \dfrac{\sqrt{5}}{2}\sin\theta \;\Rightarrow\; \sqrt{5-4x^2} = \sqrt{5 - 5\sin^2\theta} = \sqrt{5}\cos\theta$ Only the one expression here.

$1 = \dfrac{\sqrt{5}}{2}\cos\theta\dfrac{d\theta}{dx} \;\Rightarrow\; dx = \dfrac{\sqrt{5}}{2}\cos\theta\dfrac{d\theta}{dx}dx$ To be able to complete the transformation of the integral to the new variable.

$\displaystyle\int\frac{1}{\sqrt{5-4x^2}}\,dx = \int\frac{1}{\sqrt{5}\cos\theta}\cdot\frac{\sqrt{5}}{2}\cos\theta\,\frac{d\theta}{dx}\,dx$ Substituting for all terms.

Substituting gives:

$\displaystyle = \int\frac{1}{2}\,d\theta$ Simplifying the function and using the chain rule.

$\displaystyle = \frac{1}{2}\theta + c$ Integrating as a function of θ.

$\displaystyle = \frac{1}{2}\sin^{-1}\left(\frac{2}{\sqrt{5}}x\right) + c$ Expressing this in terms of the original variable.

Examples 24–28 are all functions you could not integrate without using a substitution. Examples 29–31 show how substitutions can also help you see exactly what you are doing with some integrals which you should recognise as exact derivatives of other functions.

Example 29

Find $\displaystyle\int \frac{1}{(5x-2)^2}\,\mathrm{d}x$ using the substitution $u = 5x - 2$.

$u = 5x - 2 \implies \dfrac{1}{(5x-2)^2} = \dfrac{1}{u^2}$

As normal – expressing the function and $\mathrm{d}x$ in terms of u.

$5 = \dfrac{\mathrm{d}u}{\mathrm{d}x} \implies \mathrm{d}x = \dfrac{1}{5}\dfrac{\mathrm{d}u}{\mathrm{d}x}\mathrm{d}x$

$\displaystyle\int \frac{1}{(5x-2)^2}\,\mathrm{d}x = \int \left(\frac{1}{u^2}\right)\frac{1}{5}\frac{\mathrm{d}u}{\mathrm{d}x}\,\mathrm{d}x$

$= \dfrac{1}{5}\displaystyle\int \left(\frac{1}{u^2}\right)\mathrm{d}u$

This is a standard integral form now.

$= -\dfrac{1}{5u} + c$

$= -\dfrac{1}{5(5x-2)} + c$

Expressing in terms of the original variable.

Writing $\dfrac{1}{(5x-2)^2}$ as $(5x-2)^{-2}$ allows this integral to be done without a formal substitution.

Example 30

Find $\displaystyle\int_0^{\frac{\pi}{4}} \cos 2x \sin^3 2x\,\mathrm{d}x$ using the substitution $u = \sin 2x$.

$u = \sin 2x \implies \sin^3 2x = u^3$

Only the one expression here.

$x = 0,\, u = 0;\quad x = \dfrac{\pi}{4},\, u = 1$

Changing the limits.

$\dfrac{\mathrm{d}u}{\mathrm{d}x} = 2\cos 2x \implies \dfrac{1}{2}\dfrac{\mathrm{d}u}{\mathrm{d}x}\mathrm{d}x = \cos 2x\,\mathrm{d}x$

This is the second part of the function to be integrated.

$\displaystyle\int_0^{\frac{\pi}{4}} \cos 2x \sin^3 2x\,\mathrm{d}x = \int_0^1 \frac{1}{2}u^3\,\mathrm{d}u$

Substitution makes the roles of the multiplying constants easier to see.

$= \left[\dfrac{1}{8}u^4\right]_0^1$

Now the integration and evaluation are almost trivial.

$= \dfrac{1}{8} - 0 = \dfrac{1}{8}$

Here you would be expected to notice that $\cos 2x$ is the derivative of $\sin 2x$ and integrate this without being given a formal substitution. If you find difficulty in getting the constant part of these type of integral right consistently then you might try using the substitution method yourself.

Example 31

Find $\int_{0}^{\frac{1}{12}\pi} \left(\dfrac{\cos 2x}{1+\sin 2x} \right) dx$ using the substitution $u = 1 + \sin 2x$.

$u = 1 + \sin 2x$ ←   Nothing more to do here.

$x = 0, u = 1; x = \dfrac{\pi}{12}, u = 1.5$ ←   Calculating the limits for the new variable.

$\dfrac{du}{dx} = 2\cos 2x \implies \cos 2x\,dx = \dfrac{1}{2}\dfrac{du}{dx}dx$ ←   To be able to complete the transformation of the integral to the new variable.

$\int_{0}^{\frac{1}{12}\pi} \left(\dfrac{\cos 2x}{1+\sin 2x} \right) \cdot dx = \int_{1}^{1.5} \left(\dfrac{1}{u} \right) \dfrac{1}{2} \dfrac{du}{dx} dx$ ←   Substituting throughout.

$= \int_{1}^{1.5} \left(\dfrac{1}{2u} \right) du$ ←   This is the standard logarithmic integral form.

$= \left[\dfrac{1}{2} \ln u \right]_{1}^{1.5}$ ←   Integrating and evaluating at the limits.

$= \dfrac{1}{2} \ln 1.5 - 0 = \dfrac{1}{2} \ln 1.5$

This integral was in the form $\dfrac{kf'(x)}{f(x)}$ and could have been integrated without the formal substitution.

Exercise 8.4

1. Find these indefinite integrals using the substitutions given.

   a) $\displaystyle\int \dfrac{6x}{\sqrt{2x+1}}\,dx; \quad u^2 = 2x + 1$

   b) $\displaystyle\int x^2\sqrt{1+x^3}\,dx; \quad u = 1 + x^3$

   c) $\displaystyle\int \dfrac{1}{\sqrt{9-x^2}}\,dx; \quad x = 3\sin\theta$

   d) $\displaystyle\int \dfrac{x}{\sqrt{x+2}}\,dx; \quad u = x + 2$

   e) $\displaystyle\int \dfrac{1}{25+4x^2}\,dx; \quad x = \dfrac{5}{2}\tan\theta$

   f) $\displaystyle\int \dfrac{2e^{2x}+1}{e^{2x}+x}\,dx; \quad u = e^{2x} + x$

2. Calculate the following definite integrals using the substitutions given. Note that the first three are integrating the same function as parts (a) to (c) of question 1.

a) $\displaystyle\int_0^1 \frac{6x}{\sqrt{2x+1}}\,dx; \quad u^2 = 2x+1$

b) $\displaystyle\int_0^2 x^2\sqrt{1+x^3}\,dx; \quad u = 1+x^3$

c) $\displaystyle\int_0^{\frac{\pi}{4}} \frac{1}{\sqrt{9-x^2}}\,dx; \quad x = 3\sin\theta$

d) $\displaystyle\int_0^{\frac{\pi}{6}} \frac{\cos x}{\sqrt{1+2\sin x}}\,dx; \quad u = 1+2\sin x$

e) $\displaystyle\int_0^1 (1+x)\sqrt{2x+x^2}\,dx; \quad u = 2x+x^2$

f) $\displaystyle\int_0^1 \frac{2e^{2x}+1}{e^{2x}+x}\,dx; \quad u = e^{2x}+x$

g) $\displaystyle\int_0^{\frac{\pi}{3}} \sec^2 x\,\tan^5 x\,dx; \quad u = \tan x$

h) $\displaystyle\int_0^{\sqrt{\frac{\pi}{3}}} 4x\sin(x^2)\,dx; \quad u = \sin x^2$

3. Find $\displaystyle\int_2^3 \frac{1}{x-\sqrt{x}}\,dx$ using the substitution $x = u^2$.

4. Using the substitution $u = \ln x$, find the area bounded by the graph of $y = \dfrac{(\ln x)^4}{x}$, the x-axis and the lines $x = 1$ and $x = e$.

5. Using the substitution $u = 2x + 1$, find the area bounded by the graphs of $y = \dfrac{x}{\sqrt{2x+1}}$, the x-axis and the line $x = 2$.

(shown in the diagram)

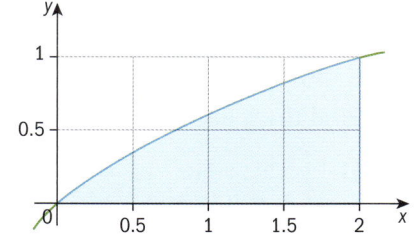

6. Find the area bounded by the graph of $y = 2x\cos\left(x^2 + \dfrac{\pi}{6}\right)$, the x-axis and the lines $x = 0$ and $x = \dfrac{\pi}{6}$.

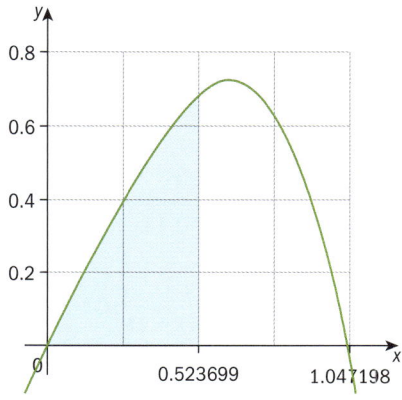

Summary exercise 8

1. By first expressing each rational expression using partial fractions, find

    a) $\displaystyle\int\left(\frac{5}{(x-2)(x+3)}\right)dx$

    b) $\displaystyle\int\left(\frac{5x+2}{(2x-1)(x+4)}\right)dx$

    c) $\displaystyle\int\left(\frac{x-2}{x(x+1)^2}\right)dx$

    d) $\displaystyle\int\left(\frac{x^2-3x-1}{(2x-1)(x+1)^2}\right)dx$

    e) $\displaystyle\int\left(\frac{3x^2-5x}{(x^2-1)}\right)dx$

EXAM-STYLE QUESTION

2. Show that $\displaystyle\frac{4}{x(x^2+4)}=\frac{1}{x}-\frac{x}{x^2+4}$. Using the

    substitution $u=x^2+4$, find $\displaystyle\int\frac{4}{x(x^2+4)}dx$

3. Find the following integrals.

    a) $\displaystyle\int\frac{4x^3}{x^4+6}dx$

    b) $\displaystyle\int\frac{\cos 3x}{3-2\sin 3x}dx$

    c) $\displaystyle\int\frac{3x-2}{3x^2-4x+7}dx$

    d) $\displaystyle\int\frac{e^{5x}}{6-5e^{5x}}dx$

    e) $\displaystyle\int\frac{2x}{x^2+9}dx$

    f) $\displaystyle\int\frac{x^2}{x^3+10}dx$

    g) $\displaystyle\int_0^{\frac{1}{6}\pi}\left(\frac{\cos 3x}{2+\sin 3x}\right)dx$

    h) $\displaystyle\int_0^1\left(\frac{e^{3x}}{2+e^{3x}}\right)dx$

    i) $\displaystyle\int_0^1\left(\frac{2x+3}{(x+1)(x+2)}\right)dx$

4. Find these integrals using integration by parts.

    a) $\displaystyle\int xe^{3x}\,dx$

    b) $\displaystyle\int 4x\cos 2x\,dx$

    c) $\displaystyle\int(2x-3)e^x\,dx$

    d) $\displaystyle\int(x^2-x+3)\cos 3x\,dx$

    e) $\displaystyle\int_0^1 x^2e^{x-1}\,dx$

    f) $\displaystyle\int_1^e(x^2\ln 4x)$

5. Find these integrals using the substitutions given.

    a) $\displaystyle\int\frac{6x}{\sqrt{3x-1}}dx;\quad u^2=3x-1$

    b) $\displaystyle\int x^3\sqrt{4+x^4}\,dx;\quad u=4+x^4$

    c) $\displaystyle\int\frac{1}{\sqrt{25-16x^2}}dx;\quad x=\frac{5}{4}\sin\theta$

    d) $\displaystyle\int\frac{x}{\sqrt{2x+7}}dx;\quad u=2x+7$

    e) $\displaystyle\int_0^1(1+6x)\sqrt{x+3x^2}\,dx;\quad u=x+3x^2$

    f) $\displaystyle\int_0^1\frac{e^{2x}+x}{e^{2x}+x^2}dx;\quad u=e^{2x}+x^2$

    g) $\displaystyle\int_0^{\frac{\pi}{4}}\sec^2 x\tan^6 x\,dx;\quad u=\tan x$

    h) $\displaystyle\int_0^{\sqrt{\frac{\pi}{6}}}4x\cos\left(x^2+\frac{\pi}{6}\right)dx;u=\sin\left(x^2+\frac{\pi}{6}\right)$

6. The diagram shows the area trapped between the graphs of $y = \dfrac{x}{(x-4)^2}$ and $y = x$. Show that the graphs intersect at $(0, 0)$ and $(3, 3)$ and that the area is $\dfrac{3}{2} + \ln 4$ (use the substitution $u = x - 4$ when dealing with the curve).

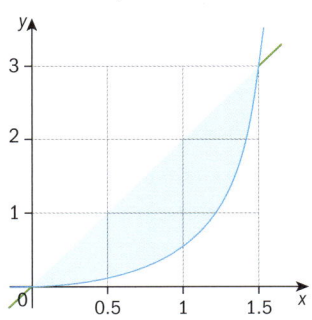

7. i) Express $\cos \theta + \left(\sqrt{3}\right) \sin \theta$ in the form $R \cos(\theta - \alpha)$, where $R > 0$ and $0 < \alpha < \dfrac{1}{2}\pi$, giving the exact values of R and α. [3]

ii) Hence show that

$$\int_0^{\frac{1}{2}\pi} \frac{1}{\left(\cos\theta + \left(\sqrt{3}\right)\sin\theta\right)^2} \, \mathrm{d}\theta = \frac{1}{\sqrt{3}}.$$ [4]

Cambridge International AS and A Level Mathematics 9709 Paper 3 Q5 May/June 2007

Key points:

Integrating rational functions

To integrate difficult algebraic fractions, first look to see if you can split the fraction into partial fractions, where each partial fraction is integrable.

Integrating $\dfrac{kf'(x)}{f(x)}$

- $\displaystyle\int \left(\frac{kf'(x)}{f(x)}\right) \mathrm{d}x = k \ln(f(x)) + c$

Integration by parts

- $\displaystyle\int \left(u \frac{\mathrm{d}v}{\mathrm{d}x}\right) \mathrm{d}x = uv - \int \left(v \frac{\mathrm{d}u}{\mathrm{d}x}\right) \mathrm{d}x$

Integration using substitution

- Before making a variable substitution, first write down everything in the integral in terms of the new variable, u, (including the dx term) and then rewrite the integral completely in one go.
- When dealing with definite integrals, make sure you express the limits in terms of the new variable, u.
- For indefinite integrals if your original integral was in terms of x, you need to express your answer also in terms of x.

Review exercise C – Pure 3

1. Expand $\dfrac{1+3x}{\sqrt{(1+2x)}}$ in ascending powers of x up to and including the term in x^2, simplyfying the coefficients. [4]

Cambridge International AS and A Level Mathematics 9709 Paper 31 Q2 May/June 2013

2. Express $\dfrac{7x^2 - 3x + 2}{x(x^2 + 1)}$ in partial fractions. [5]

Cambridge International AS and A Level Mathematics 9709 Paper 31 Q3 May/June 2013

3. When $(1 + ax)^{-2}$, where a is a positive constant, is expanded in ascending powers of x, the coefficients of x and x^3 are equal.

   i) Find the exact value of a. [4]

   ii) when a has this value, obtain the expansion up to and including the term in x^2, simplifying the coefficients. [3]

Cambridge International AS and A Level Mathematics 9709 Paper 31 Q4 October/November 2012

4. i) Expand $\dfrac{1}{\sqrt{(1-4x)}}$ in ascending powers of x, up to and including the term in x^2, simplifying the coefficients. [3]

   ii) Hence find the coefficient of x^2 in the expansion of $\dfrac{1+2x}{\sqrt{(4-16x)}}$. [2]

Cambridge International AS and A Level Mathematics 9709 Paper 31 Q2 May/June 2012

5. Using the substitution $u = e^x$, or otherwise, solve the equation
$$e^x = 1 + 6e^{-x},$$
giving your answer correct to 3 significant figures. [4]

Cambridge International AS and A Level Mathematics 9709 Paper 31 Q1 October/November 2011

6. Let $f(x) = \dfrac{3x}{(1 + x)(1 + 2x^2)}$.

   i) Express $f(x)$ in partial fractions. [5]

   ii) Hence obtain the expansion of $f(x)$ in ascending powers of x, up to and including the term in x^3. [5]

Cambridge International AS and A Level Mathematics 9709 Paper 31 Q8 October/November 2010

7. When $(1 + 2x)(1 + ax)^{\frac{2}{3}}$, where a is a constant, is expanded in ascending powers of x, the coefficient of the term in x is zero.

   i) Find the value of a. [3]

   ii) When a has this value, find the term in x^3 in the expansion of $(1 + 2x)(1 + ax)^{\frac{2}{3}}$, simplifying the coefficient. [4]

Cambridge International AS and A Level Mathematics 9709 Paper 3 Q5 May/June 2009

8. i) Show that $\displaystyle\int_2^4 4x \ln x \, dx = 56 \ln 2 - 12$. [5]

   ii) Use the substitution $u = \sin 4x$ to find the exact value of $\displaystyle\int_0^{\frac{1}{24}\pi} \cos^3 4x \, dx$. [5]

Cambridge International AS and A Level Mathematics 9709 Paper 31 Q8 May/June 2013

9. **i)** By differentiating $\dfrac{1}{\cos x}$, show that

    if $y = \sec x$ then $\dfrac{dy}{dx} = \sec x \tan x$. **[2]**

    ii) Show that $\dfrac{1}{\sec x - \tan x} \equiv \sec x + \tan x$. **[1]**

    iii) Deduce that $\dfrac{1}{(\sec x - \tan x)^2}$

    $\equiv \sec^2 x - 1 + 2 \sec x \tan x$. **[2]**

    iv) Hence show that

    $\displaystyle\int_0^{\frac{1}{4}\pi} \dfrac{1}{(\sec x - \tan x)^2}\,dx = \dfrac{1}{4}\left(8\sqrt{2} - \pi\right)$. **[3]**

    Cambridge International AS and
    A Level Mathematics 9709
    Paper 31 Q5 October/November 2012

10. By first expressing $\dfrac{4x^2 + 5x + 3}{2x^2 + 5x + 2}$ in partial

    fractions, show that

    $\displaystyle\int_0^4 \dfrac{4x^2 + 5x + 3}{2x^2 + 5x + 2}\,dx = 8 - \ln 9$. **[10]**

    Cambridge International AS and
    A Level Mathematics 9709
    Paper 31 Q9 May/June 2012

11. Let $f(x) = \dfrac{12 + 8x - x^2}{(2 - x)(4 + x^2)}$.

    i) Express $f(x)$ in the form $\dfrac{A}{2 - x} + \dfrac{Bx + C}{4 + x^2}$. **[4]**

    ii) Show that $\displaystyle\int_0^1 f(x)\,dx = \ln\left(\dfrac{25}{2}\right)$. **[5]**

    Cambridge International AS and
    A Level Mathematics 9709
    Paper 31 Q8 October/November 2011

12. Let $I = \displaystyle\int_0^1 \dfrac{x^2}{\sqrt{4 - x^2}}\,dx$.

    Using the substitution $x = 2 \sin \theta$,
    show that

    $I = \displaystyle\int_0^{\frac{1}{6}\pi} 4 \sin^2 \theta\, d\theta$. **[3]**

    Cambridge International AS and
    A Level Mathematics 9709
    Paper 31 Q5i October/November 2010

13.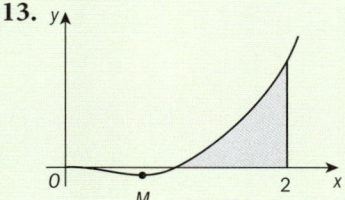

    The diagram shows the curve $y = x^3 \ln x$
    and its minimum point M.

    i) Find the exact coordinates of M. **[5]**

    ii) Find the exact area of the shaded region
    bounded by the curve, the x-axis and
    the line $x = 2$. **[5]**

    Cambridge International AS and
    A Level Mathematics 9709
    Paper 31 Q9 October/November 2010

14. **i)** Express $\dfrac{2}{(x + 1)(x + 3)}$ in partial

    fractions. **[2]**

    ii) Using your answer to part (**i**), show that

    $\left(\dfrac{2}{(x+1)(x+3)}\right)^2 \equiv \dfrac{1}{(x+1)^2} - \dfrac{1}{x+1}$

    $+ \dfrac{1}{x+3} + \dfrac{1}{(x+3)^2}$. **[2]**

    iii) Hence show that

    $\displaystyle\int_0^1 \dfrac{4}{(x+1)^2(x+3)^2}\,dx = \dfrac{7}{12} - \ln\dfrac{3}{2}$ **[5]**

    Cambridge International AS and
    A Level Mathematics 9709
    Paper 31 Q8 May/June 2010

15. i) Prove the identity
$$\cos 4\theta - 4\cos 2\theta + 3 \equiv 8\sin^4 \theta.$$ [4]

ii) Using this result find, in simplified form, the exact value of $\int_{\frac{1}{6}\pi}^{\frac{1}{3}\pi} \sin^4 \theta \, d\theta.$ [4]

Cambridge International AS and A Level Mathematics 9709 Paper 31 Q5 October/November 2009

16.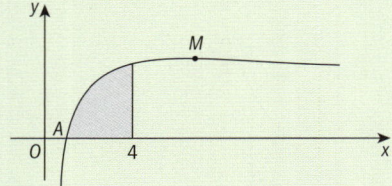

The diagram shows the curve $y = \dfrac{\ln x}{\sqrt{x}}$ and its maximum point M. The curve cuts the x-axis at the point A.

i) State the coordinates of A. [1]

ii) Find the exact value of the x-coordinate of M. [4]

iii) Using integration by parts, show that the area of the shaded region bounded by the curve, the x-axis and the line $x = 4$ is equal to $8\ln 2 - 4$. [5]

Cambridge International AS and A Level Mathematics 9709 Paper 31 Q9 October/November 2009

17.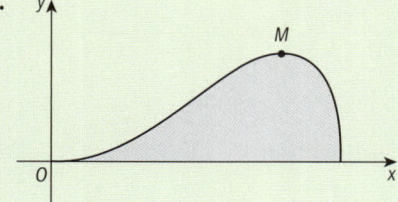

The diagram shows the curve $y = x^2\sqrt{(1-x^2)}$ for $x \geq 0$ and its maximum point M.

i) Find the exact value of the x-coordinate of M. [4]

ii) Show, by means of the substitution $x = \sin \theta$. that the area A of the shaded region between the curve and the x-axis is given by
$$A = \frac{1}{4}\int_0^{\frac{1}{2}\pi} \sin^2 \theta \, d\theta.$$ [3]

iii) Hence obtain the exact value of A. [4]

Cambridge International AS and A Level Mathematics 9709 Paper 3 Q10 May/June 2009

18.

The diagram shows the curve $y = e^{-\frac{1}{2}x}\sqrt{(1+2x)}$ and its maximum point M. The shaded region between the curve and the axes is denoted by R.

i) Find the x-coordinate of M. [4]

ii) Find by integration the volume of the solid obtained when R is rotated completely about the x-axis. Give your answer in terms of π and e. [6]

Cambridge International AS and A Level Mathematics 9709 Paper 3 Q9 May/June 2008

19. Use integration by parts to show that
$$\int_2^4 \ln x \, dx = 6\ln 2 - 2$$ [4]

Cambridge International AS and A Level Mathematics 9709 Paper 3 Q3 October/November 2007

20. Let $I = \int_1^4 \dfrac{1}{x\left(4 - \sqrt{x}\right)}\,dx$.

   i) Use the substitution $u = \sqrt{x}$ to show that

$$I = \int_1^2 \dfrac{2}{u(4 - u)}\,du. \qquad [3]$$

   ii) Hence show that $I = \dfrac{1}{2}\ln 3$.   [6]

**Cambridge International AS and
A Level Mathematics 9709 Paper 3 Q7
May/June 2007**

21. In a certain industrial process, a substance is being produced in a container. The mass of the substance in the container t minutes after the start of the process is x grams. At any time, the rate of formation of the substance is proportional to its mass. Also, throughout the process, the substance is removed from the container at a constant rate of 25 grams per minute. When $t = 0$, $x = 1000$ and $\dfrac{dx}{dt} = 75$.

Show that x and t satisfy the differential

equation $\dfrac{dx}{dt} = 0.1(x - 250)$.   [2]

**Cambridge International AS and
A Level Mathematics 9709
Paper 3 Q5i May/June 2006**

22. i) Express $\dfrac{10}{(2 - x)(1 - x^2)}$ in partial fractions.   [5]

   ii) Hence, given that $|x| < 1$, obtain the expansion of $\dfrac{10}{(2 - x)(1 + x^2)}$ in ascending powers of x, up to and including the term in x^3, simplifying the coefficients.   [5]

**Cambridge International AS and
A Level Mathematics 9709
Paper 3 Q9 May/June 2006**

9 Vectors

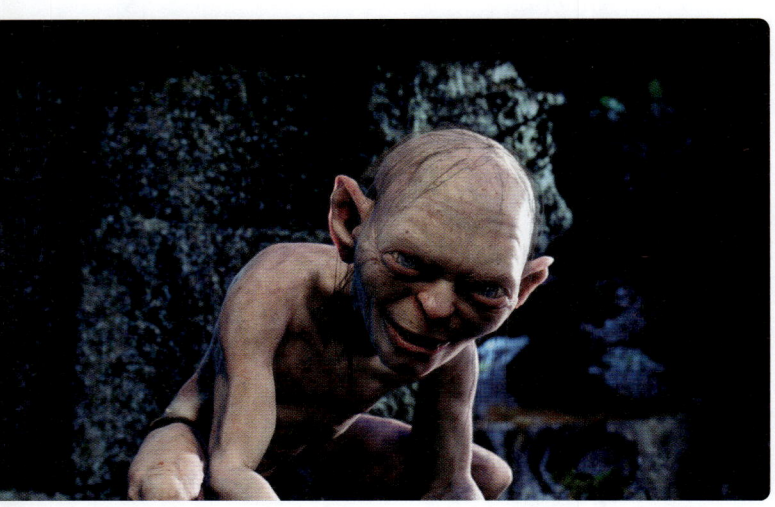

Computer generated images play an important part in modern film-making. This technique would not be possible without mathematics and, in particular, vectors. For example, the character of Gollum (from Lord of the Rings) is modelled as a wire skeleton composed of polygons. Vectors are used to describe the planes which make up the character.

Objectives

- Understand the significance of all the symbols used when the equation of a straight line is expressed in the form $\mathbf{r} = \mathbf{a} + t\mathbf{b}$.

- Determine whether two lines are parallel, intersect, or are skew.

- Find the angle between two lines, and the point of intersection of two lines when it exists.

- Understand the significance of all the symbols used when the equation of a plane is expressed in either of the forms $ax + by + cz = d$ or $(\mathbf{r} - \mathbf{a}).\mathbf{n} = 0$.

- Use equations of lines and planes to solve problems concerning distances, angles and intersections, and in particular:

  - find the equation of a line or a plane, given sufficient information,

  - determine whether a line lies in a plane, is parallel to a plane, or intersects a plane, and find the point of intersection of a line and a plane when it exists,

  - find the line of intersection of two non-parallel planes,

  - find the perpendicular distance from a point to a plane, and from a point to a line,

  - find the angle between two planes, and the angle between a line and a plane.

Before you start

You should know how to:

1. Carry out and use addition and subtraction of vectors and multiplication of a vector by a scalar.

e.g. $\begin{pmatrix} 4 \\ 3 \end{pmatrix} - 2\begin{pmatrix} 1 \\ -2 \end{pmatrix} = \begin{pmatrix} 2 \\ 7 \end{pmatrix}$

2. Use unit vectors, displacement vectors and position vectors.

e.g. If $\overrightarrow{OA} = 2\mathbf{i} - 3\mathbf{j}$ and $\overrightarrow{OB} = \mathbf{i} + \mathbf{j}$,

then $\overrightarrow{AB} = \overrightarrow{OB} - \overrightarrow{OA} = -\mathbf{i} + 4\mathbf{j}$

3. Calculate the scalar product of two vectors and use it to determine the angle between two directions.

e.g. the angle between $\begin{pmatrix} 3 \\ 1 \\ -2 \end{pmatrix}$ and $\begin{pmatrix} 2 \\ 0 \\ 1 \end{pmatrix}$ is given

by $\cos\theta = \dfrac{3 \times 2 + 1 \times 0 + (-2) \times 1}{\sqrt{3^2 + 1^2 + (-2)^2}\sqrt{2^2 + 0^2 + 1^2}}$

so $\cos\theta = \dfrac{4}{\sqrt{14}\,\sqrt{5}}$

$\theta = 61.4°$

Skills check:

1. Work out

a) $\begin{pmatrix} -3 \\ 4 \end{pmatrix} - \begin{pmatrix} 2 \\ 1 \end{pmatrix}$ **b)** $3\begin{pmatrix} 2 \\ 1 \end{pmatrix} + 2\begin{pmatrix} 0 \\ -4 \end{pmatrix}$.

2. Given that $\overrightarrow{OP} = 2\mathbf{i} + 5\mathbf{j} + \mathbf{k}$ and $\overrightarrow{OQ} = -3\mathbf{i} + \mathbf{j} + 2\mathbf{k}$, find

a) $\overrightarrow{PQ}$ **b)** $\overrightarrow{QP}$.

3. Find the angle between the vectors

a) $\begin{pmatrix} 1 \\ 2 \\ 1 \end{pmatrix}$ and $\begin{pmatrix} 1 \\ 4 \\ 3 \end{pmatrix}$

b) $\mathbf{i} - \mathbf{j} + 2\mathbf{k}$ and $2\mathbf{i} + 5\mathbf{j} - 3\mathbf{k}$.

9.1 The equation of a straight line

We are familiar with writing the equation of a two-dimensional straight line in one of the forms $y = mx + c$ or $ax + by = c$. These equations are said to be written in **cartesian form**.

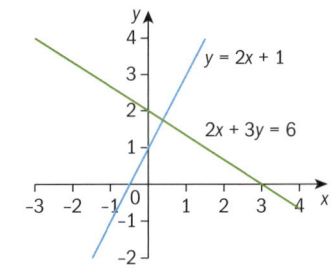

We can also write the equation of a line in **vector form**. The vector equation of a line is a way of expressing the position vector of any point on the line.

To find the vector equation of a line we need to know:

    i)   the **position vector**, say **a**, of any point A which lies on the line,

    ii) any vector, say **b**, which runs in the **direction** of the line.

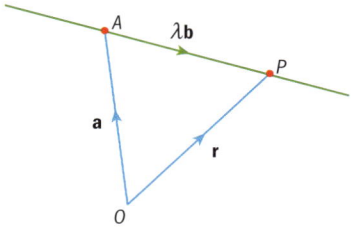

The position vector (**r**) of any point P on the line can then be expressed in the form

        $\mathbf{r} = \mathbf{a} + \lambda\mathbf{b}$, where λ is a scalar.

> The vector equation of a straight line passing through a point with position vector **a** and with direction vector **b** is
> $$\mathbf{r} = \mathbf{a} + \lambda\mathbf{b}$$
> where λ is a scalar.

Note: i) The direction vector (**b**) can have any magnitude but it must be parallel to the line.

        ii) The value of λ will be positive for points on one side of A and negative for points on the other side of A.

Example 1

a) Write down a vector equation for the straight line through the point (3, 1) with direction vector $\begin{pmatrix} 4 \\ -2 \end{pmatrix}$.

b) Find a vector equation for the line through the points $L(1, 2)$ and $M(5, 3)$.

· ·

a) $\mathbf{a} = \begin{pmatrix} 3 \\ 1 \end{pmatrix}$ ←   This is the position vector of a point on the line

       $\mathbf{b} = \begin{pmatrix} 4 \\ -2 \end{pmatrix}$ ←   This gives the direction of the line

A vector equation of the line is therefore $\mathbf{r} = \begin{pmatrix} 3 \\ 1 \end{pmatrix} + \lambda \begin{pmatrix} 4 \\ -2 \end{pmatrix}$

▶ Continued on the next page

b) $\mathbf{a} = \overrightarrow{OL} = \begin{pmatrix} 1 \\ 2 \end{pmatrix}$

Alternatively we could use $\overrightarrow{OM} = \begin{pmatrix} 5 \\ 3 \end{pmatrix}$

$\mathbf{b} = \overrightarrow{LM} = \begin{pmatrix} 5 \\ 3 \end{pmatrix} - \begin{pmatrix} 1 \\ 2 \end{pmatrix} = \begin{pmatrix} 4 \\ 1 \end{pmatrix}$

Alternatively we could use $\overrightarrow{ML} = \begin{pmatrix} -4 \\ -1 \end{pmatrix}$ or any multiple of $\overrightarrow{LM}$ or $\overrightarrow{ML}$.

A vector equation of the line is therefore $\mathbf{r} = \begin{pmatrix} 1 \\ 2 \end{pmatrix} + \lambda \begin{pmatrix} 4 \\ 1 \end{pmatrix}$

Note: Remember to put '$\mathbf{r} =$' as part of the equation.

We can also use vectors to write the equation of a line in three dimensions, as Example 2 shows.

Example 2

Points A and B have position vectors $\begin{pmatrix} 3 \\ 1 \\ -4 \end{pmatrix}$ and $\begin{pmatrix} -2 \\ 5 \\ 0 \end{pmatrix}$ respectively.

a) Find, in vector form, an equation of the straight line that passes through A and B.

b) Show that the point $C(-12, 13, 8)$ lies on this line.

a) $\overrightarrow{AB} = \begin{pmatrix} -2 \\ 5 \\ 0 \end{pmatrix} - \begin{pmatrix} 3 \\ 1 \\ -4 \end{pmatrix} = \begin{pmatrix} -5 \\ 4 \\ 4 \end{pmatrix}$

$\overrightarrow{AB} = \mathbf{b} - \mathbf{a}$

A vector equation of the line is $\mathbf{r} = \begin{pmatrix} 3 \\ 1 \\ -4 \end{pmatrix} + \lambda \begin{pmatrix} -5 \\ 4 \\ 4 \end{pmatrix}$

b) $\begin{pmatrix} 3 \\ 1 \\ -4 \end{pmatrix} + \lambda \begin{pmatrix} -5 \\ 4 \\ 4 \end{pmatrix} = \begin{pmatrix} -12 \\ 13 \\ 8 \end{pmatrix}$

For C to lie on the line there must be a value of λ which, when substituted into the equation, gives the position vector of C.

$3 - 5\lambda = -12$

$1 + 4\lambda = 13$

$-4 + 4\lambda = 8$

Write down an equation from each component.

Solving each equation gives $\lambda = 3$

This shows $\lambda = 3$ gives the position vector of C.

So $C(-12, 13, 8)$ lies on the line that passes through A and B.

Note that we must show that the three equations are consistent and the same value of λ satisfies each equation. If this is not true, then the point does not lie on the line.

Example 3

A line has a vector equation given by $\mathbf{r} = (3\mathbf{i} + \mathbf{j} + \mathbf{k}) + \mu(2\mathbf{i} - \mathbf{j} - \mathbf{k})$.

a) Show that this line intersects the x axis.

b) The point P lies on this line and has coordinates $(-5, a, b)$. Find the values of a and b.

a) If the line intersects the x axis then there will be a value of μ which gives a position vector of the form $c\mathbf{i} + 0\mathbf{j} + 0\mathbf{k}$.

> Any point on the x-axis has position vector with y and z components equal to zero.

$(3\mathbf{i} + \mathbf{j} + \mathbf{k}) + \mu(2\mathbf{i} - \mathbf{j} - \mathbf{k}) = c\mathbf{i} + 0\mathbf{j} + 0\mathbf{k}$

$(3 + 2\mu)\mathbf{i} + (1 - \mu)\mathbf{j} + (1 - \mu)\mathbf{k} = c\mathbf{i} - 0\mathbf{j} - 0\mathbf{k}$

> Rewrite the equation so we can compare the $\mathbf{i}$, $\mathbf{j}$, and $\mathbf{k}$ components.

$3 + 2\mu = c$

> Compare the $\mathbf{i}$ components.

$1 - \mu = 0$ so $\mu = 1$

> Compare the $\mathbf{j}$ and $\mathbf{k}$ components.

$\Rightarrow 3 + 2 \times 1 = c$

> Substitute $\mu = 1$ into $3 + 2\mu = c$

$\Rightarrow c = 5$.

> Find the value of c.

When $\mu = 1$ and $c = 5$, then $\mathbf{r} = 5\mathbf{i} - 0\mathbf{j} - 0\mathbf{k}$

so the point $(5, 0, 0)$ lies on the line and on the x-axis.

> Substitute $\mu = 1$ and $c = 5$ into the equation of the line.

Therefore the line intersects the x-axis.

b) $(3\mathbf{i} + \mathbf{j} + \mathbf{k}) + \mu(2\mathbf{i} - \mathbf{j} - \mathbf{k}) = -5\mathbf{i} + a\mathbf{j} + b\mathbf{k}$

> Equate the RHS of the equation of the line with the position vector of P.

$(3 + 2\mu)\mathbf{i} + (1 - \mu)\mathbf{j} + (1 - \mu)\mathbf{k} = -5\mathbf{i} + a\mathbf{j} + b\mathbf{k}$

$3 + 2\mu = -5$

> Collect terms in $\mathbf{i}$, $\mathbf{j}$ and $\mathbf{k}$ and compare components to find the value of μ.

$\mu = -4$.

When $\mu = -4$,

> Substitute $\mu = -4$ into the equation of the line.

$\mathbf{r} = (3\mathbf{i} + \mathbf{j} + \mathbf{k}) + (-4)(2\mathbf{i} - \mathbf{j} - \mathbf{k})$

$= -5\mathbf{i} + 5\mathbf{j} + 5\mathbf{k}$

$a = 5, b = 5$.

> State the values of a and b.

Exercise 9.1

1. a) The vector equation of the line l_1 is $\mathbf{r} = \begin{pmatrix} 2 \\ 3 \\ -1 \end{pmatrix} + \lambda \begin{pmatrix} 4 \\ -5 \\ 10 \end{pmatrix}$.

Write down the coordinates of the points A, B, and C on this line which correspond to the following values of λ.

i) $\lambda = 0$ **ii)** $\lambda = 1$ **iii)** $\lambda = -3$

b) The vector equation of the line l_2 is $\mathbf{r} = (2\mathbf{i} + \mathbf{k}) + \mu(-3\mathbf{i} + 3\mathbf{j} - 2\mathbf{k})$

Write down the coordinates of the points P, Q, and R on this line which correspond to the following values of μ.

    i) $\mu = 2$ **ii)** $\mu = \dfrac{1}{2}$ **iii)** $\mu = -4$

2. In two dimensions find, using column vectors, a vector equation for

  a) The line through the point $(4, -1)$ in the direction of the vector $\begin{pmatrix} -3 \\ 5 \end{pmatrix}$

  b) The line joining the points $(2, 5)$ and $(6, -1)$

  c) The x-axis

  d) The line joining the points with position vectors $\begin{pmatrix} 2 \\ -5 \end{pmatrix}$ and $\begin{pmatrix} 4 \\ 3 \end{pmatrix}$

  e) The line $y = 3x$

  f) The line $x + y = 1$.

3. In three dimensions find, using unit vectors, a vector equation for

  a) The line through $A(2, 1, 5)$ which is parallel to the vector $-\mathbf{i} - \mathbf{j} + \mathbf{k}$

  b) The line through $B(5, 2, -1)$ which is parallel to CD where $\overrightarrow{OC} = 6\mathbf{i} - 2\mathbf{j} + 3\mathbf{k}$ and $\overrightarrow{OD} = 5\mathbf{i} + 2\mathbf{j} - \mathbf{k}$

  c) The y-axis

  d) The line joining the points $(-3, 7\ 0)$ and $(-5, 10, 2)$

  e) The line through the origin in the direction $\mathbf{i} + \mathbf{j} + \mathbf{k}$

  f) The line through the point with position vector $-3\mathbf{i} + 2\mathbf{j} + 6\mathbf{k}$, in the direction of the z-axis.

4. Points A, B, and C have position vectors

$$\mathbf{a} = \begin{pmatrix} 1 \\ 2 \\ 3 \end{pmatrix}, \qquad \mathbf{b} = \begin{pmatrix} -1 \\ 2 \\ -1 \end{pmatrix}, \qquad \mathbf{c} = \begin{pmatrix} 4 \\ 2 \\ 5 \end{pmatrix}.$$

Find

  a) A vector equation for the line AB

  b) A vector equation for the line through A parallel to BC

  c) A vector equation for the line through the mid-point of AB and the mid-point of AC.

5. Find which of the points $P(6, 7, -13)$, $Q(-3, -11, 14)$, and $R(0, -6, 5)$ lie

on the straight line with equation $\mathbf{r} = \begin{pmatrix} 1 \\ -3 \\ 2 \end{pmatrix} + \lambda \begin{pmatrix} 2 \\ 4 \\ -6 \end{pmatrix}$.

6. Points A and B have position vectors $-2\mathbf{i} + \mathbf{j}$ and $4\mathbf{i} + 4\mathbf{k}$.

  a) Find an equation for the line that passes through A and B.

  b) Show that this line does not intersect the x-axis.

7. Points A and B have position vectors $\mathbf{a} = \begin{pmatrix} 2 \\ -1 \\ 3 \end{pmatrix}$ and $\mathbf{b} = \begin{pmatrix} 1 \\ 1 \\ -1 \end{pmatrix}$.

   a) Find an equation for the line AB.

   b) Find the values of α and β if the point $C(3, \alpha, \beta)$ lies on the line AB.

8. **a)** Find the vector equation of the line joining the points $(2, 3, 4)$ and $(1, 4, 3)$.

   b) Find the x and y-coordinates on this line where the z-coordinate is 0.

9.2 Intersecting lines

In two dimensions, there are 3 possible relationships between two straight lines.

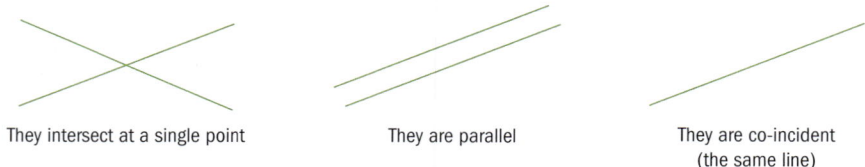

| They intersect at a single point | They are parallel | They are co-incident (the same line) |

Two lines **intersect** if they have one point in common.
Two lines are **parallel** if they have the same direction but no point in common.
Two lines are **coincident** if they have the same direction and an infinite number of points in common.

In three dimensions there is a fourth possible relationship between two lines.

Lines that are not parallel, do not intersect, and are not co-incident are called **skew** lines.

For example, in the cuboid shown below, the straight lines AC and GD are skew.

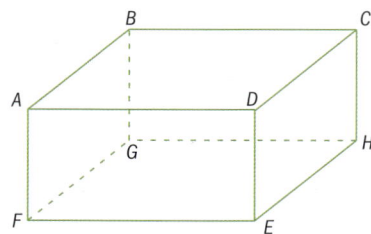

Two lines are **skew** if they have different directions and have no point in common.

Example 4

Two lines l_1 and l_2 are given by the vector equations

$$\mathbf{r} = \begin{pmatrix} 5 \\ 1 \\ -1 \end{pmatrix} + \lambda \begin{pmatrix} 2 \\ 1 \\ 5 \end{pmatrix} \text{ and } \mathbf{r} = \begin{pmatrix} 13 \\ -6 \\ 2 \end{pmatrix} + \mu \begin{pmatrix} -3 \\ 4 \\ 1 \end{pmatrix} \text{ respectively.}$$

Show that the lines intersect and find the position vector of the point of intersection.

We can see that either the lines must intersect, or else they are skew. Their directions given by $\begin{pmatrix} 2 \\ 1 \\ 5 \end{pmatrix}$ and $\begin{pmatrix} -3 \\ 4 \\ 1 \end{pmatrix}$

are not the same and are not multiples of each other so the lines are not parallel or coincident.

$$\begin{pmatrix} 5 \\ 1 \\ -1 \end{pmatrix} + \lambda \begin{pmatrix} 2 \\ 1 \\ 5 \end{pmatrix} = \begin{pmatrix} 13 \\ -6 \\ 2 \end{pmatrix} + \mu \begin{pmatrix} -3 \\ 4 \\ 1 \end{pmatrix}$$

For the lines to intersect there must be a value of λ and a value of μ which give the same position vector from each equation.

$5 + 2\lambda = 13 - 3\mu$ (i)

$1 + \lambda = -6 + 4\mu$ (ii)

$-1 + 5\lambda = 2 + \mu$ (iii)

Equation (i) $-$ 2 × equation (ii) gives

Solve two of the equations to find a pair of values which satisfies those two equations.

$3 = 25 - 11\mu$

$\mu = 2$

Substituting $\mu = 2$ into equation (ii)

$1 + \lambda = -6 + 4 \times 2$

$\lambda = 1$

Checking equation (iii)

Check that these values satisfy the third equation.

When $\lambda = 1, \mu = 2, -1 + 5\lambda = 4$

$2 + \mu = 4$

So this pair of values also satisfies the third equation and we have shown that the lines intersect.

If we solve two of the equations to get a value for each of the scalars but then, on checking, these values do not satisfy the third equation then the lines do not intersect.

$$\mathbf{r} = \begin{pmatrix} 5 \\ 1 \\ -1 \end{pmatrix} + 1 \begin{pmatrix} 2 \\ 1 \\ 5 \end{pmatrix} = \begin{pmatrix} 7 \\ 2 \\ 4 \end{pmatrix}$$

Find the position vector of the point of intersection by substituting the value of λ into the equation of l_1 (or alternatively the value of μ into the equation of l_2).

If we are asked to find the **coordinates** of the point of intersection, we give the answer in the form (7, 2, 4).

Example 5

Find whether the following pairs of lines intersect. If they do not intersect, find whether they are parallel or skew.

a) $\quad \mathbf{r} = \begin{pmatrix} 4 \\ 5 \\ 2 \end{pmatrix} + \lambda \begin{pmatrix} 2 \\ -1 \\ 3 \end{pmatrix} \qquad \mathbf{r} = \begin{pmatrix} 5 \\ 0 \\ 4 \end{pmatrix} + \mu \begin{pmatrix} -4 \\ 2 \\ -6 \end{pmatrix}$

b) $\quad \mathbf{r} = \begin{pmatrix} 4 \\ 3 \\ 1 \end{pmatrix} + \lambda \begin{pmatrix} 3 \\ 1 \\ 2 \end{pmatrix} \qquad \mathbf{r} = \begin{pmatrix} 2 \\ 1 \\ 6 \end{pmatrix} + \mu \begin{pmatrix} 5 \\ -2 \\ 4 \end{pmatrix}$

a) By inspection we see that

$$\begin{pmatrix} -4 \\ 2 \\ -6 \end{pmatrix} = -2 \begin{pmatrix} 2 \\ -1 \\ 3 \end{pmatrix},$$

> Firstly check to see if the lines are parallel.

> It is easy to check whether the lines are parallel by looking at the direction vectors so it may save work to do this first.

The lines are therefore parallel and do not intersect.

b) The lines are not parallel.

> Check the direction vectors in order to conclude this.

For intersection

$$4 + 3\lambda = 2 + 5\mu \qquad \text{(i)}$$
$$3 + 1\lambda = 1 - 2\mu \qquad \text{(ii)}$$
$$1 + 2\lambda = 6 + 4\mu \qquad \text{(iii)}$$

Equation (i) – 3 × equation (ii) gives

$$-5 = -1 + 11\mu$$

$$\mu = -\frac{4}{11}$$

> Solve two of the equations for λ and μ vectors.

Substituting $\mu = -\frac{4}{11}$ into equation (i)

$$4 + 3\lambda = 2 + 5 \times -\frac{4}{11}$$

$$\lambda = -\frac{14}{11}$$

Checking equation (iii)

$$1 + 2\lambda = 1 + 2 \times -\frac{14}{11} = -\frac{17}{11}$$

$$6 + 4\mu = 6 + 4 \times -\frac{4}{11} = \frac{50}{11}$$

$$1 + 2\lambda \neq 6 + 4\mu$$

> Check to see if the values found also satisfy the third equation.

There are no values which satisfy all three equations so there is no point in common for these two lines.

The lines are not parallel so the lines must be skew.

Example 6

Prove that the lines $\mathbf{r}_1 = (2\lambda + 1)\mathbf{i} + (1 - \lambda)\mathbf{j} + (1 - \lambda)\mathbf{k}$ and $\mathbf{r}_2 = (3 + \mu)\mathbf{i} - (2\mu + 1)\mathbf{j} + (4\mu + 2)\mathbf{k}$ are perpendicular.

The equation of the lines can be written as

$\mathbf{r}_1 = \mathbf{i} + \mathbf{j} + \mathbf{k} + \lambda(2\mathbf{i} - \mathbf{j} - \mathbf{k})$ ← Write each equation in the form $\mathbf{r} = \mathbf{a} + \lambda\mathbf{b}$.
$\mathbf{r}_2 = 3\mathbf{i} - \mathbf{j} + 2\mathbf{k} + \mu(\mathbf{i} - 2\mathbf{j} + 4\mathbf{k})$

But $(2\mathbf{i} - \mathbf{j} - \mathbf{k}).(\mathbf{i} - 2\mathbf{j} + 4\mathbf{k})$

$= (2 \times 1) + (-1 \times -2) + (-1 \times 4)$ ← Find the scalar product of the direction vectors.
$= 2 + 2 - 4$
$= 0$

Note: These two lines are perpendicular but happen to be skew.

The scalar product is 0 so the lines are perpendicular.

Exercise 9.2

1. State whether the following pairs of lines are parallel or not.
   You may assume that the lines are not coincident.

   a) $\mathbf{r} = \begin{pmatrix} 2 \\ -5 \\ 3 \end{pmatrix} + \lambda \begin{pmatrix} -1 \\ 3 \\ 2 \end{pmatrix}$, $\mathbf{r} = \begin{pmatrix} 3 \\ -6 \\ 5 \end{pmatrix} + \mu \begin{pmatrix} 4 \\ -12 \\ -8 \end{pmatrix}$

   b) $\mathbf{r} = \begin{pmatrix} -1 \\ 3 \\ 1 \end{pmatrix} + \lambda \begin{pmatrix} 3 \\ 8 \\ 6 \end{pmatrix}$, $\mathbf{r} = \begin{pmatrix} 4 \\ -7 \\ -4 \end{pmatrix} + \mu \begin{pmatrix} 1.5 \\ 4 \\ 3 \end{pmatrix}$

   c) $\mathbf{r} = \begin{pmatrix} 0 \\ 0 \\ 1 \end{pmatrix} + s \begin{pmatrix} 1 \\ -1 \\ -3 \end{pmatrix}$, $\mathbf{r} = \begin{pmatrix} 2 \\ 1 \\ 0 \end{pmatrix} + t \begin{pmatrix} 0 \\ 3 \\ 5 \end{pmatrix}$

   d) $\mathbf{r} = \begin{pmatrix} -3 \\ -2 \\ 1 \end{pmatrix} + s \begin{pmatrix} 2 \\ 1 \\ 1 \end{pmatrix}$, $\mathbf{r} = \begin{pmatrix} 1 \\ 2 \\ -5 \end{pmatrix} + \mu \begin{pmatrix} 1 \\ -1 \\ 3 \end{pmatrix}$

   e) $\mathbf{r} = (2\mathbf{i} + \mathbf{k}) + \lambda(-3\mathbf{i} + 3\mathbf{j} - 2\mathbf{k})$, $\mathbf{r} = (-5\mathbf{i} - \mathbf{j} + 2\mathbf{k}) + \mu(6\mathbf{i} + 5\mathbf{j} + 2\mathbf{k})$

   f) $\mathbf{r} = (\mathbf{i} - \mathbf{j} + 2\mathbf{k}) + \lambda(2\mathbf{i} + 5\mathbf{j} - \mathbf{k})$, $\mathbf{r} = (2\mathbf{i} - 2\mathbf{j} + 4\mathbf{k}) + \mu(2\mathbf{i} - 5\mathbf{j} + \mathbf{k})$

   g) $\mathbf{r} = (5 + 2s)\mathbf{i} + (1 + s)\mathbf{j} + (-1 + 5s)\mathbf{k}$, $\mathbf{r} = (13 - 4t)\mathbf{i} + (-6 - 2t)\mathbf{j} + (2 - 10t)\mathbf{k}$

   h) $\mathbf{r} = (s - 1)\mathbf{i} + (2s - 1)\mathbf{j} + (3s + 1)\mathbf{k}$, $\mathbf{r} = \left(1 - \frac{1}{2}s\right)\mathbf{i} + (1 - s)\mathbf{j} + \left(3 - \frac{3}{2}s\right)\mathbf{k}$

2. In each case find out whether the two lines intersect and find the coordinates of any points of intersection.

a) $\mathbf{r} = \begin{pmatrix} 2 \\ 1 \\ 0 \end{pmatrix} + \lambda \begin{pmatrix} 1 \\ 1 \\ -2 \end{pmatrix}$, $\mathbf{r} = \begin{pmatrix} 5 \\ 2 \\ -1 \end{pmatrix} + \mu \begin{pmatrix} 2 \\ 0 \\ 1 \end{pmatrix}$

b) $\mathbf{r} = \begin{pmatrix} 1 \\ 2 \\ 0 \end{pmatrix} + \lambda \begin{pmatrix} -1 \\ 5 \\ 2 \end{pmatrix}$, $\mathbf{r} = \begin{pmatrix} 2 \\ -1 \\ 5 \end{pmatrix} + \mu \begin{pmatrix} 3 \\ -2 \\ 6 \end{pmatrix}$

c) $\mathbf{r} = (2\mathbf{i} + 3\mathbf{j} + 5\mathbf{k}) + s(4\mathbf{i} - \mathbf{j} + 3\mathbf{k})$, $\mathbf{r} = (4\mathbf{i} + 7\mathbf{j} + 2\mathbf{k}) + t(2\mathbf{i} - 2\mathbf{j} + 3\mathbf{k})$

d) $\mathbf{r} = 6s\mathbf{i} + (2 - s)\mathbf{j} + (-3 - s)\mathbf{k}$, $\mathbf{r} = (-4 + 2t)\mathbf{i} + (6 + t)\mathbf{j} - (4 + t)\mathbf{k}$

3. Find the position vector of the common point for each of the following pairs of lines.

a) $\mathbf{r} = \begin{pmatrix} 1 \\ 1 \\ 0 \end{pmatrix} + \lambda \begin{pmatrix} 2 \\ -1 \\ 0 \end{pmatrix}$, $\mathbf{r} = \begin{pmatrix} 2 \\ 1 \\ 0 \end{pmatrix} + \mu \begin{pmatrix} -1 \\ 1 \\ 0 \end{pmatrix}$

b) $\mathbf{r} = (2\mathbf{i} - \mathbf{j} + \mathbf{k}) + s(3\mathbf{i} + \mathbf{j})$, $\mathbf{r} = (-\mathbf{i} + \mathbf{j} + \mathbf{k}) + \mu(-3\mathbf{i} + 2\mathbf{j})$

4. Points A, B, C, and D have position vectors $2\mathbf{i} + 3\mathbf{j}$, $3\mathbf{i} + 2\mathbf{j}$, $4\mathbf{i} + 6\mathbf{j}$, and $9\mathbf{i} + 6\mathbf{j}$ respectively. Find the position vector of the point of intersection of lines AB and CD.

5. Points P, Q, R, and S have coordinates $(2, 0, 1)$, $(-1, 3, -1)$, $(-5, -1, 2)$, and $(1, 4, 4)$ respectively. Show that the lines PQ and RS are skew.

6. Show that the lines $\mathbf{r} = \begin{pmatrix} 2 \\ 1 \\ 4 \end{pmatrix} + \lambda \begin{pmatrix} 3 \\ 6 \\ -2 \end{pmatrix}$ and $\mathbf{r} = \begin{pmatrix} 7 \\ 21 \\ 3 \end{pmatrix} + \mu \begin{pmatrix} 2 \\ -1 \\ 0 \end{pmatrix}$ are perpendicular and skew.

7. Two lines have equations $\mathbf{r} = (3\mathbf{i} + 2\mathbf{j} - \mathbf{k}) + \lambda(2\mathbf{i} - 3\mathbf{j} + 2\mathbf{k})$ and $\mathbf{r} = (-\mathbf{i} + \mathbf{j} + \mathbf{k}) + \mu(\mathbf{i} + 2\mathbf{j} + c\mathbf{k})$. Find the value of c for which the lines intersect and find the coordinates of the point of intersection.

8. Show that the two lines $\mathbf{r} = \begin{pmatrix} 5 \\ 7 \\ 1 \end{pmatrix} + \lambda \begin{pmatrix} -1 \\ 3 \\ 2 \end{pmatrix}$ and $\mathbf{r} = \begin{pmatrix} 7 \\ 1 \\ -3 \end{pmatrix} + \mu \begin{pmatrix} 1 \\ -3 \\ -2 \end{pmatrix}$ are coincident.

9.3 The angle between two straight lines

In P1 we learned how to find the angle between two vectors $\mathbf{a} = \begin{pmatrix} a_1 \\ a_2 \\ a_3 \end{pmatrix}$ and $\mathbf{b} = \begin{pmatrix} b_1 \\ b_2 \\ b_3 \end{pmatrix}$ using the result $\cos\theta = \dfrac{a_1 b_1 + a_2 b_2 + a_3 b_3}{|\mathbf{a}||\mathbf{b}|}$

We now use this result to find the angle between two lines.

> The angle between the two straight lines is defined as the angle between their direction vectors.

Example 7

Two lines have equations $\mathbf{r} = \begin{pmatrix} 2 \\ 0 \\ 1 \end{pmatrix} + \lambda \begin{pmatrix} 5 \\ 3 \\ -1 \end{pmatrix}$ and $\mathbf{r} = \begin{pmatrix} 6 \\ 7 \\ -1 \end{pmatrix} + \mu \begin{pmatrix} 1 \\ -5 \\ -3 \end{pmatrix}$.

Find the acute angle between the two lines.

$\begin{pmatrix} 5 \\ 3 \\ -1 \end{pmatrix}$ and $\begin{pmatrix} 1 \\ -5 \\ -3 \end{pmatrix}$ give the direction of each line.

$\cos\theta = \dfrac{(5 \times 1) + (3 \times -5) + (-1 \times -3)}{\sqrt{5^2 + 3^2 + (-1)^2}\ \sqrt{1^2 + (-5)^2 + (-3)^2}}$ ← Use the standard result to find $\cos\theta$.

$\qquad = -0.2$ ← A negative cosine results in an obtuse angle.

$\quad \theta = 101.5°$ (to 1 d.p.)

Acute angle between these lines $= 180 - 101.5°$

$\qquad\qquad\qquad\qquad = 78.5°$ (1 d.p.)   ← This is the acute angle required.

> Note that it can be shown that these two lines do not intersect. The angle between two skew lines is, nevertheless, still defined as the angle between their directions.

Exercise 9.3

1. Find the acute angle between the two straight lines.

   a) $\mathbf{r} = \begin{pmatrix} 5 \\ 1 \\ -1 \end{pmatrix} + \lambda \begin{pmatrix} 2 \\ 1 \\ 5 \end{pmatrix}$, $\mathbf{r} = \begin{pmatrix} 13 \\ -6 \\ 2 \end{pmatrix} + \mu \begin{pmatrix} -3 \\ 4 \\ 1 \end{pmatrix}$

   b) $\mathbf{r} = \begin{pmatrix} 1 \\ 2 \\ 0 \end{pmatrix} + s \begin{pmatrix} -2 \\ -1 \\ 3 \end{pmatrix}$, $\mathbf{r} = \begin{pmatrix} 1 \\ 0 \\ -1 \end{pmatrix} + t \begin{pmatrix} 1.5 \\ 4 \\ 3 \end{pmatrix}$

   c) $\mathbf{r} = \begin{pmatrix} 0 \\ 5 \\ 2 \end{pmatrix} + \lambda \begin{pmatrix} 1 \\ 2 \\ 3 \end{pmatrix}$, $\mathbf{r} = \begin{pmatrix} -2 \\ 3 \\ 1 \end{pmatrix} + \mu \begin{pmatrix} 2 \\ -1 \\ 4 \end{pmatrix}$

   d) $\mathbf{r} = s(\mathbf{i} - \mathbf{j} + \mathbf{k})$, $\mathbf{r} = (4\mathbf{i} + \mathbf{j}) + t(\mathbf{i} + 2\mathbf{j} + 3\mathbf{k})$

   e) $\mathbf{r} = (\mathbf{i} - \mathbf{j} + \mathbf{k}) + \lambda(\mathbf{i} - \mathbf{j} + \mathbf{k})$, $\mathbf{r} = \mu(2\mathbf{i} - 3\mathbf{j} + \mathbf{k})$

2. Show that the following two lines are perpendicular.

   $\mathbf{r} = \begin{pmatrix} 2 \\ -1 \\ 6 \end{pmatrix} + \lambda \begin{pmatrix} -2 \\ -4 \\ 5 \end{pmatrix}$ and $\mathbf{r} = \begin{pmatrix} 6 \\ 1 \\ -1 \end{pmatrix} + \mu \begin{pmatrix} -11 \\ 3 \\ -2 \end{pmatrix}$

   Hint: Show that
   $a_1 b_1 + a_2 b_2 + a_3 b_3 = 0$

3. *OABCD* is a square based pyramid.

   Find the angles between the following lines.

   a) *OD* and *AD*   b) *CD* and *AD*   c) *OB* and *BD*

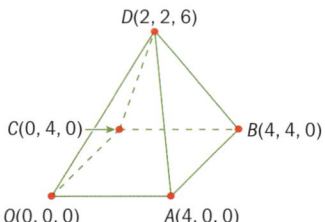

9.4 The equation of a plane

We have seen how to find the direction and equation of a line. We now turn our attention to considering the direction and equation of a plane.

For a line, a single vector is all that is needed to define its direction. However there are an infinite number of planes which could contain a single vector. It is convenient to describe the direction of a plane by specifying a direction perpendicular to it. This is usually called a normal vector to the plane and is denoted by **n**.

We can define a plane if we have a vector normal to the plane and a known point in the plane.

In the diagram, the point *A* represents a known point with position vector **a**. *P* represents a general point in the plane with position vector **r**.

As any line in the plane will be perpendicular to the normal vector,

$$\overrightarrow{AP} \cdot \mathbf{n} = 0$$

$(\mathbf{r} - \mathbf{a}) \cdot \mathbf{n} = 0$ or $\mathbf{r} \cdot \mathbf{n} = \mathbf{a} \cdot \mathbf{n}$

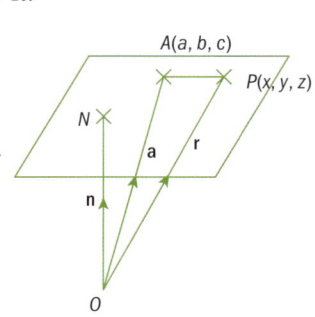

We can show that the two equations $(\mathbf{r} - \mathbf{a}).\mathbf{n} = 0$ and $\mathbf{r}.\mathbf{n} = \mathbf{a}.\mathbf{n}$ are equivalent.

Let $\mathbf{r} = \begin{pmatrix} x \\ y \\ z \end{pmatrix}$, $\quad \mathbf{a} = \begin{pmatrix} a \\ b \\ c \end{pmatrix}$, $\quad \mathbf{n} = \begin{pmatrix} n_1 \\ n_2 \\ n_3 \end{pmatrix}$

Suppose the equation of a plane is $(\mathbf{r} - \mathbf{a}).\mathbf{n} = 0$. Then

$$\left[\begin{pmatrix} x \\ y \\ z \end{pmatrix} - \begin{pmatrix} a \\ b \\ c \end{pmatrix} \right] . \begin{pmatrix} n_1 \\ n_2 \\ n_3 \end{pmatrix} = 0$$

$$(x - a) \times n_1 + (y - b) \times n_2 + (z - c) \times n_3 = 0$$

$$(xn_1 + yn_2 + zn_3) - (an_1 + bn_2 + cn_3) = 0$$

$$\mathbf{r}.\mathbf{n} - \mathbf{a}.\mathbf{n} = 0$$

And so $\mathbf{r}.\mathbf{n} = \mathbf{a}.\mathbf{n}$

The equation of a plane may be written in the form $(\mathbf{r} - \mathbf{a}).\mathbf{n} = \mathbf{0}$ or $\mathbf{r}.\mathbf{n} = \mathbf{a}.\mathbf{n}$

where $\mathbf{r}$ is the general position vector of a point on the plane, $\mathbf{r} = \begin{pmatrix} x \\ y \\ z \end{pmatrix}$ or $x\mathbf{i} + y\mathbf{j} + z\mathbf{k}$

$\mathbf{a}$ is the position vector of a known point on the plane, $\mathbf{a} = \begin{pmatrix} a \\ b \\ c \end{pmatrix}$ or $a\mathbf{i} + b\mathbf{j} + c\mathbf{k}$

and $\mathbf{n}$ is a vector normal to the plane, $\mathbf{n} = \begin{pmatrix} n_1 \\ n_2 \\ n_3 \end{pmatrix}$ or $n_1\mathbf{i} + n_2\mathbf{j} + n_3\mathbf{k}$.

This form of the equation of a plane is known as the scalar product form of the equation of a plane.

Example 8

A plane is perpendicular to the line $2\mathbf{i} + \mathbf{j} - \mathbf{k}$ and passes through the point $(-3, 2, 5)$. Find the vector equation of the plane in the form $(\mathbf{r} - \mathbf{a}).\mathbf{n} = 0$.

$\mathbf{n} = 2\mathbf{i} + \mathbf{j} - \mathbf{k}$

$\mathbf{a} = -3\mathbf{i} + 2\mathbf{j} + 5\mathbf{k}$

The equation of the plane is therefore $[\mathbf{r} - (-3\mathbf{i} + 2\mathbf{j} + 5\mathbf{k})].(2\mathbf{i} + \mathbf{j} - \mathbf{k}) = 0$

or $(\mathbf{r} + 3\mathbf{i} - 2\mathbf{j} - 5\mathbf{k}).(2\mathbf{i} + \mathbf{j} - \mathbf{k}) = 0$

Alternatively, this may be written in the form $\mathbf{r}.\mathbf{n} = \mathbf{a}.\mathbf{n}$ in the following way:

$$\mathbf{r}.(2\mathbf{i} + \mathbf{j} - \mathbf{k}) = (-3\mathbf{i} + 2\mathbf{j} + 5\mathbf{k}).(2\mathbf{i} + \mathbf{j} - \mathbf{k})$$

$$\Rightarrow \mathbf{r}.(2\mathbf{i} + \mathbf{j} - \mathbf{k}) = (-3 \times 2) + (2 \times 1) + (5 \times -1)$$

$$\Rightarrow \mathbf{r}.(2\mathbf{i} + \mathbf{j} - \mathbf{k}) = -9.$$

If we write the position vector **r** in the form $x\mathbf{i} + y\mathbf{j} + z\mathbf{k}$, then the equation $\mathbf{r}.(2\mathbf{i} + \mathbf{j} - \mathbf{k}) = -9$ (in Example 8) becomes

$$(x\mathbf{i} + y\mathbf{j} + z\mathbf{k}).(2\mathbf{i} + \mathbf{j} - \mathbf{k}) = -9$$
$$2x + y - z = -9.$$

In general, the term **a.n** in the equation $\mathbf{r.n} = \mathbf{a.n}$ is a constant. If we let d represent this constant, then $\mathbf{r.n} = d$ or $\mathbf{n}_1 x + \mathbf{n}_2 y + \mathbf{n}_3 z = d$.

Here $\mathbf{n} = a\mathbf{i} + b\mathbf{j} + c\mathbf{k}$ is a vector normal to the plane and $\mathbf{r} = x\mathbf{i} + y\mathbf{j} + z\mathbf{k}$.

> The equation of a plane may be written in the form $ax + by + cz = d$, where $a\mathbf{i} + b\mathbf{j} + c\mathbf{k}$ is a normal vector to the plane.
>
> This is called the cartesian form of the equation of a plane.

Example 9

The plane p has equation $2x - 4y - z = 3$.

a) Write down a vector which is perpendicular to plane p.

b) Find the equation of a plane q which is parallel to p and which passes through the point $(1, -2, 5)$.

Give your answer in the form $ax + by + cz = d$.

a) A normal vector to the plane is $\begin{pmatrix} 2 \\ -4 \\ -1 \end{pmatrix}$ ← Since the plane is expressed in the form $\mathbf{r.n} = d$, we can pick out the coefficients of x, y, and z.

b) Plane q has the same 'direction' as p so its equation is of the form $2x - 4y - z = d$ for some constant, d.

The plane contains the point $(1, -2, 5)$

so $(2 \times 1) - (4 \times -2) - 5 = d$ ← Put the coordinates into the equation for q and find the value of d.

$$2 + 8 - 5 = d$$
$$d = 5$$

The plane q has equation $2x - 4y - z = 5$ (or $6x - y - 5z = -7$)

Example 10

Find the equation of the plane that contains the lines

$$\mathbf{r} = \begin{pmatrix} 2 \\ 1 \\ 3 \end{pmatrix} + \lambda \begin{pmatrix} 3 \\ -2 \\ 4 \end{pmatrix} \text{ and } \mathbf{r} = \begin{pmatrix} 0 \\ -1 \\ 1 \end{pmatrix} + \mu \begin{pmatrix} 1 \\ 1 \\ 1 \end{pmatrix}, \text{ and which passes through the point } (1, 3, 2).$$

Give your answer in the form $ax + by + cz = d$.

- -

We need to find a normal vector $\mathbf{n} = \begin{pmatrix} a \\ b \\ c \end{pmatrix}$ for the plane.

As $\begin{pmatrix} 3 \\ -2 \\ 4 \end{pmatrix}$ and $\begin{pmatrix} 1 \\ 1 \\ 1 \end{pmatrix}$ are in the plane, then $\begin{pmatrix} 3 \\ -2 \\ 4 \end{pmatrix} . \begin{pmatrix} a \\ b \\ c \end{pmatrix} = 0$ and $\begin{pmatrix} 1 \\ 1 \\ 1 \end{pmatrix} . \begin{pmatrix} a \\ b \\ c \end{pmatrix} = 0$.

So $3a - 2b + 4c = 0$ (i)

$\quad\quad a + b + c = 0$ (ii)

We are unable to find unique numerical values for a, b, and c, but we can find the ratio $a : b : c$.

Equation (i) + 2 × equation (ii) gives   ⟵   Eliminate one variable, b, from the two equations.

$5a + 6c = 0$

so $a = -\dfrac{6}{5}c$ ⟵   Find one of the remaining variables, a, in terms of the other remaining variable, c.

Substituting $a = -\dfrac{6}{5}c$ into equation (ii),

$-\dfrac{6}{5}c + b + c = 0$

$b = \dfrac{1}{5}c$ ⟵   Find b in terms of c.

So $a : b : c = -\dfrac{6}{5}c : \dfrac{1}{5}c : c$

$\quad\quad\quad\quad = -\dfrac{6}{5} : \dfrac{1}{5} : 1$

$\quad\quad\quad\quad = -6 : 1 : 5$

A vector normal to the plane is $\begin{pmatrix} -6 \\ 1 \\ 5 \end{pmatrix}$ ⟵   Use the ratio to write down the normal vector.

The equation of the plane is therefore $-6x + y + 5z = d$

The plane passes through the point $(1, 3, 2)$

so $(-6 \times 1) + (1 \times 3) + (5 \times 2) = d$ ⟵   Substitute the point in the plane to find the value of d.

$\quad\quad\quad\quad\quad\quad d = 7$

The equation is $-6x + y + 5z = 7$.

Exercise 9.4

1. Find a vector equation for each of these planes. Give your answer both in the form $\mathbf{r.n} = d$ and in the form $ax + by + cz = d$.

   a) Normal vector $\mathbf{i} + 4\mathbf{j} - 3\mathbf{k}$, passing through the origin.

   b) Normal vector $\mathbf{i} - 3\mathbf{j} + 5\mathbf{k}$, passing through the point $(1, 4, 2)$.

   c) Normal vector $\begin{pmatrix} -1 \\ -3 \\ 4 \end{pmatrix}$, passing through the point $(5, 1, 2)$.

   d) Normal vector $\begin{pmatrix} 2 \\ -3 \\ 1 \end{pmatrix}$, passing through the point $(-2, 0, 3)$.

2. a) Write down a vector perpendicular to the plane $2x - y + 3z = 7$.

   b) Write down the equation of the plane which is parallel to the plane $2x - y + 3z = 7$ and which passes through the origin.

   c) Find the equation of the plane which is parallel to the plane $2x - y + 3z = 0$ and which passes through the point $(0, -2, 5)$.

3. The perpendicular from the origin to a plane intersects the plane at the point $(2, -1, 2)$.

   a) Write down a vector which is normal to the plane.

   b) Find the equation of the plane. Give your answer in the form $ax + by + cz = d$.

4. Find two direction vectors contained in the plane which passes through the points $(0, 1, 1)$, $(2, 1, -3)$, and $(1, 3, 2)$. Hence find a vector perpendicular to the plane and find the equation of the plane in the form $\mathbf{r.n} = d$.

5. Find the equation of the plane which passes through the point $(3, 1, -1)$ and is parallel to the vectors $\begin{pmatrix} 1 \\ 1 \\ 1 \end{pmatrix}$ and $\begin{pmatrix} 2 \\ -1 \\ -3 \end{pmatrix}$. Give your answer in the form $ax + by + cz = d$.

6. *OABC* is a tetrahedron. *OA, OB,* and *OC* are 3, 2, and 4 units from the origin respectively.

   a) Write down the vectors $\overrightarrow{AB}$ and $\overrightarrow{BC}$.

   b) Find a vector perpendicular to the plane *ABC*.

   c) Find an equation of the plane *ABC*.

      Give your answer in the form $ax + by + cz = d$.

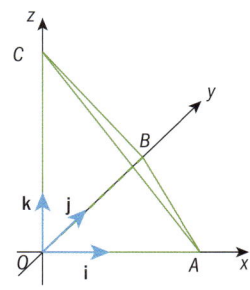

7. *OABCDEFG* is a cube with edges of length 3 units. For each of the following planes, write down a normal vector and find the Cartesian equation of the plane.

   a) *DEFG*   **b)** *ACDF*   **c)** *OBGE*

   d) *ACE*   **e)** *FDB*

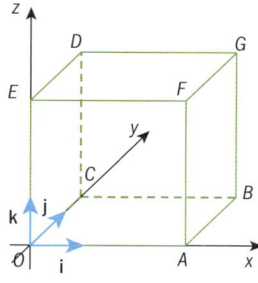

9.5 Configurations of a line and a plane

There are three possible relationships between a line and a plane.

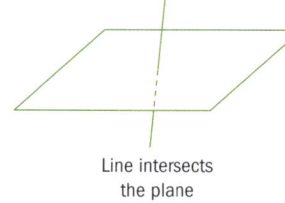

Line intersects
the plane

Line is parallel
to the plane

Line lies in
the plane

A line intersects a plane if the line and the plane have a point in common.
A line is parallel to a plane if the line and the plane have no points in common.
A line lies in a plane if the line and the plane have an infinite number of points in common.

Example 11

Find the position vector of the point where the line $\mathbf{r} = \begin{pmatrix} 1 \\ 1 \\ 1 \end{pmatrix} + \lambda \begin{pmatrix} 2 \\ -1 \\ 0 \end{pmatrix}$ intersects the plane $x - y + z = 2$.

From the equation of the line,

$x = 1 + 2\lambda$

$y = 1 - \lambda$ ← Write down the coordinates of a general point on the line in terms of λ.

$z = 1$

$(1 + 2\lambda) - (1 - \lambda) + 1 = 2$ ← Substitute the coordinates into the equation of the plane.

$1 + 2\lambda - 1 + \lambda + 1 = 2$

$3\lambda = 1$

$\lambda = \dfrac{1}{3}$ ← Work out the value of λ which gives a point on both the line and the plane.

When $\lambda = \dfrac{1}{3}$ $\mathbf{r} = \begin{pmatrix} 1 \\ 1 \\ 1 \end{pmatrix} + \dfrac{1}{3} \begin{pmatrix} 2 \\ -1 \\ 0 \end{pmatrix}$ ← Find the point on the line that corresponds to $\lambda = \dfrac{1}{3}$

$= \begin{pmatrix} \frac{5}{3} \\ \frac{2}{3} \\ 1 \end{pmatrix}$

Note that you can check that the point lies on the plane by substituting the point you found into $x - y + z = 2$, the equation of the plane.

When $x = \dfrac{5}{3}, y = \dfrac{2}{3}$ and $z = 1$ then $x - y + z = \dfrac{5}{3} - \dfrac{2}{3} + 1 = 2,$

so the point lies on the plane.

Example 12

Find whether the line $\mathbf{r} = (3\mathbf{i} - \mathbf{j}) + \lambda(2\mathbf{i} + \mathbf{j} + 4\mathbf{k})$ intersects, is parallel to, or lies in the plane $\mathbf{r}.(5\mathbf{i} - 2\mathbf{j} - 2\mathbf{k}) = 17$.

The direction of the line is given by $2\mathbf{i} + \mathbf{j} + 4\mathbf{k}$.

The direction of the normal to the plane is given by $5\mathbf{i} - 2\mathbf{j} - 2\mathbf{k}$.

> Use the scalar product as a quick test to find out if the two vectors are at right angles.

$(2\mathbf{i} + \mathbf{j} + 4\mathbf{k}).(5\mathbf{i} - 2\mathbf{j} - 2\mathbf{k}) = (2 \times 5) + (1 \times -2) + (4 \times -2) = 0$

> If the scalar product was not equal to 0 we could conclude that the line must intersect the plane at a single point.

The line is perpendicular to the normal vector so the line is either in the plane or parallel to the plane.

To find whether the line lies in the plane or not we investigate whether there are any common points.

$$\mathbf{r} = (3 + 2\lambda)\mathbf{i} + (-1 + \lambda)\mathbf{j} + 4\lambda\mathbf{k}$$

> Re-arrange the equation of the line as a general position vector, in terms $\mathbf{i}$, $\mathbf{j}$, and $\mathbf{k}$.

$[(3 + 2\lambda)\mathbf{i} + (-1 + \lambda)\mathbf{j} + 4\lambda\mathbf{k}].(5\mathbf{i} - 2\mathbf{j} - 2\mathbf{k})$

$= (3 + 2\lambda) \times 5 + (-1 + \lambda) \times -2 + 4\lambda \times -2$

$= 15 + 10\lambda + 2 - 2\lambda - 8\lambda = 17$

> Substitute $(3 + 2\lambda)\mathbf{i} + (-1 + \lambda)\mathbf{j} + 4\lambda\mathbf{k}$ for $\mathbf{r}$ in the left-hand side of the equation of the plane.

As the right-hand side of the equation of the plane is 17, the equation is satisfied whatever the value of λ. Hence there are an infinite number of points lying on both the line and the plane. Therefore the line lies in the plane.

> Note that if the resulting equation in λ had no solutions, this would show that the line is parallel to the plane.

Alternatively, to find whether the line lies in the plane or is parallel to the plane, we could assign a value of λ and find the position vector of one point on the line then check whether this position vector satisfies the equation of the plane.

For example,

$\lambda = 0$, $\mathbf{r} = 3\mathbf{i} - \mathbf{j}$

$(3\mathbf{i} - \mathbf{j}).(5\mathbf{i} - 2\mathbf{j} - 2\mathbf{k}) = (3 \times 5) + (-1 \times -2) + (0 \times -2) = 17$

Hence, the same conclusion is reached.

We can also find the **angle between a line and a plane**. We do this by firstly finding the acute angle θ between the line l and a vector normal to the plane, P. This angle is then subtracted from 90° to find the angle between the line and the plane.

To find the angle between a line and a plane, first find the angle between the direction vector of the line and the normal vector to the plane.

Example 13

Find the angle between the line $\mathbf{r} = \begin{pmatrix} 1 \\ -2 \\ 1 \end{pmatrix} + \lambda \begin{pmatrix} 2 \\ 1 \\ 1 \end{pmatrix}$ and the plane $2x + 3y - z = 6$.

Consider the vectors $\begin{pmatrix} 2 \\ 1 \\ 1 \end{pmatrix}$ and $\begin{pmatrix} 2 \\ 3 \\ -1 \end{pmatrix}$.

First find the angle between the line and the normal to the plane.

$$\cos\theta = \frac{a_1 b_1 + a_2 b_2 + a_3 b_3}{|\mathbf{a}||\mathbf{b}|}$$

Find θ by using the standard result for $\cos\theta$.

$$= \frac{(2 \times 2) + (1 \times 3) + (1 \times -1)}{\sqrt{2^2 + 1^2 + 1^2}\sqrt{2^2 + 3^2 + (-1)^2}}$$

$$= \frac{6}{\sqrt{6}\sqrt{14}}$$

$\theta = 49.1°$ (to 1 d.p.)

The angle between the line and the plane $= 90° - 49.1°$

$= 40.9°$

Subtract the angle found from 90° to give the angle between the line and the plane.

Exercise 9.5

1. Find the coordinates of the point of intersection of the plane $z = 0$ and the line $\mathbf{r} = (2 + \lambda)\mathbf{i} + (1 - \lambda)\mathbf{j} + (3 - 2\lambda)\mathbf{k}$.

2. Find the angle between the line $\mathbf{r} = \begin{pmatrix} 2 \\ 0 \\ -1 \end{pmatrix} + \lambda \begin{pmatrix} 1 \\ 2 \\ -2 \end{pmatrix}$ and the plane $9x + 12y + 8z = 10$.

3. Show that the line with equation $\mathbf{r} = \begin{pmatrix} 3 \\ -2 \\ 8 \end{pmatrix} + \lambda \begin{pmatrix} 1 \\ -1 \\ 4 \end{pmatrix}$ is parallel to the plane whose equation is $x + 5y + z = 8$.

4. In each case, find the acute angle between the line and the plane.

a) $\mathbf{r} = \begin{pmatrix} 1 \\ 2 \\ 0 \end{pmatrix} + \lambda \begin{pmatrix} 3 \\ 5 \\ 9 \end{pmatrix}$, $\qquad 2x - 3y + 2z = 7$

b) $\mathbf{r} = \begin{pmatrix} 2 \\ -\frac{7}{3} \\ \frac{3}{2} \end{pmatrix} + \lambda \begin{pmatrix} 10 \\ 5 \\ -4 \end{pmatrix}$, $\qquad 2x - 4y - 9z = 0$

c) $\mathbf{r} = (\lambda - 1)\mathbf{i} + \left(\frac{1}{2}\lambda + 3\right)\mathbf{j} + (2 + 3\lambda)\mathbf{k}$, $\quad \mathbf{r}.(4\mathbf{i} - 5\mathbf{k}) = -6$

d) $\mathbf{r} = \left(-4\mathbf{i} + \frac{1}{2}\mathbf{j} - 2\mathbf{k}\right) + \lambda\,(14\mathbf{i} + \mathbf{j})$, $\quad \mathbf{r}. \begin{pmatrix} 4 \\ -5 \\ 1 \end{pmatrix} = 13$

5. In each case, find whether the given line is parallel to, lies within or intersects the plane $2x + y - 3z = 5$.

a) $\mathbf{r} = \begin{pmatrix} 1 \\ 2 \\ 0 \end{pmatrix} + \lambda \begin{pmatrix} 3 \\ 5 \\ 9 \end{pmatrix}$, $\qquad$ **b)** $\mathbf{r} = (2 + 3\lambda)\mathbf{i} + 3\mathbf{j} + 2\lambda\mathbf{k}$

c) $\mathbf{r} = \begin{pmatrix} 3 \\ -2 \\ 7 \end{pmatrix} + \lambda \begin{pmatrix} -2 \\ 3 \\ -3 \end{pmatrix}$ $\qquad$ **d)** $\mathbf{r} = \left(\mathbf{i} - \frac{3}{2}\mathbf{j} + \frac{1}{6}\mathbf{k}\right) + \lambda(2\mathbf{i} + 2\mathbf{j} + 2\mathbf{k})$

6. Show that the line $\mathbf{r} = (6\mathbf{i} - 2\mathbf{j} + 4\mathbf{k}) + \lambda(\mathbf{i} - 3\mathbf{j} + 3\mathbf{k})$ is parallel to the plane $\mathbf{r}. \begin{pmatrix} 3 \\ 3 \\ 2 \end{pmatrix} = 5$, but does not lie in the plane.

7. The plane P has equation $z - y = 5$. The line l has equation $\mathbf{r} = \begin{pmatrix} 1 \\ -1 \\ 2 \end{pmatrix} + \lambda \begin{pmatrix} 0 \\ 4 \\ -1 \end{pmatrix}$.

   a) Find the coordinates of the point of intersection of P and l.

   b) Find the acute angle between P and l.

8. Find the value of p such that the line $\mathbf{r} = \begin{pmatrix} 1 \\ 2 \\ -1 \end{pmatrix} + \lambda \begin{pmatrix} 1 \\ p \\ -2 \end{pmatrix}$ does not intersect the plane $x - 3y - 2z = 5$.

9.6 Configurations of two planes

There are three possible relationships between two planes.

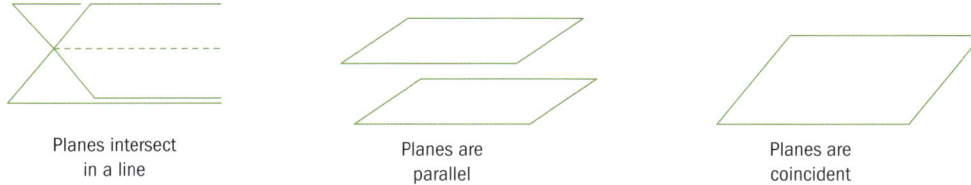

Planes intersect in a line

Planes are parallel

Planes are coincident

> Two planes intersect if they have a line, and hence an infinite number of points, in common.
> Two planes are parallel if they have no points in common.
> Two planes are coincident if they are the same plane.

Example 14

Identify which of the following planes **(i)** to **(iv)**

a) are coincident **b)** are parallel **c)** intersect.

   i) $x + y - z = 0$ **ii)** $x - y + z = 5$
   iii) $2y - 2z = 2x - 5$ **iv)** $3z = 15 - 3x + 3y$

   i) $x + y - z = 0$ **ii)** $x - y + z = 5$
   iii) $2x - 2y + 2z = 5$ **iv)** $3x - 3y + 3z = 15$

Re-write each equation in the form $ax + by + cz = d$.

Since equation (iv) = 3 × equation (ii) the equations represent the same plane.
a) Planes (ii) and (iv) are coincident.

Look to see if any one equation is a multiple of any other.

Since planes (ii), (iii), and (iv) have normal vectors in the same direction, these planes are either parallel to each other or coincident.
b) Plane (iii) is parallel to the plane represented by (ii) and (iv).

Look to see if any normal vector is a multiple of any other, meaning planes are parallel or co-incident. However, since constant terms do not share the same multiplier, (iii) must be parlallel to (ii) & (iv).

Since the normal vector for plane (i) is in a different direction to (ii), (iii), and (iv)
c) Plane (i) intersects with each of planes (ii), (iii), and (iv) to form a line in each case.

In summary, the diagram shows the relationship between the planes.

Example 15

Find the equation of the line of intersection of the planes $3x + 2y - z = 1$ and $x - 4y + 2z = 5$.

To find the equation of the line we need to find its direction and the position vector of a particular point on the line.

The line of intersection lies in each plane and therefore will have a direction perpendicular to the direction of the normal vector of each plane.

$$\begin{pmatrix} 3 \\ 2 \\ -1 \end{pmatrix} \cdot \begin{pmatrix} a \\ b \\ c \end{pmatrix} = 0, \quad 3a + 2b - c = 0 \quad \text{(i)}$$

> Let the direction of the line be $\begin{pmatrix} a \\ b \\ c \end{pmatrix}$ and use scalar product with normal vector = 0

$$\text{and} \quad \begin{pmatrix} 1 \\ -4 \\ 2 \end{pmatrix} \cdot \begin{pmatrix} a \\ b \\ c \end{pmatrix} = 0, \quad a - 4b + 2c = 0 \quad \text{(ii)}$$

Equation (i) $- 3 \times$ equation (ii)

> Find the ratio of $a : b : c$ by expressing a and b in terms of c.

$$14b - 7c = 0$$
$$b = \frac{1}{2}c$$

Substituting in (i)

$$3a = 0$$
$$a = 0$$

Direction vector of the line is $\begin{pmatrix} 0 \\ \frac{1}{2}c \\ c \end{pmatrix} = \begin{pmatrix} 0 \\ c \\ 2c \end{pmatrix} = \begin{pmatrix} 0 \\ 1 \\ 2 \end{pmatrix}$

> Multiply each component by 2 and then and divide by c to simplify the direction vector.

To find a point on the line, assign a value to z in the equation of each plane.

$z = 1$, $3x + 2y - 1 = 1$
$x - 4y + 2 = 5$

> Give z any value and substitute this value into the equations.

$$3x + 2y = 2 \quad \text{(iii)}$$
$$x - 4y = 3 \quad \text{(iv)}$$

Equation (iii) $-3 \times$ equation (iv)

> Solve to find x and y.

$$14y = -7$$
$$y = -\frac{1}{2}$$

From equation (iii), $x = 1$

The position vector of a point on both planes, i.e. on the line of intersection, is $\begin{pmatrix} 1 \\ -\frac{1}{2} \\ 1 \end{pmatrix}$

An equation of the line is $\mathbf{r} = \begin{pmatrix} 1 \\ -\frac{1}{2} \\ 1 \end{pmatrix} + \lambda \begin{pmatrix} 0 \\ 1 \\ 2 \end{pmatrix}$

> Write down the equation of the line in the form $\mathbf{r} = \mathbf{a} + \lambda\mathbf{b}$

Note that this equation is not unique. Assigning different values for z will lead to the position vector of different points on the line being found.

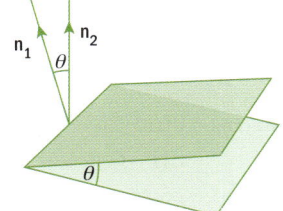

To find **the angle between 2 planes**, we use the fact that the angle between the planes will be the same as the angle between their normal vectors.

> To find the angle between two planes, find the angle between their normal vectors.

Example 16

Find the angle between the planes $x + y + z = 8$ and $2x - 7y + z = 0$.

Consider the normal vectors $\begin{pmatrix} 1 \\ 1 \\ 1 \end{pmatrix}$ and $\begin{pmatrix} 2 \\ -7 \\ 1 \end{pmatrix}$.

$$\cos\theta = \frac{a_1 b_1 + a_2 b_2 + a_3 b_3}{|\mathbf{a}||\mathbf{b}|}$$

Use the standard result for the angle between 2 vectors.

$$= \frac{(1 \times 2) + (1 \times -7) + (1 \times 1)}{\sqrt{1^2 + 1^2 + 1^2}\,\sqrt{2^2 + (-7)^2 + 1^2}}$$

$$= \frac{-4}{\sqrt{3}\sqrt{54}}$$

$$\theta = 108.3° \text{ (to 1 d.p.)}$$

The angle $= 180° - 108.3°$

Calculate the acute angle.

$= 71.7°$

Exercise 9.6

1. The plane π has equation $3x - y + 2z = 6$.
   a) Write down an equation of a coincident plane.
   b) Write down an equation of a parallel plane.
   c) Write down an equation of plane which is perpendicular to π.

2. Show that the planes $2x + 3y - 2z = 1$ and $4x - 2y + z = 1$ are perpendicular.

3. Find the angle between the planes $6x - 3y + z = 2$ and $2x + 2y - z = 5$.

4. Find a vector which is perpendicular to both the vectors $8\mathbf{i} + 2\mathbf{j} - \mathbf{k}$ and $3\mathbf{i} - \mathbf{j} - 3\mathbf{k}$.

5. **a)** Find the angle between the plane $3x - 6y - 2z = 7$ and the plane $2x + y - 2z = 5$.

    b) Find an equation of the line of intersection of these planes.

6. **a)** Find the equation of the line of intersection of the planes $\mathbf{r}.(\mathbf{i} - \mathbf{j}) = 1$ and $\mathbf{r}.(2\mathbf{i} + \mathbf{j} - 3\mathbf{k}) = 2$.

    b) Find the acute angle between these planes.

7. Show that the line of intersection of the planes $x + 2y - 2z = 5$ and $5x - 2y - z = 0$ is in the same direction as the line $\mathbf{r} = (-3\mathbf{i} + \mathbf{j}) + \lambda(2\mathbf{i} + 3\mathbf{j} + 4\mathbf{k})$.

8. Find an equation of the plane which passes through the point $(-1, 2, 1)$ and is parallel to the plane $4x - y + 3z = 0$. Give your answer in the form $\mathbf{r}.\mathbf{n} = d$.

9.7 The distance from a point to a plane or line

The **distance of a point from a plane** is defined as the shortest distance, that is, the distance from a point along a direction which is perpendicular to the plane.

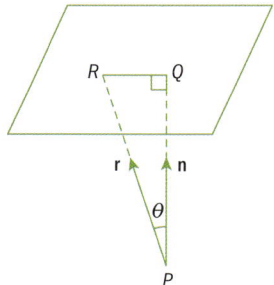

Let Q be the point on the plane at which the perpendicular from the point P meets the plane, i.e. $\overrightarrow{PQ} = \mathbf{n}$.

Let R be any other point on the plane, with $\overrightarrow{PR} = \mathbf{r}$
Angle $PQR = 90°$ and angle $QPR = \theta$.

$$\text{Consider } \overrightarrow{PR}.\overrightarrow{PQ} = \mathbf{r}.\mathbf{n}$$
$$= |\mathbf{r}|\,|\mathbf{n}|\cos\theta$$

$|\mathbf{r}|\cos\theta$ gives the distance of P from the plane, that is, PQ.

If we ensure that the length (magnitude) of $\mathbf{n}$ is 1 then $\mathbf{r}.\mathbf{n} = |\mathbf{r}| \times 1 \times \cos\theta = |\mathbf{r}| \times \cos\theta$ gives us the distance of P from the plane.

We denote a vector of unit length in the direction of $\mathbf{n}$ by $\hat{\mathbf{n}}$.

The distance PQ can now be expressed in the form $\mathbf{r}.\hat{\mathbf{n}} = d$.

> The distance, d, from point P to a plane is given by $\mathbf{r}.\hat{\mathbf{n}} = d$, where $\mathbf{r}$ is the vector from P to any point on the plane and $\hat{\mathbf{n}}$ is the unit normal vector.

Example 17

Find the distance of the point $P(2, 1, -3)$ from the plane $3x - y + 4z = 10$

Firstly find $\hat{\mathbf{n}}$

$$\hat{\mathbf{n}} = \frac{1}{\sqrt{3^2 + (-1)^2 + 4^2}} \begin{pmatrix} 3 \\ -1 \\ 4 \end{pmatrix}$$

Divide **n** by its magnitude to find $\hat{\mathbf{n}}$.

$$= \frac{1}{\sqrt{26}} \begin{pmatrix} 3 \\ -1 \\ 4 \end{pmatrix}$$

Now, find any point which lies in the plane, for example $(3, -1, 0)$

Find any set of values for x, y and z which satisfy the equation $3x - y + 4z = 10$

$$\text{Then } \mathbf{r} = \begin{pmatrix} 3 \\ -1 \\ 0 \end{pmatrix} - \begin{pmatrix} 2 \\ 1 \\ -3 \end{pmatrix} = \begin{pmatrix} 1 \\ -2 \\ 3 \end{pmatrix}$$

Find the vector from P to any point on the plane.

$$\mathbf{r}.\hat{\mathbf{n}} = \begin{pmatrix} 1 \\ -2 \\ 3 \end{pmatrix} . \frac{1}{\sqrt{26}} \begin{pmatrix} 3 \\ -1 \\ 4 \end{pmatrix}$$

Substitute **r** and $\hat{\mathbf{n}}$ into $r.\hat{\mathbf{n}}$ to find the perpendicular distance, d.

$$= \frac{1}{\sqrt{26}} ((1 \times 3) + (-2 \times -1) + (3 \times 4))$$

$$d = \frac{17}{\sqrt{26}}$$

Notes:

i) You may get a negative value for d. This happens in cases where the normal vector points away from the plane rather than towards the plane. When stating the distance, ignore the negative sign as a distance cannot be negative.

ii) To find the shortest distance between two parallel planes, we calculate the distance between them by subtracting their distances from the origin.

We can also find the distance from a point P to a line l. This is demonstrated in Example 18.

Exercise 9.7

1. Find the distance between the point $A(2, -3, 4)$ and the plane $x + 2y + 2z = 13$.

2. Find the distance between the point with position vector $\begin{pmatrix} 2 \\ 2 \\ 3 \end{pmatrix}$ and the plane $2x + 3y + 5z = 0$.

3. Find the distance between the origin and the plane $3x - 6y + 7z = 22$.

4. Find the distance between the origin and the line $\mathbf{r} = \begin{pmatrix} 2 \\ 1 \\ 2 \end{pmatrix} + \lambda \begin{pmatrix} 3 \\ 4 \\ -5 \end{pmatrix}$.

5. Find the distance between the point $(2, 2, 0)$ and the line of intersection of the planes $x + y = 0$ and $y - z = 1$.

6. Find the distance of the point $(3, -1, -2)$ from the line $\mathbf{r} = (-4\mathbf{i} + 3\mathbf{j}) + \lambda(\mathbf{i} + 2\mathbf{j} - 4\mathbf{k})$.

7. The point $(3, 2, -1)$ is an equal distance from two parallel planes. The equation of one of these planes is $3x - y + 2z = -4$. Find the equation of the other plane.

8. Find the distance between the parallel planes $2x - y + 2z - 6 = 0$ and $4x - 2y + 4z + 1 = 0$.

9. Show that the distance of the point (x_0, y_0, z_0) from the plane $ax + by + cz = d$ is given by $\left| \dfrac{ax_0 + by_0 + cz_0 - d}{\sqrt{a^2 + b^2 + c^2}} \right|$.

1. Find the vector equation of the line joining $P(1, 2, -1)$ and $Q(-1, 0, -1)$.

2. Find the cosine of the acute angle between the line $\mathbf{r} = \begin{pmatrix} 1 \\ 0 \\ 0 \end{pmatrix} + \lambda \begin{pmatrix} -4 \\ 3 \\ -5 \end{pmatrix}$ and the vector $\begin{pmatrix} 1 \\ 1 \\ 0 \end{pmatrix}$.

EXAM-STYLE QUESTIONS

3. The three lines l_1, l_2 and l_3 have equations

$$\mathbf{r} = \begin{pmatrix} 3 \\ -2 \\ 1 \end{pmatrix} + \lambda \begin{pmatrix} 7 \\ -2 \\ 4 \end{pmatrix}, \mathbf{r} = \begin{pmatrix} -2 \\ 1 \\ -1 \end{pmatrix} + \mu \begin{pmatrix} 1 \\ 0 \\ 0 \end{pmatrix} \text{ and}$$

$$\mathbf{r} = \begin{pmatrix} 0 \\ 1 \\ -1 \end{pmatrix} + v \begin{pmatrix} -1 \\ 0 \\ 3 \end{pmatrix}$$

   a) Show that l_1 and l_2 intersect and find the point of intersection.

   b) Show that l_1 and l_3 do not intersect and find the acute angle between them.

4. Find the acute angle between the line
$$\mathbf{r} = \begin{pmatrix} -1 \\ 0 \\ 3 \end{pmatrix} + \lambda \begin{pmatrix} 2 \\ 3 \\ 6 \end{pmatrix} \text{ and the plane}$$
$10x + 2y - 11z = 3$.

5. a) Find the point of intersection of the line through the points $(2, 0, 1)$ and $(-1, 3, 4)$ with the line through the points $(-1, 3, 0)$ and $(4, -2, 5)$.

   b) Calculate the acute angle between the lines.

6. Find the equation of a plane which passes through the point $(1, -1, 3)$ and is parallel to the plane $3x + y + z = 7$.

7. Show that the line with equation
$$\mathbf{r} = \begin{pmatrix} 1 \\ -1 \\ 2 \end{pmatrix} + \lambda \begin{pmatrix} 2 \\ 3 \\ 4 \end{pmatrix} \text{ is parallel to the plane}$$
$x - 2y + z = 6$.

EXAM-STYLE QUESTIONS

8. Find the position vector of the point at which the line $\mathbf{r} = (\mathbf{i} - \mathbf{j}) + \mu(2\mathbf{i} - \mathbf{j} + 3\mathbf{k})$ intersects with the plane $\mathbf{r}.(3\mathbf{i} + 2\mathbf{j} - \mathbf{k}) = 5$.

9. The line through the point $(3, 2, 1)$ which is perpendicular to the plane $2x - y + 2z = -2$ intersects the plane at point P. Find the position vector of P.

10. Find an equation of the plane which passes through the points $D(1, 1, -1)$, $E(2, 0, 2)$ and $F(0, -2, 1)$. Give your answer in the form $\mathbf{r}.\mathbf{n} = d$.

11. a) Show that the lines $\mathbf{r} = \begin{pmatrix} 2 \\ 2 \\ 3 \end{pmatrix} + \lambda \begin{pmatrix} 1 \\ 3 \\ 1 \end{pmatrix}$ and
$$\mathbf{r} = \begin{pmatrix} 2 \\ 3 \\ 4 \end{pmatrix} + \mu \begin{pmatrix} 1 \\ 4 \\ 2 \end{pmatrix} \text{ intersect.}$$

   b) Find an equation for the plane which contains these two lines.
   Give your answer in the form $ax + by + cz = d$

12. Find the position vector of the point P at which the line l, with equation $\mathbf{r} = (3\mathbf{i} + 5\mathbf{j} + \mathbf{k}) + \lambda(2\mathbf{i} + 6\mathbf{j} - \mathbf{k})$ meets the plane α with equation $\mathbf{r}.(2\mathbf{i} - 3\mathbf{j} + \mathbf{k}) = 7$

13. Find the equation of the plane which passes through the point with position vector $\begin{pmatrix} 2 \\ 1 \\ -1 \end{pmatrix}$ and which is perpendicular to the line of intersection of the planes $2x + y - z = 3$ and $x + 2y + z = 2$.

14. A plane passes through $P(2, 2, 1)$ and is perpendicular to the line joining the origin to P.

    a) Find a vector equation of the plane in the form $\mathbf{r}.\hat{\boldsymbol{n}} = d$, where $\hat{\boldsymbol{n}}$ is a unit vector.

    b) State the distance from the origin to the plane.

15. a) Find a vector equation for the plane which passes through the points $A(1, -1, 2)$, $B(2, 1, 1)$ and $C(3, -2, 1)$.

    b) Find the distance from this plane to the point $D(4, -2, 5)$

16. Find the equations of the following planes, giving your answer in the form $\mathbf{r}.\mathbf{n} = d$

    a) Direction of normal vector $\begin{pmatrix} 3 \\ 4 \\ 5 \end{pmatrix}$, distance from plane to the origin is 5.

    b) Direction of normal vector to the plane is $2\mathbf{i} + 3\mathbf{j} + 4\mathbf{k}$, plane passes through the point with position vector $2\mathbf{i} + 3\mathbf{j} + 4\mathbf{k}$.

17. $A(-2, 0, -3)$, $B(1, -2, 1)$, $C\left(-2, \frac{-13}{5}, \frac{26}{5},\right)$ and $D\left(\frac{16}{5}, \frac{-13}{5}, 0\right)$ are four points.

    Find the equation of the plane which contains the line AB and which is parallel to the line CD.

EXAM-STYLE QUESTIONS

18. Find an equation of the plane through the three points which have position vectors $3\mathbf{i} - \mathbf{j} + \mathbf{k}$, $\mathbf{i} + 2\mathbf{j} - \mathbf{k}$ and $\mathbf{i} + \mathbf{j} + \mathbf{k}$.

19. The equation of a plane is $2x - 3y + z = 4$

    a) Find the coordinates of the point where the perpendicular from the point $(4, 5, 1)$ meets the plane.

    b) Find the image of the point $(4, 5, 1)$ when it is reflected in the plane.

20. Find a vector of magnitude 2 units which is parallel to the line of intersection of the two planes $x + 2y + z = 1$ and $x - y + 2z = -7$

EXAM-STYLE QUESTIONS

21. Find the angle which the line of intersection of the two planes $\mathbf{r}.(2\mathbf{i} + \mathbf{j} - \mathbf{k}) = 0$ and $\mathbf{r}.(\mathbf{i} + \mathbf{j} + 2\mathbf{k}) = 0$ makes with the x-axis.

22. The line l is the line of intersection of the two planes $x - 2z = 3$ and $y = 2z$. This line intersects the plane $x + 3y - z = -4$ at the point P. Find the coordinates of P.

23. Three planes π_1, π_2 and π_3 have equations $x - y + z = 5$, $x - y + z = 3$ and $2x - 3y + z = 5$ respectively.

    a) Find the angle between the planes π_1 and π_3.

    b) Find the equation of the line of intersection of π_2 and π_3.

    c) Find the distance between the planes π_1 and π_2.

    d) Find an equation of the plane through $(5, 1, -1)$ which is also parallel to π_1.

24. The point A lies on the line which is parallel to the vector $2\mathbf{i} + \mathbf{j} - \mathbf{k}$ and which passes through the point $(1, 1, 2)$. The point B lies on a different line which is parallel to the vector $\mathbf{i} + \mathbf{j} - 2\mathbf{k}$ and which passes through the point $(1, 1, 4)$. The line AB is perpendicular to both of these lines. Find the equation of the line AB.

25. The points A and B have position vectors, relative to the origin O, given by

$$\overrightarrow{OA} = \begin{pmatrix} -1 \\ 3 \\ 5 \end{pmatrix} \text{ and } \overrightarrow{OB} = \begin{pmatrix} 3 \\ -1 \\ -4 \end{pmatrix}.$$

The line l passes through A and is parallel to OB. The point N is the foot of the perpendicular from B to l.

i) State a vector equation for the line l. [1]

ii) Find the position vector of N and show that $BN = 3$. [6]

iii) Find the equation of the plane containing A, B and N, giving your answer in the form $ax + by + cz = d$.

Cambridge International AS and A Level Mathematics 9709 Paper 3 Q10 May/June 2006

26. The straight line l passes through the points A and B with position vectors $2\mathbf{i} + 2\mathbf{j} + \mathbf{k}$ and $\mathbf{i} + 4\mathbf{j} + 2\mathbf{k}$ respectively. This line intersects the plane p with equation $x - 2y + 2z = 6$ at the point C.

i) Find the position vector of C. [4]

ii) Find the acute angle between l and p. [4]

iii) Show that the perpendicular distance from A to p is equal to 2. [3]

Cambridge International AS and A Level Mathematics 9709 Paper 3 Q10 October/November 2005

27. With respect to the origin O, the points A and B have position vectors given by $\overrightarrow{OA} = 2\mathbf{i} + 2\mathbf{j} + \mathbf{k}$ and $\overrightarrow{OB} = \mathbf{i} + 4\mathbf{j} + 3\mathbf{k}$.

The line l has vector equation
$$\mathbf{r} = 4\mathbf{i} - 2\mathbf{j} + 2\mathbf{k} + s(\mathbf{i} + 2\mathbf{j} + \mathbf{k}).$$

i) Prove that the line l does not intersect the line through A and B. [5]

ii) Find the equation of the plane containing l and the point A, giving your answer in the form $ax + by + cz = d$. [6]

Cambridge International AS and A Level Mathematics 9709 Paper 3 Q10 May/June 2005

Chapter summary

Straight lines

- The vector equation of a straight line passing through a point with position vector $\mathbf{a}$ and with direction vector $\mathbf{b}$ is $\mathbf{r} = \mathbf{a} + \lambda\mathbf{b}$ where λ is a scalar.

- Two lines intersect if they have one point in common.

- Two lines are parallel if they have the same direction but no point in common.

- Two lines are coincident if they have the same direction and an infinite number of points in common.

- In three dimensions, skew lines are lines that are not parallel but do not intersect.

- The angle between two straight lines is defined as the angle between their direction vectors.

Planes

- The equation of a plane can be written in the form $\mathbf{n}_1 x + \mathbf{n}_2 y + \mathbf{n}_3 z = d$ where $(\mathbf{n}_1\ \mathbf{n}_2\ \mathbf{n}_3)$ is the normal to the plane. This is called Cartesian form.

- An alternative form for the equation of a plane is $\mathbf{r}.\mathbf{n} = d$, where $\mathbf{n}$, a normal to the plane, is given by $\mathbf{n} = a\mathbf{i} + b\mathbf{j} + c\mathbf{k}$. This is called scalar product form.

- Two planes intersect if they have a line, and hence, an infinite number of points in common.

- Two planes are parallel if they have no points in common.

- Two planes are coincident if they are the same plane.

To find the angle between two planes, find the angle between their normal vectors

Lines and Planes

- A line intersects a plane if the line and the plane have one point in common

- A line is parallel to a plane if the line and the plane have no points in common.

- A line lies in a plane if the line and the plane have an infinite number of points in common.

- To find the angle between a line and a plane, first find the angle between the line and the normal vector to the plane.

Distance of a point from a line or plane

- The distance, d, from point P to a plane is given by $\mathbf{r}.\hat{\mathbf{n}} = d$ where $\mathbf{r}$ is the vector from P to any known point on the plane and $\hat{\mathbf{n}}$ is the unit normal vector.

10 Differential equations

A differential equation is an equation which relates a variable, y, to its derivatives, $\frac{dy}{dx}$ or sometimes $\frac{d^2 y}{dx^2}$. The behaviour of many real-life systems in both nature and technology can be modelled by differential equations. Finding ways to produce a function which satisfies the relationship in the differential equation has resulted in many scientific advances in recent times. For example, the Navier-Stokes non-linear partial differential equations are used to describe the motion of liquids and gases. As such, these equations are incredibly useful when predicting how a liquid or gas will behave in a wide range of fields – from improving the aerodynamic design of a car, to improving the effective delivery of drugs through the bloodstream.

Objectives

- Formulate a simple statement involving a rate of change as a differential equation, including the introduction if necessary of a constant of proportionality.
- Find by integration a general form of solution for a first order differential equation in which the variables are separable.
- Use an initial condition to find a particular solution.
- Interpret the solution of a differential equation in the context of a problem being modelled by the equation.

Before you start

You should know how to:

1. Work out the value of c, given a curve and point on the curve.

    e.g. a) $y = 2x + c$ $(-1, 4)$

           $4 = 2(-1) + c$ $c = 6$

       b) $y = e^x + \tan x + c$ $(0, -1)$

           $-1 = 1 + 0 + c$ $c = -2$

Skills check:

1. Work out the value of c, given a curve and point on the curve.

    a) $y = -2x + c$ $(3, -2)$

    b) $y = 6x^2 - x + c$ $(-4, -3)$

    c) $y = e^x + c$ $(0, -2)$

    d) $y = \ln x + c$ $(e, 3)$

2. Use trigonometric identities.

    e.g. Express $\sin^3 \theta \cos^2 \theta$ in terms of $\sin \theta$ only.

    $\sin^3 \theta \cos^2 \theta = \sin^3 \theta \, (1 - \sin^2 \theta) = \sin^3 \theta - \sin^5 \theta$

2. Express $\cos 2\theta$ in terms of $\sin \theta$ only.

3. Decompose rational functions into partial fractions.

    e.g. $\dfrac{x+3}{x^2+2x}$

    $\dfrac{x+3}{x^2+2x} = \dfrac{A}{x} + \dfrac{B}{x+2} \Rightarrow A + B = 1 \Rightarrow 2A = 3$

    $\Rightarrow A = \dfrac{3}{2}, \; B = -\dfrac{1}{2}$

3. Express $\dfrac{2}{x^2-1}$ in partial fractions.

10.1 Forming simple differential equations (DEs)

> A **differential equation** is an equation in which at least one differential expression appears.

In Chapter 5, and previously in P1, you met the simplest differential equation, $\dfrac{dy}{dx} = f(x)$. This has a solution $y = \displaystyle\int f(x)dx$, and progressively you learned how to integrate more functions, all of which were in the form $\dfrac{dy}{dx} = f(x)$.

$\dfrac{dy}{dx} = f(x)$ is a **first order differential equation** because it only contains the first derivative $\dfrac{dy}{dx}$.

With these skills, you will be able to solve first order DEs in which the variables are separable, i.e. DEs where the equation can be written the form $g(y)\dfrac{dy}{dx} = f(x)$.

Note that sometimes one of the functions $f(x)$ or $g(y)$ may be 1.

Since you know that $\dfrac{dy}{dx}$ gives the gradient of the function which satisfies the DE, you can get a feel for the behaviour of the functions which satisfy the DE by sketching a 'needle diagram': we sketch a number of short line segments at certain points on the domain, each with (approximately) the gradient (given by the value of $\dfrac{dy}{dx}$ at each point). Doing this by hand is somewhat tedious so thinking about patterns is helpful.

Consider $\dfrac{dy}{dx} = 2x$. The gradient is only a function of x, so the needles lying on the same vertical line (i.e. those with the same x-value) will have the same gradient. On the y-axis (where $x = 0$) we see that the gradient is 0, so the needles that cross the y-axis will be horizontal. When the value of x is positive, so is the value of the gradient, and thus the needles have a positive gradient which gets steeper as x increases. When the value of x is negative, so is the value of the gradient, and thus the needles have a negative gradient which gets steeper as x decreases. The needle diagram to the right shows approximately what the solutions to $\dfrac{dy}{dx} = 2x$ will look like.

Solving the DE $\dfrac{dy}{dx} = 2x$ by integrating, as you already know how to do, gives a general solution $y = \displaystyle\int 2x \, dx = x^2 + c$. The solution is a quadratic which is symmetric about the y-axis, with an undetermined constant (which is why there is actually a family of functions which satisfy the differential relationship).

Completing a number of the paths for different values of the constant c lets you see how the needle diagram gives an impression of what the solutions look like. Assigning a specific value to c determines one particular solution. We are often given an initial condition by which we calculate c, the constant of integration.

The graph shows 3 particular solutions: $y = x^2 + 1$, $y = x^2$, and $y = x^2 - 2$. However, there are an infinite number of particular solutions within the family that comprises the general solution. For first order differential equations, each particular solution is distinct (it does not cross with another particular solution at any point).

Often the mathematical description of a physical situation can be captured in a DE.

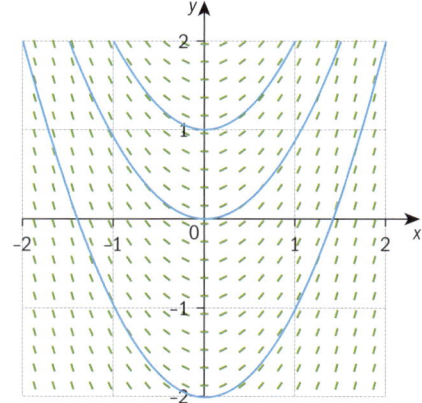

Example 1

A body falling through a fluid experiences resistances causing it to lose speed at a rate proportional to its speed at that instant. Write down a differential equation satisfied by the speed v at time t, and sketch a needle diagram in the first quadrant to illustrate the solution to this differential equation.

$$\frac{\mathrm{d}v}{\mathrm{d}t} = -kv$$

Because the speed is decreasing it makes sense to put a negative in the DE which assign k to be positive, although it is not essential.

Because the derivative is just a function of v, here the needle diagram has rows which all have the same slope and get steeper as v increases.

The question does not specifically ask us to sketch any particular solutions, but the diagram to the left shows that the shape of particular solutions follow the needles in the needle diagram.

Example 2

Water is leaking from a tank. The rate at which the depth of water is decreasing is proportional to the square root of the current depth. Write down a differential equation satisfied by the depth h at time t, and sketch a needle diagram in the first quadrant to illustrate the solution to this differential equation.

$$\frac{dh}{dt} = -k\sqrt{h}$$

This is similar to the equation from Example 1, however but the power raised to the variable h is different.

Because the derivative is a function of h, again the needle diagram has rows which all have the same slope. However, now the increase in gradient is much slower as h increases, since $\frac{dh}{dt}$ is only proportional to $\sqrt{h}$, not h.

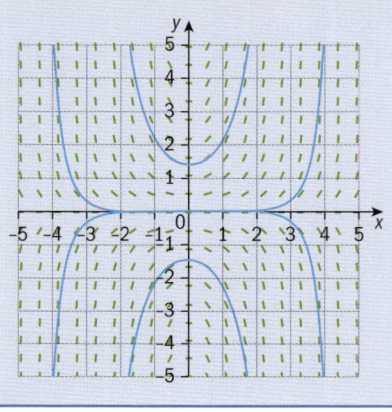

This is reflected in the particular solutions the needle diagram points towards.

Example 3

A differential equation has the form $\frac{dy}{dx} = xy$. Sketch a needle diagram in all four quadrants to illustrate the solution of this differential equation.

In quadrant 1, we see that the gradient is zero when either x or y is zero, and then increases as either x or y increase.

Also, the function xy has reflective symmetry about both the x- and y-axes, so we can easily draw in all 4 quadrants.

Example 4

A rectangular tank has a base of $5\,m^2$. Water is flowing into the tank at a constant rate of 10 litres per second $(= 0.01\,m^3s^{-1})$. At the same time, water is leaking out at a rate proportional to $\sqrt{h}$ where h is the depth of the water at time t. When the depth is 1 metre the water is leaking out at 2 litres per second. Find the differential equation describing the rate of change of h with time t.

If water comes in at rate $0.01\,m^3s^{-1}$ and the base area is $5\,m^2$, the water coming in is increasing the depth at a rate of $0.002\,ms^{-1}$.

So the DE is $\dfrac{dh}{dt} = 0.002t - k\sqrt{h}$.

When $h = 1$, $k\sqrt{h} = 0.0004 \Rightarrow k = 0.0004$ giving $\dfrac{dh}{dt} = 0.002t - 0.0004\sqrt{h}$.

Exercise 10.1

1. The rate at which body temperature, T, falls is proportional to the difference between the body temperature T and the temperature T_0 of the surroundings. Find a differential equation relating body temperature, T, and time t.

2. A certain substance is formed in a chemical reaction. The mass of the substance formed t seconds after the start of the reaction is x grams. At any time the rate of formation of the substance is proportional to $(30 - x)$. Find a differential equation relating x and t.

3. In studying the spread of a disease, a scientists thinks that the rate of infection is proportional to the product of the number of people infected and the number of people uninfected. If N is the number infected at time t and P is the total number of people in the population, form a differential equation to summarise the scientist's theory.

4. The rate of increase of a population is proportional to its size at the time. Write down a differential equation to describe this situation.

   It also known that when the population was 2 million, the rate of increase was $140\,000$ per day. Find the constant of proportionality in your DE.

5. Sketch needle diagrams for the DEs in questions 1 to 3.

10.2 Solving first-order differential equations with separable variables

We saw in **10.1** that when asked to solve $\dfrac{dy}{dx} = 2x$, we can write $y = \displaystyle\int 2x\,dx = x^2 + c$.

To be rigorous in solving this differential equation, we should write

$$\frac{dy}{dx} = 2x \implies \int 1\,dy = \int 2x\,dx \implies y = x^2 + c.$$

If $\dfrac{dy}{dx} = \dfrac{f(x)}{g(y)}$, we can write $g(y)\dfrac{dy}{dx} = f(x)$, giving $\displaystyle\int g(y)\frac{dy}{dx}\,dx = \int f(x)\,dx.$

> If $G(y) = \displaystyle\int g(y)\,dy$
>
> then $\dfrac{d}{dx}\{G(y)\} = g(y)\dfrac{dy}{dx}$

This has solution $\displaystyle\int g(y)\,dy = \int f(x)\,dx.$

> A differential equation which can be rearranged to the form
> $g(y)\dfrac{dy}{dx} = f(x)$ is said to have separable variables.

Example 5

Solve $\dfrac{dy}{dx} = 2y.$

> This DE is similar to the one in Example 1 except here the constant of proportionality is positive. The solution here is one of exponential growth, whereas the needle diagram in Example 1 showed the solution was in the form of exponential decay.

$$\frac{dy}{dx} = 2y$$

$$\frac{1}{y}\frac{dy}{dx} = 2 \qquad \longleftarrow \quad \text{Separate the variables first.}$$

$$\int \frac{1}{y}\,dy = \int 2\,dx \qquad \longleftarrow \quad \text{Show what you are intending to do.}$$

$$\ln y = 2x + c \qquad \longleftarrow \quad \text{Integrate both sides.}$$

$$y = Ae^{2x} \qquad \longleftarrow \quad \text{Write the solution in its simplest form.}$$

> The constant which is added in integration will become a multiplicative constant if there is a log integral and you take exponentials (i.e. $A = e^c$) – remember to change the letter when you do this.

Example 6

Solve $\dfrac{dy}{dx} = 2xy$.

....

$\dfrac{dy}{dx} = 2xy$

$\dfrac{1}{y}\dfrac{dy}{dx} = 2x$

$\displaystyle\int \dfrac{1}{y}\, dy = \int 2x\, dx$

$\ln y = x^2 + c$

$y = Ae^{x^2}$

This DE is very similar to Example 5. Once the variables are separated you essentially always have two integrations to do and it can require any of the integration techniques you learnt in P1 or in Chapter 6 of this book.

Example 7

Solve $\dfrac{dy}{dx} = \cos x$.

....

$\dfrac{dy}{dx} = \cos x$

$\displaystyle\int 1\, dy = \int \cos x\, dx$

$y = \sin x + c$

The needle diagram shows the gradient function, and the particular solutions with $c = 1$ is shown in blue.

Example 8

Solve $\dfrac{dy}{dx} = \cos^2 y$.

....

$\dfrac{dy}{dx} = \cos^2 y$

$\sec^2 y \dfrac{dy}{dx} = 1$

← You need to get integrable functions on both sides.

$\displaystyle\int \sec^2 y\, dy = \int 1\, dx$

$\tan y = x + c$

← Integrate both sides.

$y = \tan^{-1}(x + c)$

← If the functional form is very complicated then it is not a requirement to give the function of y explicitly, unless the question specifies this. The previous line is a perfectly acceptable form of the solution to the DE.

Example 9

Solve $\dfrac{dy}{dx} = \dfrac{3x^2 \sin^2 y}{(x^3 + 2)}$

$\dfrac{dy}{dx} = \dfrac{3x^2 \sin^2 y}{(x^3 + 2)}$

$\operatorname{cosec}^2 y \, \dfrac{dy}{dx} = \dfrac{3x^2}{(x^3 + 2)}$ — You need to get integrable functions on both sides.

$\displaystyle\int \operatorname{cosec}^2 y \, dy = \int \dfrac{3x^2}{(x^3 + 2)} \, dx$

$\cot y = -\ln A(x^3 + 2)$, where $c = \ln A$ — Integrate both sides – you need to remember standard forms of integrable functions.

Example 10

Solve $\dfrac{dy}{dx} = \dfrac{2}{x^2 - 1}$.

$\dfrac{dy}{dx} = \dfrac{2}{x^2 - 1}$

$\dfrac{dy}{dx} = \dfrac{1}{x - 1} - \dfrac{1}{x + 1}$ — You need to use partial fractions to get this into an integrable form.

$\displaystyle\int dy = \int \left(\dfrac{1}{x - 1} - \dfrac{1}{x + 1} \right) dx$

$y = \ln(x - 1) - \ln(x + 1) + c$ — Here we express c as $\ln A$ in order to combine the terms.

$y = \ln \left\{ \dfrac{A(x - 1)}{(x + 1)} \right\}$

In examinations, you will normally be asked to first express a rational function in partial fractions, and then integrate the partial fractions to solve the DE.

Example 11

Solve $\dfrac{dy}{dx} = y^2 \ln x$.

$\dfrac{dy}{dx} = y^2 \ln x$

$\dfrac{1}{y^2} \dfrac{dy}{dx} = \ln x$ — Separate the variables.

$\displaystyle\int \dfrac{1}{y^2} \, dy = \int \ln x \, dx$ — You will need to integrate this by parts.

$-\dfrac{1}{y} = x \ln x - \displaystyle\int x \dfrac{1}{x} \, dx = x \ln x - x + c$

Remember, you may be asked to use any of the integration techniques you met in Chapter 5 in solving DEs. After separating the variables you always have two integrations to perform.

Exercise 10.2

Find the general solution to the following differential equations.

1. $\dfrac{dy}{dx} = \dfrac{x}{y}$

2. $\dfrac{dy}{dx} = \dfrac{x^3}{y}$

3. $\dfrac{dy}{dx} = \dfrac{x^2 + x + 1}{y^2}$

4. $\dfrac{dy}{dx} = y(x+3)$

5. $\dfrac{dy}{dx} = \dfrac{2}{\cos y}$

6. $\dfrac{dy}{dx} = \dfrac{1}{xy}$

7. $2y\dfrac{dy}{dx} = 4x^3$

8. $x^2\dfrac{dy}{dx} = 2y^3$

9. $\dfrac{dy}{dx} = y\cos x$

10. $\dfrac{dy}{dx} = 2xe^y$

11. $\dfrac{dy}{dx} = 2x\sec y$

12. $\dfrac{dy}{dx} = \dfrac{y^2}{x(x+2)}$

13. $\dfrac{dy}{dx} = \dfrac{\cos x}{\cos y}$

14. $\dfrac{dy}{dx} = \dfrac{\cos^2 y}{x}$

15. $2\dfrac{dy}{dx} = 4x^3(y^2 - 1)$

16. $\dfrac{dy}{dx} = \dfrac{\cos^2 y}{\cos^2 x}$

17. $\dfrac{dy}{dx} = xe^x \sec y$

18. $2\dfrac{dy}{dx} = \dfrac{y^2 - 1}{x^2 + x}$

19. $\dfrac{dy}{dx} = y^2\sec^2 x$

20. $\dfrac{dy}{dx} = \dfrac{x(1 + y^4)}{y^3}$

21. $\dfrac{dy}{dx} = \left(\dfrac{x+3}{x^2+x}\right)y^2$

22. $\left(\dfrac{-2}{y^3}\right)\dfrac{dy}{dx} = \dfrac{e^x}{1 + e^{2x}}$ Use the substitution $u = e^x$.

23. $y\dfrac{dy}{dx} = \sin^3 x\cos^2 x$

24. $y\dfrac{dy}{dx} = \dfrac{27}{(9 + x^2)^2}$ Use the substitution $x = 3\tan\theta$.

10.3 Finding particular solutions to differential equations

In the last section, the only information you were given was the differential equation, and you could find the general solution: a family of solutions, each member of which is determined by the value of the constant of integration. If you also have a condition which gives a point on the solution curve, then the particular value of the constant of integration can be determined.

Essentially the DE gives the gradient function at any point in the plane. If you have a starting point and know what direction to head in then you can move along the whole function path. This extra condition is sometimes known as the **initial condition** and sometimes as a **boundary condition**.

The first couple of examples use DEs for which we have already produced the general solution, so you can see the extra work required in order to also produce a particular solution.

Example 12

Find the curve which satisfies $\dfrac{dy}{dx} = 2y$ and passes through $(0, 3)$.

$\dfrac{dy}{dx} = 2y$

$\Rightarrow y = Ae^{2x}$ ⟵ From Example 5. This is the general solution.

$x = 0,\ y = 3 \Rightarrow A = 3$ ⟵ Use the initial condition to find A.

$\Rightarrow y = 3e^{2x}$ ⟵ This is the particular solution.

Example 13

Find the curve which satisfies $\dfrac{dy}{dx} = \dfrac{2}{x^2 - 1}$ and passes through $(2, 0)$.

$\dfrac{dy}{dx} = \dfrac{2}{x^2 - 1}$ ⟵ From Example 10.

$y = \ln\left\{\dfrac{A(x - 1)}{(x + 1)}\right\}$

$x = 2,\ y = 0 \Rightarrow 0 = \ln\left(\dfrac{A}{3}\right) \Rightarrow A = 3$

$y = \ln\left\{\dfrac{3(x - 1)}{(x + 1)}\right\}$ | The needle diagram shows the DE with the particular solution through $(2, 0)$.

Example 14

Find the curve which satisfies $\dfrac{dy}{dx} = \dfrac{4xy}{(x^2 + 3)}$ and passes through $(0, 9)$.

$\dfrac{dy}{dx} = \dfrac{4xy}{(x^2 + 3)}$

$\Rightarrow \dfrac{1}{y}\dfrac{dy}{dx} = \dfrac{4x}{(x^2 + 3)}$ ⟵ Separate the variables.

$\Rightarrow \displaystyle\int \dfrac{1}{y}\,dy = \int \dfrac{4x}{(x^2 + 3)}\,dx$ ⟵ Integrate both sides.

$\Rightarrow \ln y = 2\ln(x^2 + 3) + c = \ln A(x^2 + 3)^2$

$\Rightarrow y = A(x^2 + 3)^2$

$x = 0,\ y = 9 \Rightarrow A = 1$ ⟵ Use the initial condition to find A.

$\Rightarrow y = (x^2 + 3)^2$

Finding particular solutions to differential equations

Example 15

Find the curve which satisfies $\dfrac{dy}{dx} = \sin x \sec y$ and passes through $(0, 0)$.

$$\frac{dy}{dx} = \sin x \sec y$$

$$\cos y \frac{dy}{dx} = \sin x$$

$$\int \cos y \, dy = \int \sin x \, dx$$

$$\Rightarrow \sin y = -\cos x + c$$

$$x = 0, y = 0 \Rightarrow 0 = -1 + c \Rightarrow c = 1$$

$$\Rightarrow \sin y = -\cos x + 1$$

The needle diagram shows the DE with the particular solution through $(0, 0)$. Often the solution curves for DEs are very hard to draw without using a computer package.

Example 16

Find the curve which satisfies $\dfrac{dy}{dx} = xy^2 e^{2x}$ and passes through $(0, 1)$.

$$\frac{dy}{dx} = xy^2 e^{2x}$$

$$\frac{1}{y^2} \frac{dy}{dx} = xe^{2x}$$

Separate the variables.

$$\int \frac{1}{y^2} \, dy = \int xe^{2x} \, dx$$

This will need integration by parts.

$$\Rightarrow -\frac{1}{y} = x \cdot \frac{1}{2}e^{2x} - \int \frac{1}{2}e^{2x} \, dx$$

$$\Rightarrow -\frac{1}{y} = \frac{1}{2}xe^{2x} - \frac{1}{4}e^{2x} + c$$

Remember the constant at the end here. This is the general solution.

$$x = 0, y = 1 \Rightarrow -1 = 0 - \frac{1}{4} + c \Rightarrow c = -\frac{3}{4}$$

Use the initial condition to find c.

$$\Rightarrow -\frac{1}{y} = \frac{1}{2}xe^{2x} - \frac{1}{4}e^{2x} - \frac{3}{4} = \frac{-(3 + e^{2x} - 2xe^{2x})}{4}$$

$$\Rightarrow y = \frac{4}{(3 + e^{2x} - 2xe^{2x})}$$

Reorganise to find y explicitly. This is the particular solution.

Example 17

Find the curve which satisfies $y^2 \dfrac{dy}{dx} = x^2 e^{x^3}$ and passes through the origin.

$y^2 \dfrac{dy}{dx} = x^2 e^{x^3}$

$y^2 \dfrac{dy}{dx} = x^2 e^{x^3}$ ⬅ Separate the variables.

$\Rightarrow \displaystyle\int y^2 \, dy = \int x^2 e^{x^3} \, dx$ ⬅ Since $\dfrac{d}{dx}(e^{x^3}) = 3x^2 e^{x^3}$

$\Rightarrow \dfrac{1}{3} y^3 = \dfrac{1}{3} e^{x^3} + c$

$x = 0, y = 0 \Rightarrow c = -\dfrac{1}{3}$ ⬅ Using the boundary condition to find the particular solution.

$\Rightarrow y^3 = e^{x^3} - 1$

Exercise 10.3

Find the particular solution to each of the following differential equations, given the prescribed boundary condition. The first 9 questions in this exercise use the same DEs as those used in questions 1–9 of Exercise 6.2, for which you should already have found the general solutions.

1. $\dfrac{dy}{dx} = \dfrac{x}{y}$; $x = 1, y = 2$

2. $\dfrac{dy}{dx} = \dfrac{x^3}{y}$; $x = 1, y = 2$

3. $\dfrac{dy}{dx} = \dfrac{x^2 + x + 1}{y^2}$; $x = 0, y = 3$

4. $\dfrac{dy}{dx} = y(x + 3)$; $x = 0, y = 5$

5. $\dfrac{dy}{dx} = \dfrac{2}{\cos y}$; $x = 0, y = \dfrac{\pi}{2}$

6. $\dfrac{dy}{dx} = \dfrac{1}{xy}$; $x = 1, y = 2$

7. $2y\dfrac{dy}{dx} = 4x^3$; $x = 2, y = 3$

8. $2x^2 \dfrac{dy}{dx} = y^3$; $x = 1, y = 0.5$

9. $\dfrac{dy}{dx} = y\cos x$; $x = \dfrac{\pi}{2}, y = 1$

10. $\dfrac{dy}{dx} = \dfrac{x}{y}$; $x = 2, y = 0$ This is the same DE as Q1, but with a different boundary condition.

11. $\dfrac{dy}{dx} = \dfrac{1}{xy}$; $\qquad x = e, y = 1$ $\qquad$ <mark>This is the same DE as Q6, but with a different boundary condition.</mark>

12. A curve passes through the point (1, 2) and its gradient function is $\dfrac{2x+1}{2y}$.

Find the equation of the curve.

13. A curve is such that $\dfrac{dy}{dx} = \sqrt{\dfrac{y}{x+1}}$ and the point (3, 9) lies on the curve.

Find the equation of the curve.

14. A curve is such that $x^3 \dfrac{dy}{dx} = \sec y$ and the point $\left(1, \dfrac{\pi}{6}\right)$ lies on the curve.

Find the equation of the curve.

15. A curve is such that $e^y \dfrac{dy}{dx} - 2\sec^2 x = 0$ and the point $\left(\dfrac{\pi}{4}, 0\right)$ lies on the curve.

Find the equation of the curve.

16. A curve is such that $\sqrt{xy}\, \dfrac{dy}{dx} = 1$ and the point (4, 9) lies on the curve.

Find the equation of the curve.

17. A curve is such that $\dfrac{dy}{dx} = \dfrac{2e^{-y}x}{x^2+1}$ and the point (0, 0) lies on the curve.

Find the equation of the curve.

18. A curve is such that $x(x+1)\dfrac{dy}{dx} = y$ and the point (3, 1) lies on the curve.

Find the equation of the curve.

19. A curve is such that $e^{-y}\dfrac{dy}{dx} = -e^{3x}$ and the point (0, 0) lies on the curve.

Find the equation of the curve.

20. A curve is such that $x^3 \dfrac{dy}{dx} = \cos^2 y$ and the point $\left(1, \dfrac{\pi}{4}\right)$ lies on the curve.

Find the equation of the curve.

21. A curve is such that $y\dfrac{dy}{dx} = xe^{x^2}$ and the point (0, 3) lies on the curve.

Find the equation of the curve.

10.4 Modelling with differential equations

In this chapter, you have already seen how to set up simple differential equations which describe a real-life context, and also how to find both general and particular solutions of first order DEs where the variables are separable. This section combines all these building blocks to find solutions to real-life problems (usually in order to determine when certain conditions in the future will be satisfied).

Example 18

A tank is draining in such a way that the when the height of water in the tank is h cm, it is decreasing at the rate of $0.5\sqrt{h}$ cm s^{-1}. Initially the water in the tank is at a height of 25 cm.

a) Write down a differential equation which describes this situation.

b) Solve the differential equation to find h as a function of time.

c) What is the height of the water after 10 seconds?

d) How long does it take for the water to reach a height of 5 cm.

e) Sketch a graph of the height against time for $0 \le t \le 20$.

· ·

a) $\dfrac{dh}{dt} = -0.5\sqrt{h}$

> The height is decreasing so $\dfrac{dh}{dt}$ is negative.

b) $\dfrac{1}{\sqrt{h}} \dfrac{dh}{dt} = -0.5$

> Separate the variables.

$\Rightarrow \displaystyle\int \dfrac{1}{\sqrt{h}}\, dh = \int -0.5\, dt$

$\Rightarrow 2\sqrt{h} = -0.5t + c$

> Integrating both sides. This is the general solution

$t = 0, h = 25 \Rightarrow c = 10$

> Using the initial condition to find the particular solution.

$\Rightarrow 2\sqrt{h} = -0.5t + 10$

$\Rightarrow h = \left(-0.25t + 5\right)^2; \quad t = 20 - 4\sqrt{h}$

> Writing the solution explicitly for h and for t is helpful here as parts (c) and (d) ask for a height and a time.

c) $h = \left(-0.25 \times 10 + 5\right)^2 = 6.25$ cm

d) $t = 20 - 4\sqrt{5} = 11.1$ seconds

> This is a section of the quadratic curve $h = (-0.25t + 5)^2$.

Example 19

The spread of a disease occurs at a rate proportional to the product of the number of people infected and the number not infected. Initially 50 out of a population of 1050 are infected and it is spreading at a rate of 10 new cases per day.

a) If n is the number infected after t days, show that $\dfrac{\mathrm{d}n}{\mathrm{d}t} = \dfrac{n(1050 - n)}{5000}$

b) Solve this differential equation to find the number of people infected after t days.

c) How long will it take for 250 people to be infected?

d) Explain why everyone in the population will eventually be infected.

a) $\dfrac{\mathrm{d}n}{\mathrm{d}t} = kn(1050 - n)$

n people are affected, and $(1050 - n)$ are not affected.

$t = 0, n = 50, \dfrac{\mathrm{d}n}{\mathrm{d}t} = 10 \Rightarrow k = \dfrac{1}{5000}$

The initial condition on the rate of change allows you to find the constant of proportionality.

$\Rightarrow \dfrac{\mathrm{d}n}{\mathrm{d}t} = \dfrac{n(1050 - n)}{5000}$

b) $\dfrac{1}{n(1050 - n)} \dfrac{\mathrm{d}n}{\mathrm{d}t} = \dfrac{1}{5000}$

Now: $\dfrac{1}{n(1050 - n)} = \dfrac{A}{n} + \dfrac{B}{1050 - n} \Rightarrow A = B = \dfrac{1}{1050}$

Use partial fractions to get an integrable form.

$\displaystyle\int \left(\dfrac{1}{n} + \dfrac{1}{1050 - n} \right) \mathrm{d}n = \int \dfrac{1050}{5000} \, \mathrm{d}t = \int 0.21 \, \mathrm{d}t$

It is easier to take all the constant terms into one.

$\Rightarrow \ln n - \ln(1050 - n) = 0.21t + c$

$\Rightarrow \ln\left(\dfrac{n}{1050 - n} \right) = 0.21t + c$

Solve the DE and express in simplest form.

$\Rightarrow \left(\dfrac{n}{1050 - n} \right) = A\mathrm{e}^{0.21t}$

$t = 0, n = 50 \Rightarrow A = 0.05$

Find the particular solution using the initial condition.

$\Rightarrow \left(\dfrac{n}{1050 - n} \right) = 0.05\mathrm{e}^{0.21t} \quad (*)$

c) $n = 250; \quad 0.05\mathrm{e}^{0.21t} = \dfrac{250}{800} \Rightarrow \mathrm{e}^{0.21t} = 6.25$

$\Rightarrow t = 8.73$ days

d) $\Rightarrow n = \dfrac{1050 \times 0.05\mathrm{e}^{0.21t}}{1 + 0.05\mathrm{e}^{0.21t}}$

Rearrange $(*)$ to give n explicitly.

$\Rightarrow n = 1050 \times \left(\dfrac{0.05\mathrm{e}^{0.21t}}{1 + 0.05\mathrm{e}^{0.21t}} \right)$

$n \to 1050$ as $t \to \infty$

As t increases the exponential term dominates in the bracket and this rational expression tends to 1.

Sometimes the problem will involve a constant of proportionality (given by k in Example 19) and also a constant of integration (given by c in Example 19). Be careful not to mix these up, as they have different values!

Example 20

A stone falls through the air from rest and, t seconds after it was dropped, its speed v satisfies the equation $\frac{dv}{dt} = 10 - 0.2v$.

> This is modeling motion under gravity at $10\,\text{ms}^{-2}$ with air resistance proportional to speed.

a) Show that $v = 50(1 - e^{-0.2t})$.

b) Calculate the time at which the stone reaches a speed of $20\,\text{ms}^{-1}$.

c) Sketch the graph of v against t and hence show that the velocity of the stone will never be more than $50\,\text{ms}^{-1}$.

d) Explain what the differential equation tells us would happen if the stone was thrown downwards with a speed of $60\,\text{ms}^{-1}$ instead of being dropped.

· ·

a) $\dfrac{dv}{dt} = 10 - 0.2v$

$\dfrac{1}{10 - 0.2v}\dfrac{dv}{dt} = 1$ ← Separate the variables.

$\Rightarrow \displaystyle\int \dfrac{5}{50 - v}\,dv = \int 1\,dt$

$\Rightarrow -5\ln(50 - v) = t + c$ ← Find the general solution.

$\Rightarrow \ln(50 - v) = -0.2t + d$

$\Rightarrow 50 - v = Ae^{-0.2t}$

$t = 0, v = 0 \Rightarrow A = 50$ ← The initial conditions provide the particular solution.

$\Rightarrow v = 50(1 - e^{-0.2t})$

b) $20 = 50(1 - e^{-0.2t}) \Rightarrow -0.2t = \ln 0.6$

$\Rightarrow t = 2.55$

c)

> The sketch also shows the behaviour in the 'free-fall' model with no air resistance. Until the stone starts to travel quickly the free-fall model is quite a good approximation, as air resistance is negligable at a slow speed.

The expression in the bracket $(1 - e^{-0.2t})$ approaches 1 as t increases, but $e^{-0.2t}$ cannot become negative so it will never exceed 1. Hence the speed cannot exceed $50\,\text{ms}^{-1}$

d) If v is 60 initially then there will be a negative rate of change for the speed, so the stone will slow down – again approaching a terminal velocity of $50\,\text{ms}^{-1}$, because that speed is where the air resistance exactly matches the gravitational force.

Example 21

The size of a colony of pests, n, which fluctuates during the year, is modelled by an entomologist by the differential equation $\frac{dn}{dt} = 0.2n(0.2 - \cos t)$, where t is the number of weeks from the start of the observations. There are 400 pests in the colony initially.

Did you know?

An entomologist is a scientist who studies insects.

a) Solve the differential equation to find n in terms of t.

b) Find how many pests there are after 3 weeks.

c) Show that the number of pests reaches a minimum after approximately 9.6 days, and find the number of pests at that time.

d) Find how many pests there are after 2π weeks.

e) Explain why the model predicts that the number of pests will grow infinitely large over time.

a) $\dfrac{dn}{dt} = 0.2n(0.2 - \cos t)$

$\dfrac{5}{n}\dfrac{dn}{dt} = (0.2 - \cos t)$ ← Separate the variables.

$\displaystyle\int \dfrac{5}{n}\, dn = \int (0.2 - \cos t)\, dt$

$\Rightarrow 5\ln n = 0.2t - \sin t + c$ ← Integrating both sides.

$\Rightarrow n = Ae^{0.2(0.2t - \sin t)}$

$t = 0, n = 400 \Rightarrow A = 400$ ← Using the initial condition to find the particular solution.

$\Rightarrow n = 400e^{0.2(0.2t - \sin t)}$

b) $t = 3, n = 400e^{0.2(0.6 - \sin 3)} = 438.44... = 438$ ← Substituting the given time.

c) $\dfrac{dn}{dt} = 0 \Rightarrow \cos t = 0.2 \Rightarrow t = 1.369$ weeks ≈ 9.6 days ← Solving for minimum.

$n = 347$ ← Find n, given $t = 1.369$ weeks.

d) $t = 2\pi, n = 400e^{0.2(0.4\pi - \sin 2\pi)} = 514$

e) The exponential function has a term $0.04t$ as well as the bounded periodic term, so as t increases n will grow infinitely large.

A sketch of the solution function shows the behaviour of the model. While it fluctuates over a period of 2π weeks, there is also an upward trend.

Example 22

A model of the deflation of a sphere of radius r cm assumes that at time t seconds after the start, the rate of decrease of the surface area is proportional to the volume at that time.

When $t = 0$, $r = 20$ cm and $\dfrac{dr}{dt} = -3$.

a) Show that r satisfies $\dfrac{dr}{dt} = -0.0075r^2$.

b) Solve this differential equation, obtaining an expression for r in terms of t.

c) How long will it take for the sphere to deflate to a radius of 10 cm?

d) How much longer will it be before the radius is 5 cm?

a) $A = 4\pi r^2 \Rightarrow \dfrac{dA}{dt} = 8\pi r \dfrac{dr}{dt}$; $V = \dfrac{4}{3}\pi r^3$ ← Use the formulae for a sphere.

$\Rightarrow 8\pi r \dfrac{dr}{dt} = -k\left(\dfrac{4}{3}\pi r^3\right)$ ← Express the stated relationship in terms of r.

$t = 0, r = 20, \dfrac{dr}{dt} = -3$

$\Rightarrow 8\pi \times 20 \times (-1) = -k\left(\dfrac{4}{3}\pi \times 20^3\right)$

$\Rightarrow k = 0.015$ ← Use the initial condition to find the constant of proportionality.

$\Rightarrow \dfrac{dr}{dt} = -0.0025r^2$

b) $\dfrac{-400}{3r^2}\dfrac{dr}{dt} = 1 \Rightarrow \displaystyle\int \dfrac{-400}{3r^2}\, dr = \int 1\, dt$ ← Separate the variables.

$\Rightarrow \dfrac{400}{3r} = t + c$ ← Integrating both sides.

$t = 0, r = 20 \Rightarrow c = \dfrac{20}{3} \Rightarrow \dfrac{400}{r} = t + \dfrac{20}{3}$ ← Use the initial condition to find the particular solution.

$\Rightarrow r = \dfrac{400}{\left(t + \dfrac{20}{3}\right)}$ ← Express r in terms of t.

c) $r = 10$, $t = \dfrac{400}{10} - \dfrac{20}{3} = \dfrac{100}{3}$

d) $r = 5$, $t = \dfrac{400}{5} - \dfrac{20}{3} = \dfrac{220}{3}$

so it takes another 73.3 seconds.

A sketch of the solution function shows the behaviour of the model and why it takes increasingly longer for the radius to decrease by a set amount.

Example 23

In a chemical reaction a substance X reacts with another substance Y. The masses of substances X and Y after time t seconds are x and y grams respectively.

It is given that $\dfrac{dy}{dt} = -0.5xy$ and $x = 20e^{-2t}$. Also, we know that when $t = 0$, $y = 50$.

a) Form a differential equation in y and t.

b) Solve this equation to obtain an expression for y in terms of t.

c) Find the exact mass of Y which will remain when t is very large.

----

a) $\dfrac{dy}{dt} = -0.5xy; \quad x = 20e^{-2t} \Rightarrow \dfrac{dy}{dt} = -10ye^{-2t}$ ←   Removing x from the DE.

$\Rightarrow \dfrac{1}{y}\dfrac{dy}{dt} = -10e^{-2t}$ ←   Separate the variables.

b) $\displaystyle\int \dfrac{1}{y}\,dy = \int -10e^{-2t}\,dt$ ←   Integrate both sides.

$\Rightarrow \ln y = 5e^{-2t} + c$

$t = 0, y = 50 \Rightarrow c = \ln 50 - 5$ ←   Use the initial condition to find the particular solution and simplifying the expression.

$\Rightarrow \ln y = 5e^{-2t} + \ln 50 - 5$

$\Rightarrow \ln\left(\dfrac{y}{50}\right) = 5(e^{-2t} - 1)$

$y = 50e^{5(e^{-2t}-1)}$

c) $\ln\left(\dfrac{y}{50}\right) \to -5$ as t gets very large, so $50e^{-5}$ grams will remain when t is very large.

Exercise 10.4

1. A rod has the property that the temperature gradient, $\left(\dfrac{dT}{dx}\right)$, at a distance x cm from the end of the rod being heated is proportional to the distance x. The end of the rod is heated to a temperature of $400°\,C$, and when $x = 50$ we know that $\dfrac{dT}{dx} = -8$.

   a) Write down a differential equation which is satisfied by the temperature T at a distance x from the end of the rod.

   b) Solve the differential equation to find an expression for the temperature T in terms of x.

   c) Calculate the temperature at the point $50\,cm$ from the heated end of the rod.

   d) Calculate how far from the heated end of the rod the temperature reaches $20°\,C$.

2. In a chemical reaction a substance X reacts with another substance Y. The masses of substances X and Y after time t seconds are x and y grams respectively.

It is given that $\dfrac{dy}{dt} = -0.1xy$ and $x = 30e^{-3t}$. Also, when $t = 0$, $y = 20$.

a) Form a differential equation in y and t.

b) Solve this differential to obtain an expression for y in terms of t.

c) Find the proportion of Y which will remain when t is very large.

3. Carbon-14 occurs in all living creatures at 1 part in a trillion carbon molecules, but when the creature dies, the carbon molecules are no longer exchanged with the atmosphere. Carbon-14 molecules decay at a rate proportional to the amount remaining at any given time.

a) If m is the mass of carbon-14 in 1 gram of carbon at any time t years after a creature dies, form a differential equation relating m and t.

b) Taking $m = 1$ when $t = 0$, show that the differential equation is satisfied by $m = e^{-kt}$.

c) Carbon-14 is used to date fossils. It has a half-life of 5730 years (every 5730 years the mass of carbon-14 per gram of carbon in the fossil reduces by 50%). Find the value of k in the expression for m.

d) A fossil is found to have 0.008 grams of Carbon-14 per gram of Carbon. Calculate how old the fossil is.

4. A reservoir in the shape of a cuboid has a base area of $60\,000\ m^2$. Water is seeping from the reservoir at a rate proportional to the depth and it is known that the reservoir loses water at a rate of 3000 litres per minute when the depth is 10 metres.

The depth of water in the reservoir is 8 metres when a heavy storm starts causing the reservoir to be filled at a constant rate of 1500 litres per second.

a) Write down a differential equation which is satisfied by the depth of water, d metres, at time t minutes after the storm starts.

b) Solve the differential equation.

c) If the storm lasts for two and a half hours, calculate the depth of water in the reservoir when the storm ends.

5. A colony of bacteria grows at a rate proportional to the size of the colony at any time.

a) Write down a differential equation satisfied by the number of bacteria N (in millions) after t hours.
Initially the colony has 2 million bacteria, and after 6 hours it has 2.5 million bacteria.

b) Solve the differential equation to find an expression for N in terms of t.

c) The space in which the colony is housed has room for 10 million bacteria. Calculate the time at which the space reaches saturation.

6. A body moving through a liquid experiences a resistance to motion which causes it to lose speed at a rate proportional to its speed at any time.

   a) Write down a differential equation satisfied by the speed $v\,\text{ms}^{-1}$ after time t seconds.
   Initially the body has a speed of $25\,\text{ms}^{-1}$ and after 10 seconds it has a speed of $20\,\text{ms}^{-1}$.

   b) Solve the differential equation to find an expression for v in terms of t.

   c) Calculate how long it takes the body to lose half of its initial speed.

7. A disease spreads at a rate proportional to the product of the number of people, n, infected and the number of people not yet infected. The population has size P. Initially 5% of the population is infected.

   a) Write down a differential equation which is satisfied by n and the time t.
   After 3 days it is found that 10% of the population is infected.

   b) Solve this differential equation to find the number of people infected after t days.

   c) How long will it take for half the people to be infected?

   d) Explain why everyone in the population will eventually be infected.

1. Find the general solution to the following differential equations.

   a) $\dfrac{dy}{dx} = \dfrac{x^2}{y^3}$

   b) $\dfrac{dy}{dx} = 2x\cos^2 y$

   c) $\dfrac{dy}{dx} = e^{3x}\sec y$

   d) $x\tan y\,\dfrac{dy}{dx} = 1$

   e) $\dfrac{dy}{dx} = e^{3x-y}$

   f) $\dfrac{dy}{dx} = \dfrac{xy}{x^2+1}$

   g) $(x^2+6)\dfrac{dy}{dx} = 8xy$

   h) $\sin 3\theta\,\dfrac{dx}{d\theta} = (x+2)\cos 3\theta.$

2. Find the particular solution to the following differential equations, with given boundary conditions.

   a) $\dfrac{dy}{dx} = \dfrac{3x^2+x}{4y^3};\quad x=2,\,y=1$

   b) $\dfrac{dy}{dx} = y^2\,2\sin x;\quad x=\pi,\,y=1$

   c) $x\dfrac{dy}{dx}+1 = y-2\dfrac{dy}{dx};\quad x=1,\,y=19$

   d) $\dfrac{dy}{dx} = \dfrac{x^2 y+y}{x^2-1};\quad x=0,\,y=3$

   e) $\dfrac{dy}{dx} = e^{3x-y};\quad x=0,\,y=0$

   f) $\dfrac{dv}{dt} = e^{-t}\sqrt{v};\quad t=0,\,v=9$

   g) $\dfrac{dy}{dx} = xe^{x-y};\quad x=0,\,y=0$

   h) $\ln y\,\dfrac{dy}{dx} = \dfrac{x-1}{x^2+x};\quad x=1,\,y=1.$

EXAM-STYLE QUESTIONS

3. A curve is such that $\dfrac{dy}{dx} = \dfrac{y}{\sqrt{x+1}}$ and the point $(1, 1)$ lies on the curve.

   Find the equation of the curve.

4. A curve is such that $\dfrac{dy}{dx} = xe^x\sec y$ and the point $\left(0, \dfrac{\pi}{6}\right)$ lies on the curve.

   Find the equation of the curve.

5. A curve is such that $ye^y\cos^2 x\,\dfrac{dy}{dx} = 1$ and the point $\left(\dfrac{\pi}{4}, 0\right)$ lies on the curve.

   Find the equation of the curve.

6. A curve is such that $\sqrt{y}\,\dfrac{dy}{dx} = \dfrac{2}{x^2-1}$ and the point $(2, 0)$ lies on the curve.

   Find the equation of the curve.

7. A curve is such that $y\dfrac{dy}{dx} = \cos^2 x$ and the point $\left(\dfrac{\pi}{4}, 0\right)$ lies on the curve.

   Find the equation of the curve.

8. A curve is such that $\left(\dfrac{2y+1}{y^2+1}\right)\dfrac{dy}{dx} = \dfrac{1}{x}$ and the point $(1, 0)$ lies on the curve.

   Find the equation of the curve.

9. The temperature of a hot body decreases at a rate proportional to the difference between its temperature and the air temperature around it. A body is heated to $90°\,C$ and placed in air which is at $20°\,C$.

   a) Write down a differential equation relating the temperature T of the body and the time t.

   After 4 minutes the body has cooled to $60°\,C$.

   b) Solve the differential equation to find an expression for the temperature T in terms of t.

   c) Calculate the temperature after 10 minutes.

   d) Calculate how long it takes for the body to get down to within $5°C$ of the air temperature.

10. The resistance a body experiences passing through a liquid is proportional to the square of its speed at any time. The body has a speed of 50 ms^{-1} entering the liquid.

a) Write down a differential equation relating the speed v of the body and the time t.

After 4 seconds the body has slowed to 30 ms^{-1}.

b) Solve the differential equation to find an expression for the speed v in terms of t.

c) Calculate the speed after 10 seconds.

d) Calculate how long it takes for the body to slow down to 10 ms^{-1}.

11. The spread of a rumour in a large group of people is thought to reach new people at a rate proportional to the product of the number of people who have heard the rumour and the number who have not heard it.

In a population of 2000 people, initially 50 people have heard the rumour. Two hours later, 300 people have heard it.

a) Write down a differential equation relating the number of people, N, who have heard the rumour with the time t.

b) Solve the differential equation to find an expression for N in terms of t.

c) Calculate how long it takes for there to be a majority of people who have heard the rumour.

12. a) Show that $\dfrac{dv}{dt} = v\dfrac{dv}{dx}$ where $v = \dfrac{dx}{dt}$.

The rate a body loses speed, v, is proportional to its displacement, x, from a fixed point.

b) Show that the differential equation

$v\dfrac{dv}{dx} = -kx$ describes this situation.

The body has speed 10 ms^{-1} as it passes through the fixed point and just reaches a point 5 metres from the fixed point as it comes to rest.

c) Solve the differential equation to find an equation relating v and x, showing that $k = 4$.

d) Find the speed of the body when it is 3 metres from the fixed point.

13. A tank contains a solution of a mineral in water. Initially there is 600 litres of water with 12 kg of the dissolved mineral. The mixture is drained at a rate of 30 litres per minute and simultaneously pure water is added at a rate of 30 litres per minute. The tank is stirred continuously to keep the mixture uniform at all times.

a) Write down a differential equation which is satisfied by the mass, m kg, of the mineral in the solution at time t minutes.

b) Solve the differential equation to find an expression for m in terms of t. When the solution contains 8 kg of the mineral another 4 kg is added, and the process is repeated.

c) Find the time at which the solution contains 8 kg of the mineral.

Note: Q.14 is beyond the scope of the Cambridge syllabus but may aid in your understanding.

14. A man walks along a pier pulling a toy boat by a rope of length L. The man keeps the rope taut and straight. Initially the rope is at right angles to the pier, with the boat at A and the man at B. A little while later the man is at E and the boat is at C. D is the point on the pier such that CD is perpendicular to the pier. $CD = x$, $BD = y$.

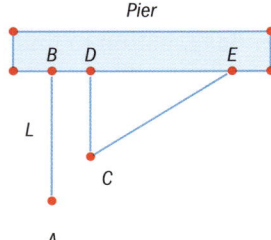

The path followed by the boat is called a tractrix, and has the property that the rope is always tangential to the path of the boat.

a) If $y = f(x)$ is the path of the boat, show that $\dfrac{dy}{dx} = \dfrac{-\sqrt{L^2 - x^2}}{x}$.

b) Use the substitution $x = L\cos\theta$ to show that the solution to the differential equation is
$f(x) = L\ln(x / L - \sqrt{L^2 - x^2}) - \sqrt{L^2 - x^2}$.

[**You** may use without derivation that $\int\operatorname{cosec}\theta \, d\theta = \ln|\operatorname{cosec}\theta - \cot\theta|$].

15. In a certain chemical reaction the amount, x grams, of a substance present is decreasing. The rate of decrease of x is proportional to the product of x and the time, t seconds, since the start of the reaction. Thus x and t satisfy the differential equation
$$\frac{dx}{dt} = -kxt,$$
where k is a positive constant. At the start of the reaction, when $t = 0$, $x = 100$.

i) Solve this differential equation, obtaining a relation between x, k and t. [5]

ii) 20 seconds after the start of the reaction the amount of substance present is 90 grams. Find the time after the start of the reaction at which the amount of substance present is 50 grams. [3]

Cambridge International AS and A Level Mathematics 9709 Paper 3 Q8 October/November 2005

16. i) Using partial fractions, find
$$\int \frac{1}{y(4-y)} \, dy. \qquad [4]$$

ii) Given that $y = 1$ when $x = 0$, solve the differential equation
$$\frac{dy}{dx} = y(4 - y),$$
obtaining an expression for y in terms of x. [4]

iii) State what happens to the value of y if x becomes very large and positive. [1]

Cambridge International AS and A Level Mathematics 9709 Paper 3 Q8 May/June 2005

Chapter summary

Differential equations

- A differential equation is an equation in which at least one differential expression appears.

- A first order differential equation only contains a first derivative $\left(\dfrac{dy}{dx} \text{ or equivalent} \right)$.

- The differential equation has separable variables if the equation can be written in the form $g(y)\dfrac{dy}{dx} = f(x)$.

- The general solution is obtained by integrating both sides of $\displaystyle\int g(y)\,dy = \int f(x)\,dx$ and including $+c$ on one side.

- A particular solution is obtained by using an initial or boundary condition to calculate the value c has to take in the general solution to allow the solution to satisfy that condition.

- All the techniques used in integration can be required to solve the two integrals obtained when the variables are separated.

11 Complex numbers

We can use computers to generate beautiful images from complex numbers, such as the Mandelbrot set, shown on the left. However, complex numbers are also widely used in real-life applications. For example, they are used in the study of electrical circuits, and the flow of fluids around objects, as well as the technology found in digital cameras and mobile phones. They are also used in the design of aeroplane wings.

Objectives

- Understand the idea of a complex number, recall the meaning of the terms real part, imaginary part, modulus, argument, conjugate, and use the fact that two complex numbers are equal if and only if both real and imaginary parts are equal.

- Carry out operations of addition, subtraction, multiplication and division of two complex numbers expressed in cartesian form $x + iy$.

- Use the result that, for a polynomial equation with real coefficients, any non-real roots occur in conjugate pairs.

- Represent complex numbers geometrically by means of an Argand diagram.

- Carry out operations of multiplication and division of two complex numbers expressed in polar form $r(\cos\theta + i\sin\theta) \equiv re^{i\theta}$.

- Find the two square roots of a complex number.

- Understand in simple terms the geometrical effects of conjugating a complex number and of adding, subtracting, multiplying and dividing two complex numbers.

- Illustrate simple equations and inequalities involving complex numbers by means of loci in an Argand diagram, e.g. $|z - a| < k$, $|z - a| = |z - b|$, $\arg(z - a) = \alpha$.

Before you start

You should know how to:

1. Use the trigonometric identities to simplify expressions.

e.g. $(1 + \cos\theta)^2 + \sin^2\theta$

$= 1 + 2\cos\theta + \cos^2\theta + \sin^2\theta$

$= 1 + 2\cos\theta + 1$

$= 2 + 2\cos\theta = 2(1 + \cos\theta)$

2. Manipulate and rationalise surds.

e.g. $\left(3\sqrt{2}\right)^2 + \left(2\sqrt{3}\right)^2$

$= \left(3\sqrt{2}\right)\left(3\sqrt{2}\right) + \left(2\sqrt{3}\right)\left(2\sqrt{3}\right)$

$= 9 \times 2 + 4 \times 3$

$= 30$

e.g. $\dfrac{2 - \sqrt{3}}{1 + \sqrt{3}} = \dfrac{2 - \sqrt{3}}{1 + \sqrt{3}} \times \dfrac{1 - \sqrt{3}}{1 - \sqrt{3}}$

$= \dfrac{2 - 3\sqrt{3} + \left(\sqrt{3}\right)^2}{1 + \sqrt{3} - \sqrt{3} - \left(\sqrt{3}\right)^2}$

$= \dfrac{5 - 3\sqrt{3}}{1 - 3}$

$= \dfrac{3\sqrt{3} - 5}{2}$

3. Work out the discriminant for a quadratic equation.

e.g. $2x^2 - 5x + 3 = 0$

$a = 2, b = -5, c = 3$

Discriminant: $b^2 - 4ac = (-5)^2 - 4 \times 2 \times 3 = 1$

Skills check:

1. Simplify

a) $(1 + \sin\theta)^2 + \cos^2\theta$

b) $\sin^2\theta - \sin^2\theta\cos^2\theta$

c) $\cos^4\theta + 2\sin^2\theta\cos^2\theta - \sin^4\theta$

2. Show that

a) $\dfrac{2\sqrt{3} - 3}{\sqrt{3}} = 2 - \sqrt{3}$

b) $\dfrac{4 + \sqrt{2}}{1 + \sqrt{2}} = -2 + 3\sqrt{2}$

c) $\dfrac{1}{2\sqrt{3} - 2} = \dfrac{\sqrt{3} + 1}{4}$

3. Work out the discriminant for

a) $x^2 - 5x + 2 = 0$

b) $2x^2 - 7x - 4 = 0$

c) $3x - 5x^2 - 4 = 0$

11.1 Introducing complex numbers

In P1 we saw that if the discriminant, $b^2 - 4ac$, of the quadratic equation $ax^2 + bx + c = 0$ is negative, then the equation has no real roots. This is because there is no real number which, when squared, gives a negative real number.

The ancient Greeks were puzzled by the fact that there was no real number which satisfies the equation $x^2 = -1$. However, it was not until 1545 that the Italian mathematician Gerolamo Cardano thought about the possibility of complex numbers when he solved the equation $x(10 - x) = 40$ to give the solutions $x = 5 \pm \sqrt{-15}$. Later in the sixteenth century Rafael Bombelli, another Italian mathematician who is generally regarded as the inventor of complex numbers, developed this concept to enable him to solve cubic and quartic equations.

Before we discuss complex numbers in more detail, we introduce the symbol i which we will use to denote the negative non-real number $\sqrt{-1}$.

It follows that $i^2 = -1$

$$\boxed{\sqrt{-1} = i \Leftrightarrow i^2 = -1}$$

As such, $\sqrt{-9} = \sqrt{9 \times -1} = \sqrt{9} \times \sqrt{-1} = 3i$.

More generally, $\sqrt{-b^2} = bi$.

> Numbers of the form bi, where b is real, are called imaginary numbers.

Example 1

Solve the equation in each case.

a) $x^2 = -25$ b) $4x^2 + 484 = 0$ c) $x^2 + 20 = 0$

. . . .

a) $x^2 = -25$

$\quad x = \pm\sqrt{-25}$ ← Take **both** square roots of -25.

$\quad x = \pm 5i$ ← Write $\sqrt{-25}$ as a multiple of i.

b) $4x^2 + 484 = 0$

$\quad x^2 = -121$ ← Rearrange to find the value of x^2.

$\quad x = \pm\sqrt{-121} = \pm 11i$

c) $x^2 + 20 = 0$

$\quad x = \pm\sqrt{-20} = \pm\sqrt{20}i$ ← Leave the value of x in exact terms.

$\quad = \pm 2\sqrt{5}i$ ← You may wish to write your answer in surd form.

Example 2

Simplify:

a) i^3 **b)** i^4 **c)** i^9 **d)** $(-i)^6$ **e)** $(2i)^5$

a) $i^3 = i^2 \times i = -1 \times i = -i$ Replace i^2 with -1.

b) $i^4 = i^2 \times i^2 = -1 \times -1 = 1$

c) $i^9 = i^4 \times i^4 \times i = 1 \times 1 \times i = i$ Write i^9 as a product of powers of i each of which you can replace with 1, -1, i or $-i$. Use $i^4 = 1$ here.

d) $(-i)^6 = i^6 = i^4 \times i^2 = 1 \times -1 = -1$ Expand the brackets and simplify.

e) $(2i)^5 = 32 \times i^5 = 32 \times i^4 \times i = 32i$

Use the ordinary laws of algebra to write the expression without brackets, then simplify.

Exercise 11.1

It is useful to remember
$i^2 = -1$, $i^3 = -i$, $i^4 = 1$

1. Which of the following equations have no real roots?

   a) $x^2 + 5x + 2 = 0$

   b) $2x^2 - 4x + 1 = 0$

   c) $x^2 + 5x + 7 = 0$

   d) $3x^2 - x + 2 = 0$

   e) $2x^2 + 5x - 1 = 0$

2. Solve the equation in each case.

   a) $x^2 = -4$ **b)** $x^2 = -100$ **c)** $x^2 = -13$ **d)** $x^2 = -75$

3. Simplify the expression in each case.

   a) i^5 **b)** i^7 **c)** i^{11} **d)** i^{25} **e)** i^{64}

4. Simplify the expression in each case.

   a) $(-i)^{14}$ **b)** $(2i)^7$ **c)** $(3i)^3$ **d)** $(-2i)^{10}$

5. Write down the square roots of

   a) 64 **b)** -64 **c)** -19 **d)** -32.

6. Solve the equation in each case.

   a) $x^2 + 49 = 0$ **b)** $x^2 + 23 = 5$ **c)** $2x^2 + 9 = 5$ **d)** $(5x)^2 = -125$

7. Simplify the following expressions.

   a) $\left(\sqrt{2}\,i\right)^2$ **b)** $\left(-\sqrt{7}\,i\right)^4$ **c)** $\left(\frac{1}{3}i\right)^3$ **d)** $\left(-\frac{1}{2}i\right)^8$

8. We wish to find two numbers, x and y, whose sum is 10 and whose product is 40.

    a) Write down two equations in x and y which fully describe this problem.

    b) Show that $x^2 - 10x + 40 = 0$.

    c) Show that this equation has no real roots.

> Cardano, an Italian mathematician, tried to solve this problem in 1545.

At the beginning of this chapter we discussed how the mathematician Cardano solved the quadratic equation $x(10 - x) = 40$ to give the solutions $x = 5 \pm \sqrt{-15}$. Our experience of solving quadratic equations by using the formula $x = \dfrac{-b \pm \sqrt{b^2 - 4ac}}{2a}$ shows us that our answer is made up of two parts, $\dfrac{-b}{2a}$ and $\dfrac{\sqrt{b^2 - 4ac}}{2a}$. The first of these expressions will always be real (provided the coefficients a, b, and c are real). However we have learned that the second expression might not be real; it might be imaginary and therefore expressed as bi (where b is real). This leads us to our definition of a complex number.

> **Note:** The 'b' used in the formula for solving a quadratic equation should not be confused with the 'b' used in the complex number $a + bi$'.

> A number of the form $a + bi$, where a and b are real and $i^2 = -1$ is called a **complex number**.

A number such as $2i$ is said to be imaginary, whereas a number such as $5 + 2i$ is said to be complex. We use the letter z to denote a complex number, where $z = x + iy$, and x and y are both real numbers.

We can also see from the quadratic formula, provided that a, b and c are real, that the roots occur in pairs, one root of the form $a + bi$ and the other of the form $a - bi$. For example, if one root is $3 + 2i$ then the other root will be $3 - 2i$. Pairs of complex numbers like this are referred to as **complex conjugates** of each other.

> **Note:** All imaginary numbers are also complex numbers as they can be written as $0 + bi$. Similarly, all real numbers are complex numbers as they can be written in the form $a + 0i$

> The complex conjugate of $a + bi$ is $a - bi$, where a and b are real numbers.
>
> The complex conjugate of a complex number z is denoted by z^*

> **Note:** In this book we use a + bi when we know the values of the real and imaginary parts (e.g. 3 + 7i) and x + iy when we do not know the values of the real and imaginary parts. However this is just a matter of preference and you may like to write it differently.

Example 3

Solve the equation $2z^2 - z + 3 = 0$

Using the quadratic formula,

$$z = \frac{-b \pm \sqrt{b^2 - 4ac}}{2a}$$

$$z = \frac{-(-1) \pm \sqrt{(-1)^2 - 4 \times 2 \times 3}}{2 \times 2}$$

Substitute $a = 2$, $b = -1$ and $c = 3$ into the formula.

$$z = \frac{1 \pm \sqrt{1 - 24}}{4}$$

Simplify the expression.

$$z = \frac{1 \pm \sqrt{-23}}{4}$$

$$z = \frac{1}{4} \pm \frac{\sqrt{23}}{4}i$$

Here we leave the answer in the form $a \pm bi$ but we could separate the two values and write $z = \frac{1}{4} + \frac{\sqrt{23}}{4}i$ and $\frac{1}{4} - \frac{\sqrt{23}}{4}i$.

or

Completing the square,

$$2z^2 - z + 3 = 0$$

$$2\left(z^2 - \frac{1}{2}z\right) + 3 = 0$$

$$2\left\{\left(z - \frac{1}{4}\right)^2 - \frac{1}{16}\right\} + 3 = 0$$

$$2\left(z - \frac{1}{4}\right)^2 - \frac{1}{8} + 3 = 0$$

$$2\left(z - \frac{1}{4}\right)^2 = -\frac{23}{8}$$

$$\left(z - \frac{1}{4}\right)^2 = -\frac{23}{16}$$

$$\left(z - \frac{1}{4}\right) = \pm\sqrt{-\frac{23}{16}}$$

$$z = \frac{1}{4} \pm \sqrt{-\frac{23}{16}}$$

$$z = \frac{1}{4} \pm \frac{\sqrt{23}}{4}i$$

Exercise 11.2

1. Express each of the following complex numbers in the form complex $a + bi$.

   a) $5 + \sqrt{-1}$ b) $4 - \sqrt{-6}$ c) $-2 + \sqrt{-40}$ d) $-1 - \sqrt{-81}$

2. Write down the complex conjugate of the following.

   a) $4 - 2i$ b) $-5 + i$ c) $-9 - 7i$ d) $3i - 6$

3. Write down the conjugate of the complex number in each case.

   a) $z = 10 + 12i$ b) $z^* = 3 - 3i$

   c) $z = 5i$ d) $z^* = 8$

4. Find whether the following equations have real or complex roots.

   a) $z^2 - 6z + 7 = 0$

   b) $5z^2 + z + 2 = 0$

   c) $z^2 + 8z = 0$

   d) $4 + 3z - 3z^2 = 0$

   e) $4z^2 + 5z + 3 = 1$

   f) $1 - z^2 = 4 + z$

5. Solve the following quadratic equations by using the quadratic formula.

   a) $z^2 - 4z + 5 = 0$

   b) $4z^2 + 2z + 1 = 0$

   c) $2z^2 - 9z - 3 = 0$

   d) $2 - z - 3z^2 = 0$

   e) $5z - 3z^2 = 1 + 4z$

   f) $10z^2 + 5z + 20 = 15$

   g) $z^2 + iz - 1 = 0$

6. Solve the following quadratic equations by completing the square.

   a) $z^2 - 2z + 3 = 0$

   b) $3z^2 + 6z + 10 = 0$

   c) $2z^2 - 8z + 7 = 0$

   d) $9z^2 - 27z + 3 = 0$

   e) $(z + 1)^2 - (2z + 2)^2 = 3$

   f) $z(z + 6) = -10$

   g) $z^2 - 2iz - 5 = 0$

7. Solve $iz^2 + z + 2i = 0$

   (You may leave your answer in the form $\dfrac{p}{q}$, where p is a real number and q is an imaginary number).

11.2 Calculating with complex numbers

We are about to see that when adding, subtracting, and multiplying complex numbers we use the normal rules of algebra. The division of complex numbers is carried out in a similar way to that of dividing surds.

It is common to use $z_1, z_2, z_3 \ldots$ if we want to use more than one complex number.

Suppose we have two complex numbers, $z_1 = a + bi$ and $z_2 = c + di$.
We add them in the following way:

$z_1 + z_2 = (a + bi) + (c + di) = (a + c) + (b + d)i$

To add two complex numbers we add their real parts and add their imaginary parts.

We subtract them in the following way:

$z_1 - z_2 = (a + bi) - (c + di) = (a - c) + (b - d)i$

To subtract two complex numbers we subtract their real parts and subtract their imaginary parts.

We multiply them in the following way:

$$z_1 z_2 = (a + bi)(c + di)$$
$$= ac + bci + adi + bdi^2$$
$$= (ac - bd) + (bc + ad)i$$

To multiply two complex numbers we use the normal rules of algebra and simplify the answer using $i^2 = -1$.

We divide them in the following way:

$$\frac{z_1}{z_2} = \frac{a + bi}{c + di}$$

$$= \frac{a + bi}{c + di} \times \frac{c - di}{c - di}$$

$$= \frac{(a + bi)(c - di)}{(c + di)(c - di)}$$

$$= \frac{ac + bci - adi - bdi^2}{c^2 + cdi - cdi - d^2 i^2}$$

$$= \frac{(ac + bd) + (bc - ad)i}{c^2 + d^2}$$

$$= \frac{(ac + bd)}{c^2 + d^2} + \frac{(bc - ad)}{c^2 + d^2}i$$

To divide two complex numbers we multiply the numerator and the denominator of the fraction by the complex conjugate of the denominator.

Example 4

The complex numbers z_1 and z_2 are given by $z_1 = 5 + 2i$ and $z_2 = 3 + 4i$.

Find $z_1 + z_2$, $z_1 - z_2$, $z_1 z_2$ and $\dfrac{z_1}{z_2}$.

$z_1 + z_2 = 5 + 2i + 3 + 4i$ ←   Add the real parts and the imaginary parts separately.

$\quad\quad = 8 + 6i$

$z_1 - z_2 = (5 + 2i) - (3 + 4i)$ ←   Use brackets as you would when subtracting one algebraic expression from another.

$\quad\quad = 5 + 2i - 3 - 4i$

$\quad\quad = 2 - 2i$

$z_1 z_2 = (5 + 2i)(3 + 4i)$ ←   Expand the brackets and simplify.

$\quad\quad = 15 + 6i + 20i + 8i^2$

$\quad\quad = 7 + 26i$ ←   Use $i^2 = -1$

$\dfrac{z_1}{z_2} = \dfrac{5 + 2i}{3 + 4i}$

$\quad\quad = \dfrac{5 + 2i}{3 + 4i} \times \dfrac{3 - 4i}{3 - 4i}$ ←   Multiply the numerator and denominator by $3 - 4i$

$\quad\quad = \dfrac{(5 + 2i)(3 - 4i)}{(3 + 4i)(3 - 4i)}$

$\quad\quad = \dfrac{15 + 6i - 20i - 8i^2}{9 + 12i - 12i - 16i^2}$ ←   Expand the brackets and simplify.

$\quad\quad = \dfrac{23 - 14i}{9 - 16}$

$\quad\quad = \dfrac{(23 - 14i)}{25}$ or $\dfrac{23}{25} - \dfrac{14}{25}i$

Exercise 11.3

1. Simplify

   a) $(2 + i) + (3 + 5i)$ b) $(7 + 6i) + (4 - 2i)$

   c) $(1 - 3i) + (-4 + 3i)$ d) $(6 - 5i) + (-3 + 8i)$

   e) $(-2 + 3i) + (3 + 2i)$ f) $(1 + i) + (-2 - 9i)$.

2. Simplify

   a) $(4 + 3i) - (2 + i)$ b) $(5 + 2i) - (3 - 4i)$

   c) $(7 - i) - (-2 + 5i)$ d) $(5 - 5i) - (-6 + 3i)$

   e) $(-3 + 6i) - (3 + 4i)$ f) $(-1 - 2i) - (-3 - 8i)$.

3. Simplify

a) $2(3 + 4i) + 3(5 + i)$

b) $2(1 - 6i) - (4 - 2i)$

c) $3(1 - 3i) + 2(-4 - 3i)$

d) $5(6 + 5i) + 2(-3 + 8i)$

e) $6(2 - 3i) - 3(3 + 5i)$

f) $4(1 + 4i) - 3(2 + 7i)$.

4. Express in the form $x + iy$, where x and y are real.

a) $(5 + i)(2 + i)$

b) $(3 + 2i)(5 - 4i)$

c) $i(7 - i)$

d) $(3 - 4i)(-1 + 2i)$

e) $(-1 + 6i)(2 + 2i)$

f) $3(1 - 3i)(-4 + 3i)$.

5. Express, in the form $x + iy$, where x and y are real.

a) $\dfrac{5 + 4i}{2 + i}$

b) $\dfrac{2 - 7i}{1 - 2i}$

c) $\dfrac{5 + 6i}{5 - 6i}$

d) $\dfrac{2i}{8i}$

e) $\dfrac{9 - 3i}{9 + 3i}$

f) $\dfrac{2i}{3 + 4i}$

g) $\dfrac{6 + 2i}{4i}$

h) $\dfrac{2 - \sqrt{3}i}{2 + \sqrt{3}i}$.

6. The complex number z is given by

$z = \sqrt{2} - i$.

a) Express z^2 in the form $x + iy$, where x and y are real.

b) The complex conjugate of z is denoted by z^*. Express the following in the form $x + iy$, where x and y are real.

i) $z + z^*$

ii) $2(z - z^*)$

iii) $5zz^*$

iv) $\dfrac{z}{iz^*}$

7. The complex numbers z_1 and z_2 are given by $z_1 = 5 - 12i$ and $z_2 = 3 - 4i$.

Express the following in the form $x + iy$ where x and y are real.

a) $z_1 + z_2$,

b) $z_2 - z_1$,

c) $z_1 z_2$

d) $\dfrac{z_2}{z_1}$

8. Express the following in the form $x + iy$, where x and y are real.

a) $(2 - i)^2$

b) $(2 - i)^3$

c) $(2 - i)^4$

d) $(2 - i)^7$.

9. If $z_1 = 1 + 2i$ and $z_2 = 2 - 3i$, express the following in the form $x + iy$ where x and y are real.

a) $\dfrac{1}{z_1 + z_2}$

b) $\dfrac{1}{z_1} + \dfrac{1}{z_2}$

10. Express in the form $x + iy$, where x and y are real.

a) $\dfrac{1 + i}{1 - i}$

b) $\dfrac{2}{3 + 5i}$

c) $\dfrac{2i - 1}{1 + 2i}$

d) $\dfrac{1}{a + ib}$

e) $\dfrac{1}{4 + 3i} + \dfrac{12}{4 - 3i}$.

11. $u = a + ib$

Find the real part and the imaginary part of

a) u^2

b) $u - u^*$

c) $u^3 - (u^*)^3$.

12. Express in the form $x + iy$ where x and y are real.

a) $(\cos\theta + i\sin\theta)^2$

b) $\dfrac{1}{\cos\theta + i\sin\theta}$

c) $\dfrac{\cos\theta + i\sin\theta}{\cos\theta - i\sin\theta}$.

11.3 Solving equations involving complex numbers

We can order real numbers by placing them on a number line, but because complex numbers have both a real part and an imaginary part, we cannot order them in the same way.

If two real numbers are equal they would occupy the same position on the number line. For two complex numbers to be equal, their real parts must be the same **and** their imaginary parts must be the same. So if for example $x + iy = 5 - 2i$, it follows that $x = 5$ and $y = -2$.

> Two complex numbers are equal if both the real parts are equal and the imaginary parts are equal.

Example 5

a) Find the value of the real number p such that $p + (2 - 3i)(1 + 5i)$ is an imaginary number.

b) Find the value of the real numbers x and y such that $4(x + iy) = -2y - 3ix - 5(3 + 2i)$

a) $p + (2 - 3i)(1 + 5i) = p + (2 - 3i + 10i - 15i^2)$ — Expand the brackets and simplify.

$$= p + 7i + 17$$

$$= (p + 17) + 7i$$ — Write in the form $x + iy$.

$$p + 17 = 0$$

$$p = -17$$ — Equate the real part to 0 and solve to find p.

b) $4x + 4iy = 2y - 3ix - 15 - 10i$

$4x + 4iy = (-2y - 15) + (-3x - 10)i$ — Collect the real parts and the imaginary parts on the RHS.

Comparing real parts, $4x = -2y - 15$

Comparing imaginary parts, $4y = -3x - 10$

$4x + 2y = -15$ (i) — Solve the simultaneous equations.

$3x + 4y = -10$ (ii)

$2 \times$ equation (i) $-$ equation (ii)

$5x = -20$

$x = -4$

Substituting $x = -4$ into equation (i)

$4 \times -4 + 2y = -15$

$y = 0.5$

$x = -4, y = 0.5$

Example 6

The complex number $3 + 2i$ is a root of the quadratic equation $z^2 - (5 + 2i)z + a + bi = 0$ where a, b are real.

a) Find the values of a and b.

b) Explain why the two roots are not complex conjugates of each other.

a) As $z = 3 + 2i$ is a root,

$(3 + 2i)^2 - (5 + 2i)(3 + 2i) + a + bi = 0$ — Substitute for z into the equation.

$(9 + 12i + 4i^2) - (15 + 16i + 4i^2) + a + bi = 0$ — Simplify.

$-6 - 4i + a + bi = 0$

$-6 + a + (-4 + b)i = 0$

$-6 + a = 0$ and $-4 + b = 0$ — Compare the real part on the LHS with the real part on the RHS, and similarly, compare the imaginary parts.

$a = 6, b = 4$

b) The coefficients of the quadratic equation $z^2 - (5 + 2i)z + a + bi = 0$ are not all real so the roots will not be complex conjugates.

We have seen that the square roots of a negative number can be expressed as imaginary numbers, that is, the solutions of $z^2 = -a$ are $z = \pm\sqrt{a}\,i$

We can use the method of comparing real parts and imaginary parts to find the square roots of a complex number, $a + bi$.

Example 7 demonstrates how we do this.

Example 7

Find the square roots of the complex number $7 - 24i$.

As with any other number we expect to get two square roots of any complex number.

To find the square roots, solve the equation

$z^2 = 7 - 24i$, where $z = x + iy$, where x and y are real.

$(x + iy)^2 = 7 - 24i$ — Substitute $x + iy$ for z

$x^2 + 2xyi + i^2y^2 = 7 - 24i$ — Expand the bracket and simplify.

$x^2 + 2xyi - y^2 = 7 - 24i$

$x^2 - y^2 = 7$ (i) — Compare real and imaginary parts to get simultaneous equations.

$2xy = -24$ (ii)

From (ii) $y = \dfrac{-12}{x}$ and substitute for y in equation (ii),

$x^2 - \left\{\dfrac{-12}{x}\right\}^2 = 7$

> Eliminate one variable to get a quartic equation in x.

$x^2 - \dfrac{144}{x^2} = 7$

$x^4 - 144 = 7x^2$

> Solve the equation to get the possible values of x.

$x^4 - 7x^2 - 144 = 0$

$(x^2 - 16)(x^2 + 9) = 0$

$x^2 = 16$ or $x^2 = -9$, but x is real so $x = \pm 4$ only.

Substitute $x = 4$ and $x = -4$ into (ii),

$y = \dfrac{-12}{\pm 4} \Rightarrow y = \mp 33$

> Find the corresponding y values.

$x = 4, y = -3$ or $x = -4, y = 3$

The square roots of $7 - 24i$ are $4 - 3i$ and $-4 + 3i$

> Note that these square roots may be written in the form $\pm (4 - 3i)$.

Exercise 11.4

1. Find the square roots of the following.
   a) -400
   b) -60
   c) $-3 + 4i$
   d) $3 + 4i$
   e) $-5 + 12i$
   f) $21 + 20i$
   g) $-21 - 20i$
   h) $35 - 12i$
   i) $36i$

2. a) Find the square roots of $3 - 4i$.
   b) Hence, solve the equation $z^2 + (2 + i)z + 2i = 0$.

3. Solve the equation $z + 2z^* = \dfrac{15}{2 + i}$.

4. Express the square roots of i and $-i$ in a form not involving i.

5. Solve the following equations.
   a) $z^2 = 25i$
   b) $z^2 = -16i$
   c) $3z^2 + 12 = 0$.

6. Find the real numbers a and b which satisfy the equation $(2 + 3i)(5 + bi) = a + 11i$.

7. The complex number z is defined by $\dfrac{z-1}{z + 1 - 2i} = 1 + i$.

   Express z in the form $x + iy$, where x and y are real.

8. Solve the equations, expressing your solutions in the form $x + iy$, where x and y are real.
   a) $z = iz + 1$
   b) $z^2 = iz + 1$

9. Solve the equation $(4 + 5i)z - (1 + i)z^* = 15 + 7i$, expressing z in the form $x + iy$, where x and y are real.

10. Given that $(a - 3i)^2 = 16 - bi$, where a and b are real, find the values of a and b.

We now turn our attention to solving polynomial equations involving solutions which may be complex numbers.

We saw earlier in this chapter that if one root of a quadratic equation with real coefficients is complex then the other root will be its conjugate. We now extend this result to any polynomial with real coefficients.

> If $z = x + iy$ is a root of a polynomial equation with real coefficients, then $z^* = x - iy$ is also a root of the polynomial equation, where z^* is the complex conjugate of z.

This result means that the number of complex roots of a polynomial equation with real coefficients will always be even.

Example 8

Given that $2 + i$ is a root of the equation $z^3 - z^2 - 7z + 15 = 0$, find the other two roots.

Since one complex root is $2 + i$ and the coefficients of the equation are all real, then the conjugate $2 - i$ is also a root.

$[z - (2 + i)] [z - (2 - i)]$ is a factor.   ← Write down the product of the known factors.

$z^2 - (2 + i)z - (2 - i)z + (2 + i)(2 - i)$

$z^2 - 4z + (4 + 2i - 2i - i^2)$ ← Multiply out and simplify.

$z^2 - 4z + 5$ is a factor.

To find the third root (which must be real),

$z^3 - z^2 - 7z + 15 = (\text{linear factor}) \times (z^2 - 4z + 5)$

By considering the term in z^3 and the constant term, we can see that the linear factor will be $(z + 3)$

The other two roots are -3 and $2 - i$.

> **Note:** In this case, we can find the other factor by inspection. An alternative method is to carry out a long division of $(z^3 - z^2 - 7z + 15)$ by $(z^2 - 4z + 5)$

Exercise 11.5

1. A cubic equation with real coefficients has 3 roots. Two of the roots are 2 and $1 + \sqrt{3}i$.

   a) Write down the third root.

   b) Find the equation which has these 3 roots. Write it in the form $az^3 + bz^2 + cz + d = 0$.

2. a) Find the roots of the equation $z^2 - 2\sqrt{2}z + 4 = 0$, giving your solutions in the form $x + iy$, where x and y are real.

   b) Showing all your working, verify that each root is also a root of the equation $z^8 - 256 = 0$.

3. Solve the following equations, giving your solutions in the form $x + iy$, where x and y are real.

   a) $(z + 2i)^2 = -9$ **b)** $z^2 = (z + i)^2$ **c)** $z^3 = (z + i)^3$

4. Solve the simultaneous equations

   $3z + w = 11$ $2iz + 5w = 8 - 9i$.

5. Solve the simultaneous equations

   $(1 - i)z + iw = 7 + 4i$ $iz + (1 - i)w = -7 - 3i$.

6. **a)** Write down one solution of the equation $z^3 = 1$.

   b) Hence write down one factor of $z^3 - 1$.   **c)** Solve the equation $z^3 - 1 = 0$.

7. Solve the equation $z^3 = 27$.

8. The complex number u is defined by $u = (1 + i)^4$.

   a) Express u in the form $x + iy$, where x and y are real.

   b) Solve the equation $z^4 = -4$.

9. The complex number $5 + 2i$ is a root of the equation $2z^3 - 15z^2 + 8z + 145 = 0$. Find the other roots.

10. **a)** Verify that $-1 + 2i$ is a root of the equation $z^3 + z - 10 = 0$.

   b) Find the other two roots of the equation.

11. **a)** Verify that $2 + 3i$ is a root of the equation $z^4 - 4z^3 + 12z^2 + 4z - 13 = 0$.

   b) Find the other roots of the equation.

12. The polynomial $x^5 + 2x^3 + 10x^2 + x + 10$ is denoted by $p(x)$.

   a) Find a real root of the equation $p(x) = 0$.

   b) Showing all your working, verify that i is a root of the equation $p(x) = 0$.

   c) Find all the roots of the equation $p(x) = 0$.

11.4 Representing complex numbers geometrically

Any complex number $z = x + iy$ can be represented by using a 2-dimensional set of coordinate axes where the horizontal axis is used as the 'real axis' and the vertical axis is used as the 'imaginary axis'.

So, for example, the complex number $3 + 2i$ is represented by the point $(3, 2)$, as shown in the diagram.

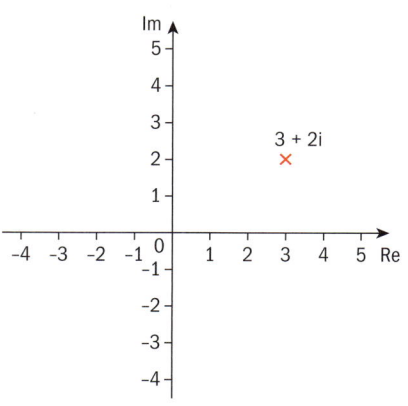

Such a diagram is called an **Argand diagram** and is named after the Swiss mathematician Jean-Robert Argand (1768 – 1822).

It follows that the point representing $z^* = x - iy$ is the reflection, in the real axis, of the point representing $z = x + iy$.

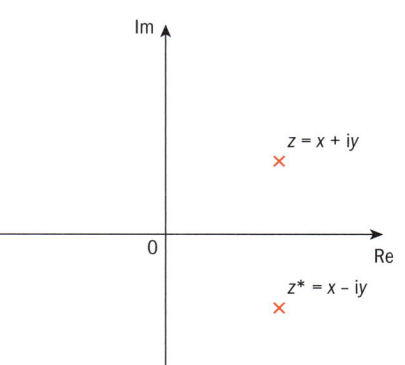

> The position of the point representing z^*, the complex conjugate of z, on the Argand diagram is found by reflecting the point representing z in the real axis.

Example 9

$u = 2 - 3i$

On a sketch of an Argand diagram, show the points representing the complex numbers $u, u^*, u^2, u + u^*, u - u^*, uu^*$ and $\dfrac{u}{u^*}$

. . . .

$u^* = 2 + 3i$ ← Calculate the complex numbers listed.

$u^2 = (2 - 3i)(2 - 3i) = -5 - 12i$

$u + u^* = (2 - 3i) + (2 + 3i) = 4$

$u - u^* = (2 - 3i) - (2 + 3i) = -6i$

$uu^* = (2 - 3i)(2 + 3i) = 13$

$\dfrac{u}{u^*} = \dfrac{2 - 3i}{2 + 3i} \times \dfrac{2 - 3i}{2 - 3i} = \dfrac{-5 - 12i}{13} = -\dfrac{5}{13} - \dfrac{12}{13}i$

Draw the axes and plot your answers. Label them clearly.

We will now consider how to represent the addition and subtraction of two complex numbers on an Argand diagram. To do this, it is helpful to consider the complex number $z = x + iy$ as a point in the Argand diagram with position vector $\begin{pmatrix} x \\ y \end{pmatrix}$.

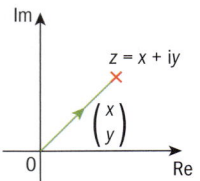

If $z_1 = a + ib$ and $z_2 = c + id$, then
$$z_1 + z_2 = (a + ib) + (c + id)$$
$$= (a + c) + i(b + d)$$

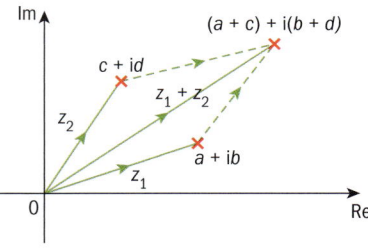

The point representing $z_1 + z_2$ can be constructed by adding the vectors representing z_1 and z_2; that is, the point representing $z_1 + z_2$ can be found by drawing the diagonal of the parallelogram formed by adding the vectors representing z_1 and z_2. This is illustrated on the diagram.

Similarly,
$$z_1 - z_2 = (a - c) + i(b - d)$$

The point representing $z_1 - z_2$ can be constructed by adding the vectors representing z_1 and $-z_2$, then drawing the vector representing $z_1 - z_2$ from the origin.

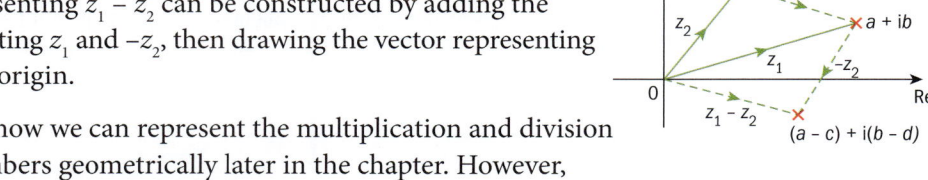

We will look at how we can represent the multiplication and division of complex numbers geometrically later in the chapter. However, at this point, it is worth considering what happens when we multiply a complex number by i.

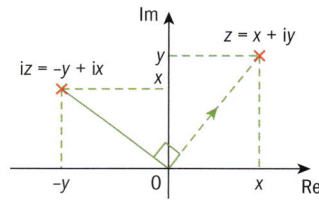

If $z = x + iy$, then $iz = i(x + iy) = ix - y$, and so the point (x, y) representing z is mapped onto the point $(-y, x)$. The diagram shows that this is equivalent to a rotation of 90° anticlockwise about the origin.

> Multiplying a complex number z by the imaginary number 'i' is geometrically equivalent to rotating the point representing z by 90° anticlockwise about the origin.

Exercise 11.6

1. Show on a sketch of an Argand diagram the points A, B, C, and D representing the complex numbers $2 + 3i$, $-1 - 4i$, $3 - i$ and $5 + 0i$ respectively.

2. The complex number u is defined by $u = 5 + 3i$.
   Show on a sketch of an Argand diagram the points P, Q, R, and S representing u, u^*, $2u$ and $-u$ respectively.

3. Write down the complex numbers z_1, z_2, z_3, z_4, z_5, z_6 and z_7 represented on the Argand diagram by the points A, B, C, D, E, F and G respectively.

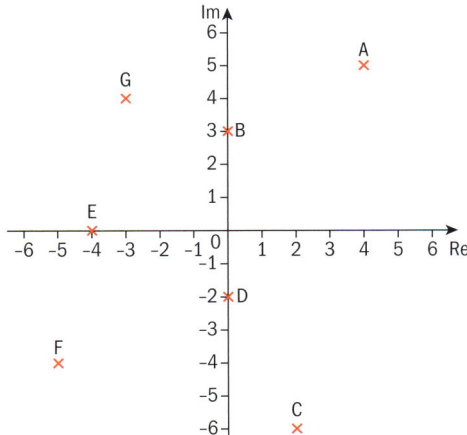

4. The complex number w is defined by $w = (2 + i)^2$

   a) Express w, iw, $\dfrac{w}{i}$ and $\dfrac{25i}{w}$ in the form $x + iy$, where x and y are real.

   b) Show on a sketch of an Argand diagram the points A, B, C, and D representing w, iw, $\dfrac{w}{i}$ and $\dfrac{25i}{w}$ respectively.

5. The complex numbers z_1 and z_2 are defined by $z_1 = 3 + i$ and $z_1 z_2 = -11 + 13i$

   a) Find z_2, expressing your answer in the form $x + iy$, where x and y are real.

   b) Show, on a sketch of an Argand diagram, the points representing z_1, z_2, $z_1 z_2$ and $iz_1 z_2$.

6. The complex numbers z and w are given below.

   In each case, sketch an Argand diagram to show the positions of z, w, $z + w$ and $z - w$.

   a) $z = 4 + 2i$, $w = 1 + 3i$ **b)** $z = 2 + 5i$, $w = -4 + i$

   c) $z = -5 - 3i$, $w = 2 + 3i$ **d)** $z = 3 - 4i$, $w = 1 + i$

7. The complex number z is defined by $z = 2 - 2i$.

   Show on a sketch of an Argand diagram the points representing z, $\dfrac{1}{z}$, and $z + \dfrac{1}{z}$

11.5 Polar form and exponential form

As we know, the position of a point can be described by using the Cartesian coordinates (x, y) and this can be plotted on a set of coordinate axes.

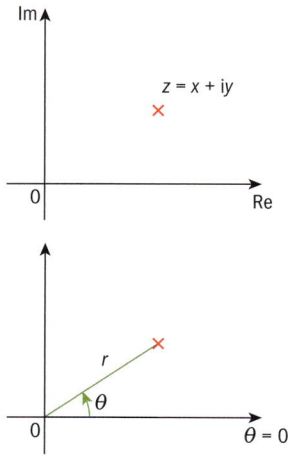

Alternatively, we can describe the position of a point by giving its distance from the origin (r) together with the angle the line joining the origin to the point makes with the positive x-axis (θ radians).

This way of describing the position of a point is called polar coordinate form. Polar coordinates are written in the form (r, θ) where $r > 0$ and $-\pi < \theta \leq \pi$.

As we saw in section **9.4**, complex numbers expressed in Cartesian form $(x + iy)$ can be plotted on an Argand diagram.

We can also represent complex numbers in **polar form**. This is more commonly called **modulus-argument form** when it is being used with complex numbers.

θ may be defined in the interval $-\pi < \theta \leq \pi$ or alternatively in the interval $0 \leq \theta < 2\pi$

The **modulus** of a complex number z, written as $|z|$, is the length of OP, where P is the point representing the complex number on an Argand diagram.

The **argument** of a complex number z, written as **arg (z),** is the angle (θ radians) between the line $\theta = 0$ and OP.

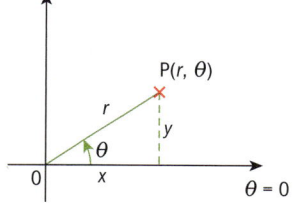

Hence, $r = \sqrt{x^2 + y^2}$ and $\tan \theta = \dfrac{y}{x}$

Also, $x = r\cos \theta$ and $y = r\sin \theta$

From the diagram we see that $x + iy = r\cos \theta + i\, r\sin \theta = r(\cos \theta + i\sin \theta)$

$\boxed{x + iy = r(\cos \theta + i\sin \theta)}$

Any complex number that can be expressed in Cartesian form as $z = x + iy$ or in modulus-argument form as $z = (r, \theta)$.

Use $r = \sqrt{x^2 + y^2}$ and $\tan \theta = \dfrac{y}{x}$ to convert from Cartesian to modulus-argument form.

Use $x = r\cos \theta$ and $y = r\sin \theta$ to convert from modulus-argument form to Cartesian form.

Notes:

i) $r > 0, -\pi < \theta \leq \pi$

ii) Care needs to be taken to make sure you give the correct argument for a complex number. Always use a diagram to check.

Example 10

a) The complex numbers z_1 and z_2 are defined by $z_1 = 1 + i$ and $z_2 = -2 - 4i$.
   Find the modulus and argument of z_1 and z_2.

b) The complex numbers z_3 and z_4 are defined by $z_3 = \left(5, \dfrac{\pi}{3}\right)$ and $z_4 = \left(2, \dfrac{5\pi}{6}\right)$.
   Write z_3 and z_4 in the form $x + iy$, where x and y are real numbers.

a) $|z_1| = \sqrt{1^2 + 1^2} = \sqrt{2} = 1.41$ (3 s.f.)

$\arg(z_1) = \tan^{-1}\left(\dfrac{1}{1}\right) = \dfrac{\pi}{4} = 0.785$ (3 s.f.)

$|z_2| = \sqrt{(2)^2 + (4)^2} = \sqrt{20} = 4.47$ (3 s.f.)

$\arg(z_2) = -\left(\pi - \tan^{-1}\left(\dfrac{4}{2}\right)\right) = -2.03$ (3 s.f.)

b) $z_3 = 5\left(\cos\dfrac{\pi}{3} + i\sin\dfrac{\pi}{3}\right) = \dfrac{5}{2} + \dfrac{5\sqrt{3}}{2}i = 2.5 + 4.33i$

$z_4 = 2\left(\cos\dfrac{5\pi}{6} + i\sin\dfrac{5\pi}{6}\right) = -\sqrt{3} + 1i = -1.73 + i$

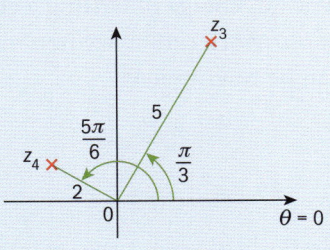

Exponential form is similar to modulus-argument form. A complex number is said to be in **exponential form** when it is expressed in the form $re^{i\theta}$, where r is the modulus and θ is the argument.

> The derivation of this form is beyond the Cambridge 9709 syllabus

For example, the following are all different forms for the same complex number:

(1) $1 + i$ Cartesian form (or real-imaginary form)

(2) $\left(\sqrt{2}, \dfrac{\pi}{4}\right) \equiv \sqrt{2}\left(\cos\dfrac{\pi}{4} + i\sin\dfrac{\pi}{4}\right)$ Polar form (or modulus-argument form)

(3) $\sqrt{2}e^{i\frac{\pi}{4}}$ Exponential form.

A complex number written in the form $re^{i\theta}$, where r is the modulus and θ is the argument, is said to be in exponential form.

Expressing a complex number in the form $r(\cos\theta + i\sin\theta) \equiv re^{i\theta}$ can help us to multiply or divide two complex numbers more easily.

To show why this is, consider the two complex numbers

$z_1 = (r_1, \theta_1) = r_1(\cos\theta_1 + i\sin\theta_1)$ and $z_2 = (r_2, \theta_2) = r_2(\cos\theta_2 + i\sin\theta_2)$

then, $z_1 \times z_2 = r_1(\cos\theta_1 + i\sin\theta_1) \times r_2(\cos\theta_2 + i\sin\theta_2)$

$$= r_1 r_2(\cos\theta_1 + i\sin\theta_1)(\cos\theta_2 + i\sin\theta_2)$$

$$= r_1 r_2(\cos\theta_1\cos\theta_2 + i\sin\theta_1\cos\theta_2 + i\sin\theta_2\cos\theta_1 + i^2\sin\theta_1\sin\theta_2)$$

$$= r_1 r_2((\cos\theta_1\cos\theta_2 - \sin\theta_1\sin\theta_2) + i(\sin\theta_1\cos\theta_2 + \sin\theta_2\cos\theta_1))$$

$$= r_1 r_2(\cos(\theta_1 + \theta_2) + i\sin(\theta_1 + \theta_2))$$

So, to multiply two complex numbers in polar form, we multiply their moduli and add their arguments.

If $z_1 = (r_1, \theta_1)$ and $z_2 = (r_2, \theta_2)$ then $z_1 z_2 = (r_1 r_2, \theta_1 + \theta_2)$

A special case of this can be used to find powers of a complex number.

If $z = (r, \theta)$ then $z^n = (r^n, n\theta)$

One advantage of writing a complex number in exponential form is that much of our work can be written more concisely. For example, the proof above can be rewritten as

$$z_1 z_2 = r_1 e^{\theta_1} \times r_2 e^{\theta_2}$$

$$= r_1 r_2 e^{\theta_1 + \theta_2}$$

Similarly, if we use this notation to divide two complex numbers,

$$\frac{z_1}{z_2} = \frac{r_1 e^{\theta_1}}{r_2 e^{\theta_2}} = \frac{r_1}{r_2}e^{\theta_1 - \theta_2}$$

So, to divide two complex numbers in polar form, we divide their moduli and subtract their arguments.

If $z_1 = (r_1, \theta_1)$ and $z_2 = (r_2, \theta_2)$ then $\dfrac{z_1}{z_2} = \left(\dfrac{r_1}{r_2}, \theta_1 - \theta_2\right)$

This proof using $z_1 = (r_1, \theta_1) = r_1(\cos\theta_1 + i\sin\theta_1)$ and $z_2 = (r_2, \theta_2) = r_2(\cos\theta_2 + i\sin\theta_2)$ is left as an exercise.

Example 11

The complex numbers z and w are defined by $z = 2 + 2i$ and $w = -1 + \sqrt{3}i$.

Find the modulus and argument of

a) z^2 **b)** z^3 **c)** z^2w^2 **d)** $\dfrac{w^2}{z^3}$.

a) $z = \left(2\sqrt{2}, \dfrac{\pi}{4}\right)$

        Write z in modulus-argument form.

$z^2 = \left\{\left(2\sqrt{2}\right)^2, 2 \times \dfrac{\pi}{4}\right\} = \left(8, \dfrac{\pi}{2}\right)$

        Use $(r, \theta)^2 = (r^2, 2\theta)$ and simplify.

$|z^2| = 8$, $\arg(z^2) = \dfrac{\pi}{2}$

b) $z^3 = \left\{\left(2\sqrt{2}\right)^3, 3 \times \dfrac{\pi}{4}\right\} = \left(16\sqrt{2}, \dfrac{3\pi}{4}\right)$

        Use $(r, \theta)^3 = (r^3, 3\theta)$ and simplify.

$|z^3| = 16\sqrt{2}$, $\arg(z^2) = \dfrac{3\pi}{4}$

> Alternatively, exponential form notation may be used.
> For example, $z = 2\sqrt{2}e^{\frac{i\pi}{4}}$
> $$z^2 = \left(2\sqrt{2}e^{\frac{i\pi}{4}}\right)^2$$
> $$= 8e^{\frac{i\pi}{2}} \text{ etc.}$$

c) $w = \left(2, \dfrac{2\pi}{3}\right)$

$w^2 = \left(2^2, 2 \times \dfrac{2\pi}{3}\right) = \left(4, \dfrac{4\pi}{3}\right)$

        Find w^2 and use your answer for z^2 from part **(a)**.

$z^2w^2 = \left(8, \dfrac{\pi}{2}\right) \times \left(4, \dfrac{4\pi}{3}\right) = \left(8 \times 4, \dfrac{\pi}{2} + \dfrac{4\pi}{3}\right)$

        Use $(r_1, \theta_1) \times (r_2, \theta_2) = (r_1 r_2, \theta_1 + \theta_2)$.

$|z^2w^2| = 32$, $\arg(z^2w^2) = \dfrac{11\pi}{6} = -\dfrac{\pi}{6}$

        Write $\arg(z)$ so it lies between $-\pi$ and π.

d) $\dfrac{w^2}{z^3} = \left\{\dfrac{4}{16\sqrt{2}}, \dfrac{4\pi}{3} - \dfrac{3\pi}{4}\right\} = \left\{\dfrac{\sqrt{2}}{8}, \dfrac{7\pi}{12}\right\}$

        Use $\dfrac{(r_1, \theta_1)}{(r_2, \theta_2)} = \left(\dfrac{r_1}{r_2}, \theta_1 - \theta_2\right)$ and simplify.

$\left|\dfrac{w^2}{z^3}\right| = \dfrac{\sqrt{2}}{8}$, $\arg\left(\dfrac{w^2}{z^3}\right) = \dfrac{7\pi}{12}$

We now turn our attention to understanding the geometrical effect of multiplying and dividing two complex numbers.

If we take a complex number, say $z_1 = (r_1, \theta_1)$ and multiply it by a second complex number $z_2 = (r_2, \theta_2)$, then as $z_2z_1 = (r_2, \theta_2) \times (r_1, \theta_1) = (r_2 r_1, \theta_2 + \theta_1)$, we see that the effect of multiplying one complex number by another is equivalent to multiplying the modulus of z_1 by the modulus of z_2. This is equivalent to an 'enlargement', centre the origin and scale factor (r_2) equal to the modulus of z_2. Using the same result, we can see that we add the argument of z_2 to the argument of z_1. This is the equivalent of a rotation (anti-clockwise) about the origin through an angle (θ_2) equal to the argument of z_2.

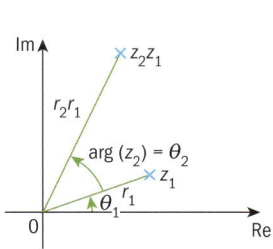

The combined effect of these two aspects results in what is often referred to as a spiral enlargement.

Dividing one complex number by another can be interpreted in a similar way, though the enlargement scale factor will be $\frac{1}{r_2}$ and the rotation will be anti-clockwise by an angle θ_2.

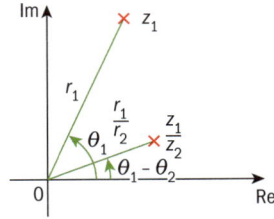

Exercise 11.7

1. Find the modulus and argument of the following complex numbers.

a) 1

b) -1

c) $1 + i$

d) $1 - i$

e) $3i$

f) $5 + 12i$

g) $\sqrt{13} - 6i$

h) $\frac{\sqrt{2}}{2} + \frac{\sqrt{2}}{2}i$

2. $z = \frac{1}{2} + \frac{1}{2}\sqrt{3}\,i$

   Find the modulus and argument in each case.

a) z

b) z^*

c) z^2

d) $(z^*)^2$

e) $(z^2)^*$

f) z^3

g) z^6

h) $z^3 + z^6$

i) $\frac{1}{z}$

j) $\frac{2}{z-1}$

3. $z = 4 + 3i$

a) Find   i) $|z|$ ii) $|z^2|$ iii) $|z^3|$

b) Find   i) $\arg z$ ii) $\arg z^2$ iii) $\arg z^3$

4. Write the following complex numbers, z, in the form $x + iy$.

a) $|z| = 2$, $\arg (z) = \frac{\pi}{2}$

b) $|z| = 5$, $\arg (z) = \pi$

c) $|z| = 4$, $\arg (z) = \frac{\pi}{3}$

d) $|z| = 3$, $\arg (z) = \frac{\pi}{6}$

e) $|z| = 10$, $\arg (z) = \frac{\pi}{4}$

f) $|z| = \frac{1}{2}$, $\arg (z) = -\frac{\pi}{2}$

g) $|z| = 1$, $\arg (z) = -\frac{3\pi}{4}$

h) $|z| = 6$, $\arg (z) = \frac{2\pi}{3}$

i) $|z| = 7$, $\arg (z) = 0$.

5. Show on a sketch of an Argand diagram the points representing the complex numbers in question 4.

6. $z = 1 + i$, $w = 1 - \sqrt{3}\,i$

   Find the modulus and argument of

a) wz

b) $\frac{w}{z}$

c) iz

d) $\frac{1}{w^*}$

7. Write the following complex numbers in the form $x + iy$.

a) $e^{\pi i}$

b) $e^{-\frac{\pi i}{2}}$

c) $4e^{\frac{\pi i}{4}}$

8. Express in the form $re^{i\theta}$

a) $1 + i$

b) -3

c) $-1 + 2i$

d) $7i$

e) $6 + 8i$

f) $6 - 8i$

g) $-\sqrt{2} - \sqrt{2}i$

h) $\sqrt{2}(5 - i)$

i) $\frac{1}{2} + \frac{\sqrt{3}}{2}i$

9. If $z_1 = 2\left(\cos\frac{\pi}{6} + i\sin\frac{\pi}{6}\right)$, $z_2 = 3\left(\cos\frac{\pi}{3} + i\sin\frac{\pi}{3}\right)$ and $z_3 = 2\left(\cos\frac{\pi}{4} + i\sin\frac{\pi}{4}\right)$,
   find, in modulus-argument form,

   a) $z_1 z_2$ b) $z_1 z_2 z_3$ c) $\dfrac{z_1}{z_2}$ d) $z_3{}^2$

10. The complex numbers z and w are defined by
    $|z| = 3$, arg $(z) = \dfrac{\pi}{4}$ and $|w| = 2$, arg $(z) = \dfrac{\pi}{6}$
    Find the modulus and argument of

    a) z^2 b) z^3 c) z^4 d) z^8 e) zw

    f) zw^2 g) $(zw)^3$ h) $\dfrac{z}{w}$ i) $\dfrac{1}{z^2}$ j) $w + w^*$

    k) $w - w^*$ l) $\dfrac{1}{(wz)^*}$

11. The complex number z is given by $z = x + iy$.
    Prove algebraically that

    a) $z + z^* = 2\text{Re}(z)$ b) $z - z^* = 2\text{Im}(z)i$ c) $|\text{Re}(z)| \le |z|$.

12. Illustrate each of the results from question 11 on a separate sketch of
    an Argand diagram.

11.6 Loci in the Argand diagram

In this section we will see how we can illustrate simple equations and inequalities
involving complex numbers as loci in an Argand diagram.

In particular we will consider the locus of z (where z is a variable complex
number) in the following three cases, taking into account how the locus
changes if the "=" sign is replaced by an inequality sign:

Case 1 $|z - z_1| = r$, where z_1 is a known complex number and r is real

$|z - z_1|$ is the modulus (or length) of $z - z_1$.
In section 4 of this chapter we saw that $z - z_1$
can be represented as the vector joining
the point z_1 to the point representing z.
So $|z - z_1| = r$ represents the locus of a point z
such that z moves so that its distance from
a fixed point z_1 is always r.

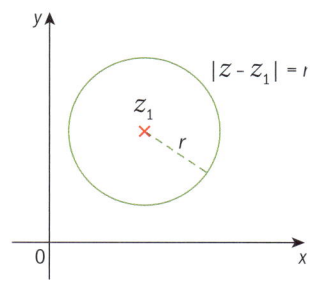

z lies on a circle, centre z_1, radius r.

> If $|z - z_1| = r$, then z lies on a circle, centre z_1, radius r.

We can also see that if $|z - z_1| < r$, then z lies anywhere inside a circle, centre z_1, radius r.

and   if $|z - z_1| > r$, then z lies anywhere outside a circle, centre z_1, radius r.

Example 12

a) Sketch an Argand diagram and show all the points which represent the complex numbers z which satisfy the equation $|z - 1 + i| = 2$.

b) The points in an Argand diagram representing $2 + 5i$ and $-6 + i$ are the ends of a diameter of a circle. Find the equation of the circle, giving your answer in the form $|z - (a + bi)| = k$.

a) $|z - (1 - i)| = 2$ ← Write the equation in the form $|z - z_1| = r$.

z lies on a circle, centre $(1, -1)$ and radius 2.

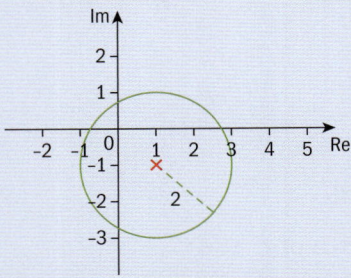

b) The centre of the circle is given by ← Use the result that the mid-point of the line joining (x_1, y_1) and (x_2, y_2) is given by

$$\left\{ \frac{(2 + -6)}{2}, \frac{(5i + i)}{2} \right\} = (-2 + 3i).$$

$$\left\{ \frac{x_1 + x_2}{2}, \frac{y_1 + y_2}{2} \right\}$$ and that the length of

The radius of the circle is given by

the line joining (x_1, y_1) and (x_2, y_2) is given by

$$\frac{1}{2} \left\{ \sqrt{(-6 - 2)^2 + (1 - 5)^2} \right\} = \frac{1}{2} \sqrt{80} = 2\sqrt{5}.$$

$$\sqrt{(x_2 - x_1)^2 + (y_2 - y_1)^2}$$

The equation of the circle is $|z - (-2 + 3i)| = 2\sqrt{5}$.

Example 13

Sketch an Argand diagram and shade the region whose points represent the complex numbers z which satisfy both the inequality $|z - (2 + 3i)| \leq 2$ and the inequality $|z - 4| > 3$.

$|z - (2 + 3i)| \leq 2$

z lies inside or on the circumference of a circle, centre $(2, 3)$ and radius 2.

$|z - 4| > 3$

z lies outside a circle, centre $(4, 0)$ and radius 3.

The set of points which satisfies both conditions lies in the shaded region and on the part of the circle, centre $(2, 3)$ which is shown in bold.

Note that the circle touches the y-axis. This should be clearly shown on your diagram.

Case 2 $|z - z_1| = |z - z_2|$ where z_1 and z_2 are known complex numbers

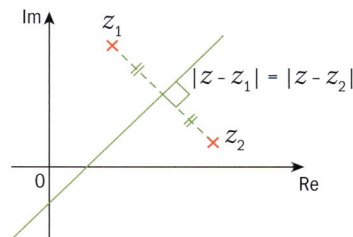

Here the locus of z is such that its distance from the fixed point z_1 is equal to its distance from the fixed point z_2. z will therefore lie on the perpendicular bisector (or mediator) of the line joining z_1 and z_2.

> If $|z - z_1| = |z - z_2|$, then z lies on the perpendicular bisector of the line joining z_1 to z_2

We can also see that if $|z - z_1| < |z - z_2|$, then z lies anywhere in the region on one side of the perpendicular bisector of the line joining z_1 to z_2 such that the distance from z to z_1 is less than the distance from z to z_2.

Example 14

a) Sketch on an Argand diagram the loci given by $|z - (1 + 2i)| = 5$ and $|z - 5 + i| = |z + 3 - 5i|$.

b) Show that these loci intersect at the point $-2 - 2i$.

....

a) $|z - (1 + 2i)| \geq 5$

z lies on or outside a circle, centre $(1, 2)$ and radius 5.

$|z - 5 + i| = |z + 3 - 5i|$

$|z - (5 - i)| = |z - (-3 + 5i)|$

z lies on the perpendicular bisector of the line joining $(5, -1)$ to $(-3, 5)$.

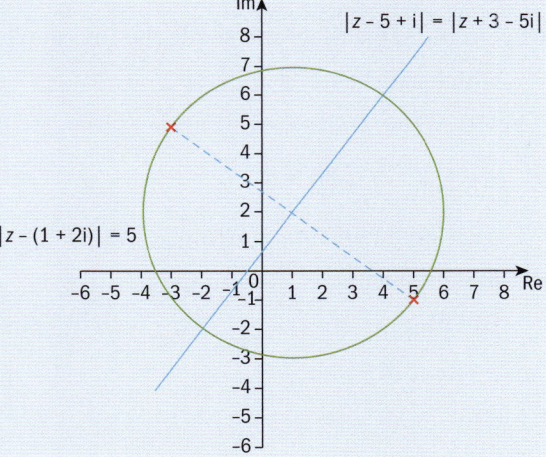

b) The loci will intersect at $-2 - 2i$ if $z = -2 - 2i$ satisfies both equations.

When $z = -2 - 2i$,

$|z - (1 + 2i)| = |-2 - 2i - (1 + 2i)|$

$\qquad\qquad\quad = |-3 - 4i|$

$\qquad\qquad\quad = 5$

so $|z - (1 + 2i)| = 5$ is satisfied. ← Substitute $z = -2 - 2i$ into the left-hand side of the inequality.

When $z = -2 - 2i$,

$|z - 5 + i| = |-2 - 2i - 5 + i|$

$\qquad\quad = |-7 - i|$

$\qquad\quad = \sqrt{50}$ ← Substitute $z = -2 - 2i$ into each side of the equation in turn.

Also $|z + 3 - 5i| = |-2 - 2i + 3 - 5i| = |1 - 7i|$

$\qquad\qquad\quad = \sqrt{50}$

so $|z - 5 + i| = |z + 3 - 5i|$ is satisfied.

Both conditions are satisfied and so these loci intersect at the point $-2 - 2i$. ← Make a clear statement in conclusion.

Case 3 arg $(z - z_1) = \theta$ where z_1 is a known complex number and θ is an angle, measured in radians

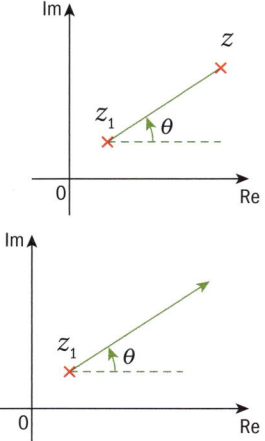

As we saw earlier in this chapter, the vector $z - z_1$ can be represented by the vector joining the point z_1 to the point representing z. We can interpret arg $(z - z_1) = \theta$ as the line joining z_1 to z having argument θ.

The locus of z is therefore a half-line, starting at z_1, at an angle θ with the positive real axis. It is called a half-line as we only want the part of the line which starts at z_1 but which has an infinite length.

> If arg $(z - z_1) = \theta$, then the locus of z is a half-line, starting at z_1, at an angle θ with the positive real axis.

Example 15

a)   Sketch on an Argand diagram the locus of the complex number z where z satisfies the equation arg $(z + 2 - i) = \dfrac{2\pi}{3}$

b)   On the same diagram, shade the region whose points represent the complex number z which satisfies $\dfrac{\pi}{2} \leq$ arg $(z + 2 - i) \leq \dfrac{2\pi}{3}$

∙∙

a)   arg $\{z - (-2 + i)\} = \dfrac{2\pi}{3}$ ← Write in the form arg $(z - z_1) = \theta$

z lies on a half-line, starting at $(-2, 1)$, making an angle $\dfrac{2\pi}{3}$ with the positive real axis.   ← Identify z_1 and θ

Draw a clear diagram. It is helpful to write a description to accompany your diagram.

b)   z may lie anywhere in the shaded region or on its boundaries.

Example 16

a) On a sketch of an Argand diagram, shade the region whose points represent the complex numbers z which satisfy the inequality $|z + 4 - 4\sqrt{3}i| \leq 4$.

b) i) Find the least value of $|z|$ in this region.

   ii) Find the greatest value of arg z in this region.

· ·

a) $|z - (-4 + 4\sqrt{3}i)| \leq 4$

   z lies on or inside a circle,

   centre $(-4, 4\sqrt{3})$ and radius 4.

b) i) The least value of $|z|$ is given by the length OP on the diagram.

   OP = OC − PC

   But $OC^2 = 4^2 + \left(4\sqrt{3}\right)^2 = 16 + 48 = 64$ ←   Use Pythagoras' theorem on triangle OBC.

   So OP = OC − PC = 8 − 4 = 4   ←   Use PC = 4 as it is a radius of the circle.

   Least value of $|z| = 4$.

   ii) The greatest value of arg z is given by the angle OA makes with the positive real axis.

   Angle $COA = \sin^{-1}\left(\dfrac{CA}{OC}\right) = \sin^{-1}\left(\dfrac{4}{8}\right) = \dfrac{\pi}{6}$

   The greatest value of arg $z = \dfrac{\pi}{6} + \dfrac{\pi}{6} + \dfrac{\pi}{2} = \dfrac{5\pi}{6}$ ←   Angle COA = Angle COB = $\dfrac{\pi}{6}$

Exercise 11.8

1. On separate sketches of an Argand diagram, show the locus of the complex number z which satisfies

   a) $|z| = 1$ b) $|z - 3| = 1$ c) $|z + 5| = 2$

   d) $|z - (2 - 2i)| \leq 2$ e) $|z + (1 + 3i)| \geq 1$

   f) $|z - a| = a$, where a is a positive real number.

2. a) Express, using modulus notation, the equations of the following circles on the Argand diagram

   i) centre 0, radius 3   ii) centre $2 + 2i$, radius 2   iii) centre $-4 - 3i$, radius 5.

   b) On separate sketches of an Argand diagram, illustrate these circles.

3. On separate sketches of an Argand diagram, show the locus of the complex number z which satisfies

a) $\arg(z - 1) = \frac{\pi}{2}$

b) $\arg(z + 1) = \frac{\pi}{3}$

c) $\arg(z + i) = -\frac{2\pi}{3}$

d) $\arg\{z - (2 + 3i)\} = \frac{3\pi}{4}$

4. On separate sketches of an Argand diagram, show the locus of the complex number z which satisfies

a) $|z| = |z - 5|$

b) $|z + i| = |z - 7i|$

c) $|z + 4| = |z - 2i|$

d) $|z - (2 + 2i)| = |z - (-2 - 2i)|$

e) $|z - (5 - 2i)| = |z - 1 - 2i|$

5. Show, by shading on separate sketches of an Argand diagram, the region whose points represent the complex number z which satisfies

a) $\frac{\pi}{6} \le \arg z \le \frac{\pi}{2}$

b) $0 \le \arg(z - i) \le \frac{\pi}{3}$

c) $\frac{\pi}{4} \le \arg(z - 2) \le \frac{2\pi}{3}$

d) $-\frac{\pi}{3} \le \arg\{z - (5 + 3i)\} \le \frac{\pi}{2}$

6. Show, by shading on separate sketches of an Argand diagram, the region whose points represent the complex number z which satisfies

a) $|z - 6| \le 3$

b) $|z - (1 + 3i)| \ge 1$

c) $|z - 3| \ge |z - 1|$

d) $|z - i| \ge |z + 1|$

e) $|z + (1 + i)| \ge |z - 3 - 3i|$

f) $|z - (2 + 3i)| \le |z - 2i|$

7. Sketch in a diagram the region described by $|z - 2| \le 2$ **and** $-\frac{\pi}{6} \le \arg z \le \frac{\pi}{6}$

8. Sketch in a diagram the region described by $|z| \le 3$ **and** $|z| \ge |z - 4|$

9. a) On a sketch of an Argand diagram, shade the region whose points represent complex numbers satisfying the inequalities $|z| \le 4$ and $\mathrm{Re}(z) \ge \mathrm{Im}(z)$, where $\mathrm{Re}(z)$ and $\mathrm{Im}(z)$ denote the real and imaginary parts of z respectively.

b) Write down the greatest and least values of $|z|$.

10. a) On a sketch of an Argand diagram, shade the region whose points represent complex numbers satisfying the inequality $|z - \sqrt{5} - 2i| \le 2$.

b) Calculate the greatest value of $|z|$.

c) Calculate the greatest value of $\arg z$.

Summary exercise 11

1. Find the values of x and y, where x and y are real, if $\dfrac{1}{x+iy} = \dfrac{4+5i}{6-3i}$

2. a) Show that the two roots of the equation $z^2 + 2z + 10 = 0$ have the same modulus.

   b) Find the argument of each root of the equation $z^2 + 2z + 10 = 0$.

3 a) Show that 1, $-\dfrac{1}{2} + \dfrac{\sqrt{3}}{2}i$, and $-\dfrac{1}{2} - \dfrac{\sqrt{3}}{2}i$ are all solutions of the equation $z^3 = 1$

   b) Hence find $\left\{-\dfrac{1}{2} + \dfrac{\sqrt{3}}{2}i\right\}^4$ and $\left\{-\dfrac{1}{2} + \dfrac{\sqrt{3}}{2}i\right\}^5$ giving your answers in the form $x + iy$.

4. The complex number z is given by $z = 1 - \sqrt{5}i$

   a) Express z^2 in the form $x + iy$ where x and y are real.

   b) Find the value of the real number p such that $z^2 - pz$ is real.

5. Two complex numbers, z and w, satisfy the inequalities $|z - 3 - 3i| \leq 2$ and $|w + 1 + i| \leq 1$.

   a) On a sketch of an Argand diagram, show the locus of the complex number z and the locus of the complex number w.

   b) Find the least and greatest possible values of $|z - w|$.

6. Show that, for any complex number z, $zz^* + 2(z + z^*)$ is real.

7. a) Solve the equation $z^2 + 5z + 4 = 0$.

   b) Hence solve the equation $z^4 + 5z^2 + 4 = 0$.

8. a) Sketch in a diagram the locus of the complex number z described by $|z + 3 + 12i| = 3\sqrt{17}$.

   b) Calculate the greatest and least values of $|z|$ for this locus.

9. a) Sketch in a diagram the locus of the complex number z described by $|z - 5i| = 3$.

   b) Calculate the greatest and least values of $|z|$ and arg z for this locus.

10. Two complex numbers z and w satisfy the inequalities $|z - (3 - 5i)| \leq 2$ and $|w - (7 - 10i)| \leq 2$. Draw an Argand diagram and find the least possible value and the greatest possible value of $|z - w|$.

11. a) Verify that $z = 1 + i$ is a root of the equation $z^4 - 6z^3 + 23z^2 - 34z + 26 = 0$.

    b) Write down another root of this equation.

    c) Find all the roots of the equation $z^4 - 6z^3 + 23z^2 - 34z + 26 = 0$.

12. $2 - \sqrt{5}i$ is a root of the equation $z^4 + 4z^3 - 20z^2 + 60z + 27 = 0$.
    Find the other roots of this equation.

13. a) Verify that $z = 1 + 2i$ is a root of the equation $z^3 - 3z^2 + 7z - 5 = 0$.

    b) Solve the equation completely.

14. An equilateral triangle has its vertices on the circle $|z| = 3$. One vertex is at the point such that arg $z = \dfrac{\pi}{6}$. Find the three vertices of the equilateral triangle expressing your answers in the form $x + iy$ where x and y are real.

15. The complex number w is given by

$$w = -\frac{1}{2} + \frac{\sqrt{3}}{2}i$$

   a) Find the value of w^2 expressing your answer in the form $x + iy$ where x and y are real.

   b) Verify that $w^4 = w$.

   c) a) Write w and w^2 in modulus argument form.

   b) Explain, using an Argand diagram, why $1 + w + w^2 = 0$.

16. a) Use a sketch of an Argand diagram illustrating the addition of the two complex numbers z_1, z_2, to show that
$$|z_1 + z_2| \le |z_1| + |z_2|.$$

   b) Use a sketch of an Argand diagram illustrating the subtraction of the two complex numbers z_1, z_2, to show that
$$|z_1 - z_2| \ge |z_1| - |z_2|.$$

17. The complex number $2 + i$ is denoted by u. Its complex conjugate is denoted by u^*.

   i) Show, on a sketch of an Argand diagram with origin O, the points A, B and C representing the complex number u, u^* and $u + u^*$ respectively. Describe in geometrical terms the relationship between the four points O, A, B and C. [4]

   ii) Express $\dfrac{u}{u^*}$ in the form $x + iy$, where x and y are real. [3]

   iii) By considering the argument of $\dfrac{u}{u^*}$, or otherwise, prove that
$$\tan^{-1}\left(\frac{4}{3}\right) = 2\tan^{-1}\left(\frac{1}{2}\right).$$
[2]

Cambridge International AS and A Level Mathematics 9709 Paper 3 Q7 May/June 2006

18. The equation $2x^3 + x^2 + 25 = 0$ has one real root and two complex roots.

   i) Verify that $1 + 2i$ is one of the complex roots. [3]

   ii) Write down the other complex root of the equation. [1]

   iii) Sketch an Argand diagram showing the point representing the complex number $1 + 2i$. Show on the same diagram the set of points representing the complex numbers z which satisfy
$$|z| = |z - 1 - 2i|.$$
[4]

Cambridge International AS and A Level Mathematics 9709 Paper 3 Q7 October/November 2005

19. i) Solve the equation $z^2 - 2iz - 5 = 0$, giving your answers in the form $x + iy$ where x and y are real. [3]

   ii) Find the modulus and argument of each root. [3]

   iii) Sketch an Argand diagram showing the points representing the roots. [1]

Cambridge International AS and A Level Mathematics 9709 Paper 3 Q3 May/June 2005

Chapter summary

Introducing complex numbers

- $\sqrt{-1} = i$ and therefore $i^2 = -1$
- A number of the form $a + bi$, where a and b are real and $i^2 = -1$, is called a complex number.
- The complex conjugate of $z = a + bi$ is $z^* = a - bi$, where a and b are real numbers.

Calculating with complex numbers.

- To add two complex numbers we add their real parts and add their imaginary parts.
- To subtract two complex numbers we subtract their real parts and subtract their imaginary parts.
- To multiply two complex numbers we use the normal rules of algebra and simplify the answer using $i^2 = -1$.
- To divide two complex numbers we multiply the numerator and the denominator of the fraction by the complex conjugate of the denominator.

Solving equations involving complex numbers

- Two complex numbers are equal if both the real parts are equal and the imaginary parts are equal.
- If $z = x + iy$ is a root of a polynomial equation with real coefficients, then $z^* = x - iy$ is also a root of the polynomial equation, where z^* is the complex conjugate of z.
- We can use the method of comparing real parts and imaginary parts to find the square roots of a complex number, $a + bi$

Representing complex numbers geometrically on an Argand diagram

- The position of z^*, the complex conjugate of z, on the Argand diagram is found by reflecting z in the real axis.

- $z_1 + z_2$ can be represented by drawing the diagonal of the parallelogram formed by adding the vectors representing z_1 and z_2.

- The points representing $z_1 - z_2$ constructed by adding the vectors representing z_1 and $-z_2$, then drawing the vector representing $z_1 - z_2$ from the origin.

- Multiplying a complex number z, by the imaginary number 'i' is geometrically equivalent to rotating the point representing z by 90° anticlockwise about the origin.

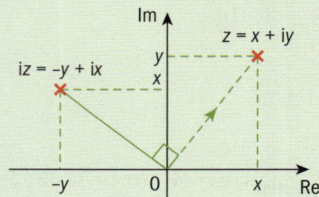

Modulus-argument and exponential forms

- Any complex number that can be expressed in Cartesian form as $z = x + iy$ or in modulus-argument (or polar) form as $z = (r, \theta)$.
- For modulus argument form $r > 0$ and $-\pi \leq \theta \leq \pi$.
- Use $r = \sqrt{x^2 + y^2}$ and $\tan \theta = \dfrac{y}{x}$ to convert from cartesian to modulus-argument form.
- Use $x = r\cos \theta$ and $y = r\sin \theta$ to convert from modulus-argument form to Cartesian form.
- $x + iy = r(\cos \theta + i \sin \theta)$

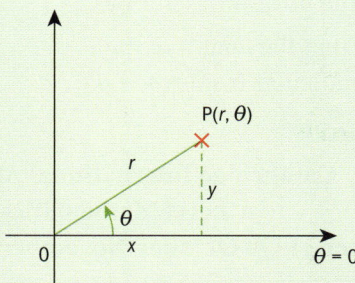

- A complex number written in the form $re^{i\theta}$, where r is the modulus and θ is the argument, is said to be in exponential form.

- If $z_1 = (r_1, \theta_1)$ and $z_2 = (r_2, \theta_2)$ then $z_1 z_2 = (r_1 r_2, \theta_1 + \theta_2)$.

- If $z_1 = (r_1, \theta_1)$ and $z_2 = (r_2, \theta_2)$ then $\dfrac{z_1}{z_2} = \left(\dfrac{r_1}{r_2}, \theta_1 - \theta_2\right)$.

Loci in the Argand diagram

- $|z - z_1| = r$, then z lies on a circle, centre z_1, radius r.

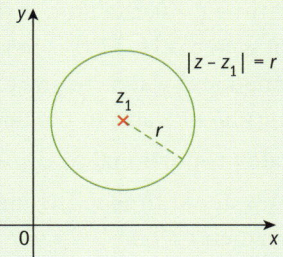

- $|z - z_1| = |z - z_2|$, then z lies on the perpendicular bisector of the line joining z_1 to z_2

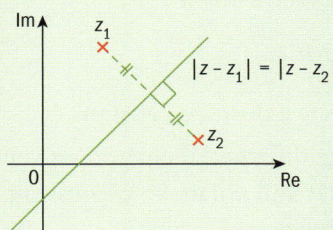

- $\arg(z - z_1) = \theta$, then the locus of z is a half-line, starting at z_1, at an angle θ with the positive real axis.

- Inequalities may be represented by shading appropriate regions on an Argand diagram.

Review exercise D – Pure 3

1. **Throughout this question the use of a calculator is not permitted.**

   i) The complex numbers u and v satisfy the equations

   $u + 2v = 2i$ and $iu + v = 3$. Solve the equations for u and v. giving both answers in the form $x + iy$, where x and y are real. **[5]**

   ii) On an Argand diagram, sketch the locus representing complex numbers z satisfying $|z + i| = 1$ and the locus representing complex numbers w satisfying $\arg(w - 2) = \frac{3}{4}\pi$. Find the least value of $|z - w|$ for points on these loci. **[5]**

Cambridge International AS and A Level Mathematics 9709 Paper 31 Q8 October/November 2013

2.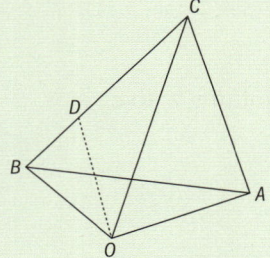

   The diagram shows three point A, B and C whose position vectors with respect to the origin O are given by $\overrightarrow{OA} = \begin{pmatrix} 2 \\ -1 \\ 2 \end{pmatrix}$,

   $\overrightarrow{OB} = \begin{pmatrix} 0 \\ 3 \\ 1 \end{pmatrix}$ and $\overrightarrow{OC} = \begin{pmatrix} 3 \\ 0 \\ 4 \end{pmatrix}$. The point D lies

   on BC, between B and C, and is such that $CD = 2DB$.

   i) Find the equation of the plane ABC, giving your answer in the form $ax + by + cz = d$. **[6]**

   ii) Find the position vector of D. **[1]**

   iii) Show that the length of the perpendicular from A to OD is $\frac{1}{3}\sqrt{(65)}$. **[4]**

Cambridge International AS and A Level Mathematics 9709 Paper 31 Q9 October/November 2013

3.

   A tank containing water is in the form of a cone with vertex C. The axis is vertical and the semi-vertical angle is 60°, as shown in the diagram. At time $t = 0$, the tank is full and the depth of water is H. At this instant, a tap at C is opened and water begins to flow out. The volume of water in the tank decreases at a rate proportional to $\sqrt{h}$ where h is the depth of water at time t. The tank becomes empty when $t = 60$.

   i) Show that h and t satisfy a differential equation of the form

   $$\frac{dh}{dt} = -Ah^{-\frac{3}{2}},$$ **[4]**

   where A is a positive constant. **[4]**

   ii) Solve the differential equation given in part (i) and obtain an expression for t in terms of h and H. **[6]**

   iii) Find the time at which the depth reaches $\frac{1}{2}H$. **[1]**

   [The volume V of a cone of vertical height h and base radius r is given by $V = \frac{1}{3}pr^2h$.]

Cambridge International AS and A Level Mathematics 9709 Paper 31 Q10 October/November 2013

4. The points P and Q have position vectors, relative to the origin O, given by

$$\overrightarrow{OP} = 7\mathbf{i} + 7\mathbf{j} - 5\mathbf{k} \text{ and } \overrightarrow{OQ} = -5\mathbf{i} + \mathbf{j} + \mathbf{k}.$$

The mid-point of PQ is the point A. The plane II is perpendicular to the line PQ and passes through A.

i) Find the equation of II, giving your answer in the form $ax + by + cz = d$. **[4]**

ii) The straight line through P parallel to the x-axis meets II at the point B. Find the distance AB, correct to 3 significant figures. **[5]**

Cambridge International AS and A Level Mathematics 9709 Paper 31 Q6 May/June 2013

5.

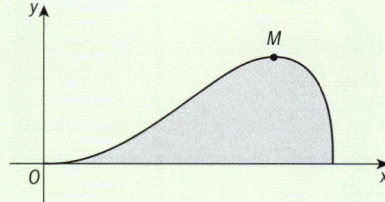

The diagram shows the curve $y = x^2\sqrt{(1-x^2)}$ for $x \geq 0$ and its maximum point M.

i) Find the exact value of the x-coordinate of M. **[4]**

ii) Show, by means of the substitution $x = \sin\theta$, that the area A of the shaded region between the curve and the x-axis is given by

$$A = \frac{1}{4}\int_0^{\frac{1}{2}\pi} \sin^2 2\theta \, d\theta.$$ **[3]**

iii) Hence obtain the exact value of A. **[4]**

Cambridge International AS and A Level Mathematics 9709 Paper 31 Q10 May/June 2009

6. **i)** Express $5\sin x + 12\cos x$ in the form $R\sin(x + \alpha)$, where $R > 0$ and $0° < \alpha < 90°$, giving the value of α correct to 2 decimal places. **[3]**

ii) Hence solve the equation

$$5\sin 2\theta + 12\cos 2\theta = 11,$$

giving all solutions in the interval $0° < \theta < 180°$. **[5]**

Cambridge International AS and A Level Mathematics 9709 Paper 3 Q6 October/November 2008

7. Two planes have equations $2x - y - 3z = 7$ and $x + 2y + 2z = 0$.

i) Find the acute angle between the planes. **[4]**

ii) Find a vector equation for their line of intersection. **[6]**

Cambridge International AS and A Level Mathematics 9709 Paper 3 Q7 October/November 2008

8. **i)** Without using a calculator, solve the equation

$$3w + 2iw^* = 17 + 8i,$$

where w^* denotes the complex conjugate of w. Give your answer in the form $a + bi$. **[4]**

ii) In an Argand diagram, the loci $\arg(z - 2i)\frac{1}{6}\pi$ and $|z - 3| = |z - 3i|$ intersect at the point P. Express the complex number represented by P in the form $re^{i\theta}$, giving the exact value of θ and the value of r correct to 3 significant figures. **[5]**

Cambridge International AS and A Level Mathematics 9709 Paper 31 Q7 May/June 2013

9. The complex number $1 + \left(\sqrt{2}\right)$i is denoted by u. The polynomial $x^4 + x^2 + 2x + 6$ is denoted by p(x).

   i) Showing your working, verify that u is a root of the equation p(x) = 0, and write down a second complex root of the equation. [4]

   ii) Find the other two roots of the equation p(x) = 0. [6]

Cambridge International AS and A Level Mathematics 9709 Paper 31 Q9 October/November 2012

10. With respect to the origin O, the points A, B and C have position vectors given by

$$\overrightarrow{OA} = \begin{pmatrix} 3 \\ -2 \\ 4 \end{pmatrix}, \overrightarrow{OB} = \begin{pmatrix} 2 \\ -1 \\ 7 \end{pmatrix} \text{ and } \overrightarrow{OC} = \begin{pmatrix} 1 \\ -5 \\ -3 \end{pmatrix}.$$

   The plane m is parallel to $\overrightarrow{OC}$ and contains A and B.

   i) Find the equation of m, giving your answer in the form $ax + by + cz = d$. [6]

   ii) Find the length of the perpendicular from C to the line through A and B. [5]

Cambridge International AS and A Level Mathematics 9709 Paper 31 Q10 October/November 2012

11. The complex number u is defined by

$$u = \frac{(1+2\text{i})^2}{2+\text{i}}.$$

   i) Without using a calculator and showing your working, express u in the form $x + \text{i}y$, where x and y are real. [4]

   ii) Sketch an Argand diagram showing the locus of the complex number z such that $|z - u| = |u|$. [3]

Cambridge International AS and A Level Mathematics 9709 Paper 31 Q4 May/June 2012

12. The variables x and y are related by the differential equation

$$\frac{\text{d}y}{\text{d}x} = \frac{6x\text{e}^{3x}}{y^2}.$$

   It is given that $y = 2$ when $x = 0$. Solve the differential equation and hence find the value of y when $x = 0.5$, giving your answer correct to 2 decimal places. [8]

Cambridge International AS and A Level Mathematics 9709 Paper 31 Q7 May/June 2012

13. The point P has coordinates $(-1, 4, 11)$ and the line l has equation

$$\mathbf{r} = \begin{pmatrix} 1 \\ 3 \\ -4 \end{pmatrix} + \lambda \begin{pmatrix} 2 \\ 1 \\ 3 \end{pmatrix}. \quad [4]$$

   i) Find the perpendicular distance from P to l [4]

   ii) Find the equation of the plane which contains P and I, giving your answer in the form $ax + by + cz = d$, where a, b, c and d are integers. [5]

Cambridge International AS and A Level Mathematics 9709 Paper 31 Q8 May/June 2012

14. The variables x and θ are related by the differential equation

$$\sin 2\theta \frac{\text{d}x}{\text{d}\theta} = (x + 1)\cos 2\theta,$$

   where $0 < \theta < \frac{1}{2}\pi$. When $\theta = \frac{1}{12}\pi$, $x = 0$. Solve the differential equation, obtaining an expression for x in terms of θ, and simplifying your answer as far as possible. [7]

Cambridge International AS and A Level Mathematics 9709 Paper 31 Q4 October/November 2011

15. With respect to the origin O, the position vectors of two points A and B are given by $\overrightarrow{OA} = \mathbf{i} + 2\mathbf{j} + 2\mathbf{k}$ and $\overrightarrow{OB} = 3\mathbf{i} + 4\mathbf{j}$. The point P lies on the line through A and B, and $\overrightarrow{AP} = \lambda\overrightarrow{AB}$.

   a) Show that $\overrightarrow{OP} = (1 + 2\lambda)\mathbf{i} + (2 + 2\lambda)\mathbf{j} + (2 - 2\lambda)\mathbf{k}$. [2]

   b) By equating expressions for cos AOP and cos BOP in terms of λ, find the value of λ for which OP bisects the angle AOB. [5]

   c) When λ has this value, verify that $AP : PB = OA : OB$. [1]

Cambridge International AS and A Level Mathematics 9709 Paper 31 Q7 October/November 2011

16. i) Showing your working, find the two square roots of the complex number $1 - (2\sqrt{6})i$. Give your answers in the form $x + iy$, where x and y are exact [5]

   ii) On a sketch of an Aigand diagram, shade the region whose points represent the complex numbers z which satisfy the inequality $|z - 3i| \leq 2$. Find the greatest value of arg z for points in this region. [5]

Cambridge International AS and A Level Mathematics 9709 Paper 31 Q10 October/November 2011

17. Points A and B have coordinates $(-1, 2, 5)$ and $(2, -2, 11)$ respectively. The plane p passes through B and is perpendicular to AB.

   i) Find an equation of p, giving your answer in the form $ax + by + cz = d$. [3]

   ii) Find the acute angle between p and the y-axis. [4]

Cambridge International AS and A Level Mathematics 9709 Paper 31 Q3 May/June 2011

18. The complex number u is defined by
$$u = \frac{6 - 3i}{1 + 2i}.$$

   i) Showing all your working, find the modulus of u and show that the argument of u is $-\frac{1}{2}\pi$. [4]

   ii) For complex numbers z satisfying $\arg(z - u) = \frac{1}{4}\pi$, find the least possible value of $|z|$. [3]

   iii) For complex numbers z satisfying $|z - (1 + i)u| = 1$, find the greatest possible value of $|z|$. [3]

Cambridge International AS and A Level Mathematics 9709 Paper 31 Q8 May/June 2011

19. The number of birds of a certain species in a forested region is recorded over several years. At time t years, the number of birds is N, where N is treated as a continuous variable. The variation in the number of birds is modelled by
$$\frac{dN}{dt} = \frac{N(1800 - N)}{3600}.$$
It is given that $N = 300$ when $t = 0$.

   i) Find an expression for N in terms of t. [9]

   ii) According to the model, how many birds will (here be after a long time? [1]

Cambridge International AS and A Level Mathematics 9709 Paper 31 Q10 May/June 2011

20. The complex number z is given by
$$z = (\sqrt{3}) + i$$

   i) Find the modulus and argument of z. [2]

   ii) The complex conjugate of z is denoted by z^*. Showing your working, express in the form $x + iy$, where x and y are real.

   a) $2z + z^*$,

   b) $\dfrac{iz^*}{z}$ [4]

c) On a sketch of an Argand diagram with origin O, show the points A and B representing the complex numbers z and iz* respectively. Prove that angle $AOB = \frac{1}{6}\pi$. [3]

Cambridge International AS and
A Level Mathematics 9709
Paper 31 Q6 October/November 2010

21. With respect to the origin O, the points A and B have position vectors given by $\overrightarrow{OA} = \mathbf{i} + 2\mathbf{j} + 2\mathbf{k}$ and $\overrightarrow{OB} = 3\mathbf{i} + 4\mathbf{j}$. The point P lies on the line AB and OP is perpendicular to AB.

i) Find a vector equation for the line AB. [1]

ii) Find the position vector of P. [4]

iii) Find the equation of the plane which contains AB and which is perpendicular to the plane OAB, giving your answer in the form $ax + by + cz = d$. [4]

Cambridge International AS and
A Level Mathematics 9709
Paper 31 Q7 October/November 2010

22. A certain substance is formed in a chemical reaction. The mass of substance farmed t seconds after the start of the reaction is x grams, At any time the rate of formation of the substance is proportional to $(20 - x)$. When $t = 0$, $x = 0$ and $\frac{dx}{dt} = 1$.

i) Show that x and t satisfy the differential equation
$$\frac{dx}{dt} = 0.05(20 - x).$$ [2]

ii) Find, in any form, the solution of this differential equation. [5]

iii) Find x when $t = 10$, giving your answer correct to 1 decimal place. [2]

iv) State what happens to the value of x as t becomes very large. [1]

Cambridge International AS and
A Level Mathematics 9709
Paper 31 Q10 October/November 2010

23. Given that $y = 0$ when $x = 1$, solve the differential equation
$$xy\frac{dy}{dx} = y^2 + 4,$$
obtaining an expression for y^2 in terms of x. [6]

Cambridge International AS and
A Level Mathematics 9709
Paper 31 Q5 May/June 2010

24. The complex number 2 + 2i is denoted by u.

i) Find the modulus and argument of u. [2]

ii) Sketch an Argand diagram showing the points representing the complex numbers 1, i and u. Shade the region whose points represent the complex numbers z which satisfy both the inequalities
$$|z - 1| \le |z - 1| \text{ and } |z - u| \le 1.$$ [4]

iii) Using your diagram, calculate the value of |z| for the point in this region for which arg z is least. [3]

Cambridge International AS and
A Level Mathematics 9709
Paper 31 Q7 May/June 2010

25. The lines l and m have vector equations
$$\mathbf{r} = \mathbf{i} + \mathbf{j} + \mathbf{k} + s(\mathbf{i} - \mathbf{j} + 2\mathbf{k}) \text{ and}$$
$$\mathbf{r} = 4\mathbf{i} + 6\mathbf{j} + \mathbf{k} + t(2\mathbf{i} + 2\mathbf{j} + \mathbf{k}) \text{ respectively.}$$

i) Show that l and m intersect. [4]

ii) Calculate the acute angle between the lines. [3]

iii) Find the equation of the plane containing l and m, giving your answer in the form $ax + by + cz = d$. [5]

Cambridge International AS and
A Level Mathematics 9709
Paper 31 Q10 May/June 2010

26. With respect to the origin O, the points A, B and C have position vectors given by $\overrightarrow{OA} = \mathbf{i} - \mathbf{k}$, $\overrightarrow{OB} = 3\mathbf{i} + 2\mathbf{j} - 3\mathbf{k}$ and $\overrightarrow{OC} = 4\mathbf{i} - 3\mathbf{j} + 2\mathbf{k}$.

The mid-point of AB is M. The point N lies on AC between A and C and is such that $AN = 2NC$.

i) Find a vector equation of the line MN. [4]

ii) It is given that MN intersects BC at the point P. Find the position vector of P. [4]

Cambridge International AS and A Level Mathematics 9709 Paper 31 Q6 October/November 2009

27. The complex number $-2 + \mathrm{i}$ is denoted by u.

i) Given that u is a root of the equation $x^3 - 11x - k = 0$, where k is real, find the value of k. [3]

ii) Write down the other complex root of this equation. [1]

iii) Find the modulus and argument of u. [2]

iv) Sketch an Argand diagram showing the point representing u. Shade the region whose points represent the complex numbers z satisfying both the inequalities

$|z| < |z - 2|$ and $0 < \arg(z - u) < \tfrac{1}{4}\pi$. [4]

Cambridge International AS and A Level Mathematics 9709 Paper 31 Q7 October/November 2009

28. In a model of the expansion of a sphere of radius r cm, it is assumed that, at time t seconds after the start, the rate of increase of the surface area of the sphere is proportional to its volume. When $t = 0$, $r = 5$ and $\dfrac{\mathrm{d}r}{\mathrm{d}t} = 2$.

i) Show that r satisfies the differential equation [4]

$$\frac{\mathrm{d}r}{\mathrm{d}t} = 0.08r^2.$$

[The surface area A and volume V of a sphere of radius r are given by the formulae $A = A\pi r^2$, $V = \frac{4}{3}\pi r^3$.]

ii) Solve this differential equation, obtaining an expression for r in terms of t. [5]

iii) Deduce from your answer to part (**ii**) the set of values that t can take, according to this model. [1]

Cambridge International AS and A Level Mathematics 9709 Paper 31 Q10 October/November 2009

29. i) Express $\dfrac{100}{x^2(10-x)}$ in partial fractions. [4]

ii) Given that $x = 1$ when $t = 0$, solve the differential equation

$$\frac{\mathrm{d}x}{\mathrm{d}t} = \frac{1}{100}x^2(10 - x),$$ [6]

obtaining an expression for t in terms of x

Cambridge International AS and A Level Mathematics 9709 Paper 31 Q8 May/June 2009

30. The line l has equation $\mathbf{r} = 4\mathbf{i} + 2\mathbf{j} - \mathbf{k} + t(2\mathbf{i} - \mathbf{j} - 2\mathbf{k})$. It is given that l lies in the plane with equation $2x + by + cz = 1$, where b and c are constants.

i) Find the values of b and c. [6]

ii) The point P has position vector $2\mathbf{j} + 4\mathbf{k}$. Show that the perpendicular distance from P to l is $\sqrt{5}$. [5]

Cambridge International AS and A Level Mathematics 9709 Paper 31 Q9 May/June 2009

31. The equation of a curve is $xy(x + y) = 2a^3$, where a is a non-zero constant. Show that there is only one point on the curve at which the tangent is parallel to the x-axis, and find the coordinates of this point. [8]

Cambridge International AS and A Level Mathematics 9709 Paper 3 Q6 May/June 2008

32.

The diagram shows the curve $y = e^{-\frac{1}{2}x}\sqrt{(1+2x)}$ and its maximum point M. The shaded region between the curve and the axes is denoted by R.

i) Find the x-coordinate of M. [4]

ii) Find by integration the volume of the solid obained when R is rotated completely about the x-axis. Give your answer in terms of π and e. [6]

Cambridge International AS and A Level Mathematics 9709 Paper 3 Q9 May/June 2008

33. The points A and B have position vectors, relative to the origin O, given by $\overrightarrow{OA} = \mathbf{i} + 2\mathbf{j} + 3\mathbf{k}$ and $\overrightarrow{OB} = \mathbf{i} + \mathbf{j} + 3\mathbf{k}$.

The line l has vector equation $r = (1 - 2t)\mathbf{i} + (5 + t)\mathbf{j} + (2 - t)\mathbf{k}$.

i) Show that l does not intersect the line passing through A and B. [4]

ii) The point P lies on l and is such that angle PAB is equal to $60°$. Given that the position vector of P is $(1 - 2t)\mathbf{i} + (5 + t)\mathbf{j} + (2 - t)\mathbf{k}$. show that $3t^2 + 7t + 2 = 0$. Hence find the only possible position vector of P. [6]

Cambridge International AS and A Level Mathematics 9709 Paper 3 Q10 May/June 2008

34. The curve with equation $y = e^{-x} \sin x$ has one stationary point for which $0 \le x \le \pi$.

i) Find the x-coordinate of this point. [4]

ii) Determine whether this point is a maximum or a minimum point. [2]

Cambridge International AS and A Level Mathematics 9709 Paper 3 Q4 October/November 2007

35. The number of insects in a population t days after the start of observation is denoted by N. The variation in the number of insects is modelled by a differential equation of the form $$\frac{dN}{dt} = kN\cos(0.02t),$$ where k is a constant and N is taken to be a continuous variable. It is given that $N = 125$ when $t = 0$.

i) Solve the differential equation, obtaining a relation between N, k and t. [5]

ii) Given also that $N = 166$ when $t = 30$, find the value of k. [2]

iii) Obtain an expression for N in terms of t, and find the least value of N predicted by this model. [3]

Cambridge International AS and A Level Mathematics 9709 Paper 3 Q7 October/November 2007

36.

The diagram shows a set of rectangular axes Ox, Oy and Oz, and three points, A, B and C with position vectors

$$\overrightarrow{OA} = \begin{pmatrix} 2 \\ 0 \\ 0 \end{pmatrix}, \overrightarrow{OB} = \begin{pmatrix} 1 \\ 2 \\ 0 \end{pmatrix} \text{ and } \overrightarrow{OC} = \begin{pmatrix} 1 \\ 1 \\ 2 \end{pmatrix}.$$

i) Find the equation of the plane ABC, giving your answer in the form $ax + by + cz = d$. [6]

ii) Calculate the acute angle between the planes ABC and OAB. **[4]**

Cambridge International AS and A Level Mathematics 9709 Paper 3 Q9 May/June 2007

37. A model for the height, h meters, of a certain type of tree at time t years after being planted assumes that, while the tree is growing, the rate of increase in height is proportional to $(9 - h)^{\frac{1}{3}}$. It is given that, when $t = 0$, $h = 1$ and $\dfrac{dh}{dt} = 0.2$.

i) Show that h and t satisfy the differential equation

$$\frac{dh}{dt} = 0.1(9 - h)^{\frac{1}{3}}$$ **[2]**

ii) Solve this differential equation, and obtain an expression for h in terms of t. **[7]**

iii) Find the maximum height of the tree and the time taken to reach this height after planting. **[2]**

iv) Calculate the time taken to reach half the maximum height. **[1]**

Cambridge International AS and A Level Mathematics 9709 Paper 3 Q10 May/June 2007

38. Given that $y = 2$ when $x = 0$, solve the differential equation

$$y\frac{dy}{dx} = 1 + y^2,$$

obtaining an expression for y^2 in terms of x. **[6]**

Cambridge International AS and A Level Mathematics 9709 Paper 3 Q4 October/November 2006

39. The line l has equation $r = \mathbf{j} + \mathbf{k} + s(\mathbf{i} - 2\mathbf{j} + \mathbf{k})$. The plane p has equation $x + 2y + 3z = 5$.

i) Show that the line l lies in the plane p. **[3]**

ii) A second plane is perpendicular to the plane p, parallel to the line l and contains the point with position vector $2\mathbf{i} + \mathbf{j} + 4\mathbf{k}$. Find the equation of this plane, giving your answer in the form $ax + by + cz = d$. **[6]**

Cambridge International AS and A Level Mathematics 9709 Paper 3 Q7 October/November 2006

40. The complex number u is given by

$$u = \frac{3 + i}{2 - i}.$$

i) Express u in the form $x + iy$, where x and y are real. **[3]**

ii) Find the modulus and argument of u. **[2]**

iii) Sketch an Argand diagram showing the point representing the complex number u. Show on the same diagram the locus of the point representing the complex number z such that $|z - u| = 1$. **[3]**

iv) Using your diagram, calculate the least value of $|z|$ for points on this locus. **[2]**

Cambridge International AS and A Level Mathematics 9709 Paper 3 Q9 October/November 2006

41. i) Express $7\cos\theta + 24\sin\theta$ in the form $R\cos(\theta - \alpha)$, where $R > 0$ and $0° < \alpha < 90°$, giving the exact value of R and the value of α correct to 2 decimal places.

ii) Hence solve the equation

$$7\cos\theta + 24\sin\theta = 15,$$

giving all solutions in the interval $0° \leq \theta \leq 360°$. **[4]**

Cambridge International AS and A Level Mathematics 9709 Paper 3 Q4 May/June 2006

Maths in real-life

Electrifying, magnetic and damp: how complex mathematics makes life simpler

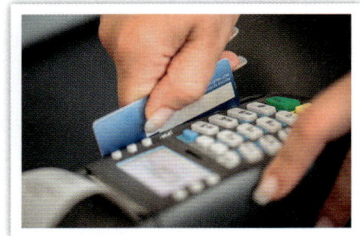

If you think of complex numbers in the Argand diagram, you can appreciate that the real and imaginary parts are both separate and interlinked, and so it is perhaps not surprising that they should have applications in physical situations where there are separate and linked entities.

In electromagnetism, the electric field and magnetic field can be represented by a single complex valued field $a + bi$, where a is normally the electric field and b the magnetic field. All the devices shown in these images depend on the use of electromagnetism.

One of the most spectacular applications is in Maglev trains. The Shanghai Maglev train (also known as the Transrapid) is the fastest commercial train in operation as of 2015. It operates by using magnets and electric current to levitate the train above a guide-rail and propel it forwards. With conventional trains there is a lot of friction in the moving parts, but with the Maglev the resistance is almost all related to air resistance, which was present in the conventional train motion as well however becomes greater at the higher speeds the Maglev train travels at. The future development of even faster trains looks possible as designers explore the practicalities of running Maglev trains in a vacuum tube where the air resistances would be removed. However, the economics of these projects means that although technically possible they may not become practical realities.

You have met simple first order differential equations in P3, however physical situations involving forces are described by second order differential equations, where the solutions sometimes require the use of complex numbers. The first model for a standard spring is that it has no resistances (like a projectile) so it continues to oscillate indefinitely once set in motion. However the reality is that the oscillations will die away. In systems like shock absorbers, the designers want the oscillations to die away quickly, and increase the friction in the system by putting it in a liquid rather than in air.

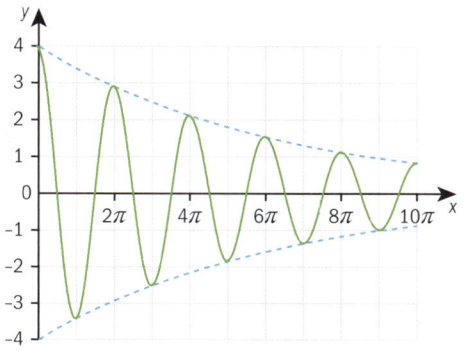

The motion of damped oscillations looks like the graph above: $x = 4e^{-0.05t} \cos t$ (shown in green) is bounded by the curves $x = \pm 4e^{-0.05t}$ shown as dotted blue lines.

Changing the physical characteristics of the system can change how quickly they die away or how fast the oscillations take place:

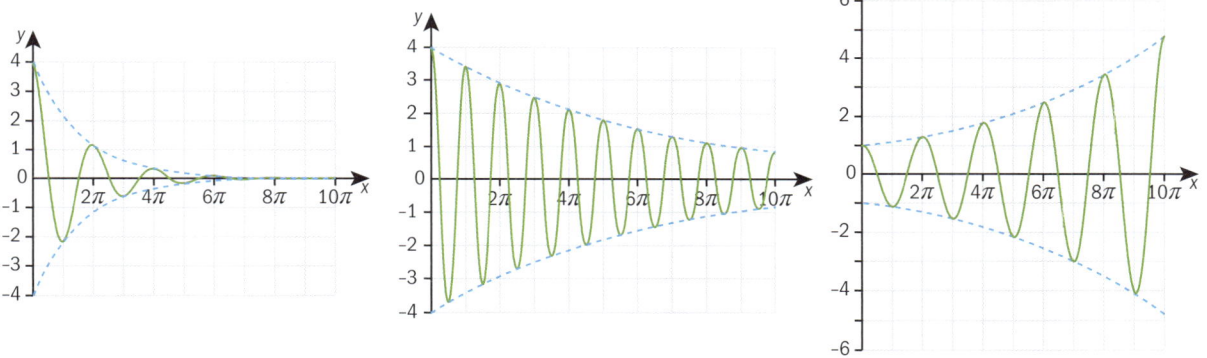

Resonance occurs where the opposite effect is seen: the amplitude of the oscillations grows rather than dies away. Sometimes this is the desired effect but more often it is destructive: feedback in amplification systems, glasses shattering if a singer exactly hits the resonant (or natural) frequency of the glass, people walking in step on a bridge, etc.

The Millennium Bridge in London had to be closed to the public for alterations shortly after it opened when it was found that people walking on the bridge naturally adjusted their walk because the bridge had a natural lateral sway – they adjusted their speed of step unconsciously to match the rhythm of the swaying motion, but this set up resonance and exaggerated the swaying motion.

Exam-style paper A – Pure 2

1. Solve the inequality $|2x - 4| < |4x|$ **(4)**

2. Solve the equation

   $\ln(64 - 3x^4) = 4\ln x$

   giving your answer correct to 3 significant figures. **(4)**

3. The polynomial $2x^4 - 3x^3 - 24x^2 + 13x + 12$ is denoted by p(x).
   i) Find the quotient when p(x) is divided by $x^2 - x - 12$ **(3)**
   ii) Hence solve the equation p(x) = 0. **(3)**

4. The curve with equation $y = \dfrac{3e^x}{x^2}$ has one stationary point.
   i) Find the exact coordinates of this point. **(4)**
   ii) Determine whether this point is a maximum or a minimum point. **(2)**

5. i) Given that $\sin(\theta + 45°) = 2\sin\theta$, show that $\tan\theta = \dfrac{1}{2\sqrt{2} - 1}$ **(2)**
   ii) Solve the equation

   $2\tan^2\theta - \sec\theta = 4$

   giving all solutions in the interval $0° < \theta < 360°$ **(5)**

6. i) Show that $(3\sin x - \cos x)^2$ can be written in the form $5 - 4\cos 2x - 3\sin 2x$ **(4)**
   ii) Hence find the exact value of $\displaystyle\int_0^{\frac{1}{2}\pi}(3\sin x - \cos x)^2\,dx$ **(4)**

7. Find the value of exact value of $\dfrac{dy}{dx}$ when $x = -3$ in each of the following cases:
   i) $y = x^2\ln(x + 13)$. Find the exact value of $\dfrac{dy}{dx}$ when $x = -3$ **(4)**

   ii) $y = \dfrac{3 - 5x}{6 + 5x}$. Find the exact value of $\dfrac{dy}{dx}$ when $x = -3$ **(3)**

8.

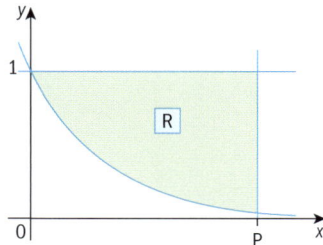

The diagram shows the curve $y = e^{-2x}$. The shaded region **R** is bounded by the curve and the lines $y = 1$

and $x = p$, where p is a constant.

i)   Find the area of **R** in terms of p.   **(4)**

ii) Show that if the area of **R** is equal to 1 then

$$p = \frac{3}{2} - \frac{1}{2}e^{-2p}$$ **(1)**

iii) Use the iterative formula

$$p_{n+1} = \frac{3}{2} - \frac{1}{2}e^{-2p_n}$$

with initial value $p_1 = 1.5$, to calculate the value of p correct to 2 decimal places. Give the result of each iteration to 4 decimal places.   **(3)**

1. Solve the inequality $|x| > |x - 1|$ **(3)**

2. Use the trapezium rule with two intervals to estimate the value of

$$\int_0^1 \frac{2}{4 + e^x}\,dx$$

   Giving your answer correct to 2 decimal places. **(3)**

3.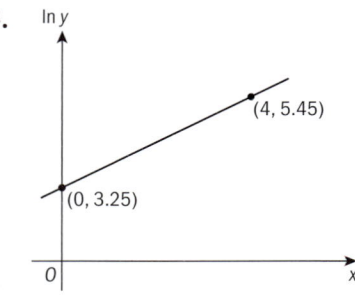

   The variables x nd y satisfy the equation $y = A(b^x)$ where A and b are constants.
   The graph of $\ln y$ against x is a straight line passing through the points $(0, 3.25)$
   and $(4, 5.45)$ as shown in the diagram.
   Find the values of A and b, correct to 1 decimal place. **(5)**

4. **i)** Find the quotient when the polynomial

   $$4x^3 + 8x^2 - 45x + 31$$

   is divided by $2x^2 + 9x - 5$, and show that the remainder is 6 **(3)**

   ii) Hence, or otherwise, factorise the polynomial

   $$4x^3 + 8x^2 - 45x + 25$$ **(2)**

5. **i)** Express $12 \cos \theta - 5 \sin\theta$ in the form $R\cos(\theta + \alpha)$, where $R > 0$ and $0° \le \alpha \le 90°$
   giving the value of α correct to 2 decimal places. **(3)**

   ii) Hence, solve the equation

   $$12 \cos \theta - 5 \cos \theta = 10,$$

   giving all solutions in the interval $0° \le \theta \le 360°$ **(3)**

   iii) State the least value of $12 \cos \theta - 5 \cos\theta$ as θ varies. **(1)**

6. **a)** Show that $\displaystyle\int_0^{\frac{1}{4}\pi}(2 - \cos^2 x)dx = \frac{3\pi - 2}{8}$ **(4)**

   b) Show that $\displaystyle\int_0^{\sqrt{15}} \frac{x}{x^2 + 5}\,dx = \ln 2$ **(5)**

7. **i)** By sketching a suitable pair of graphs show that the equation
    $$e^{3x} = 7 - 2x^2$$
    has exactly two real roots. **(3)**

    ii) Show by calculation that there is a root between 0.6 and 0.7. **(2)**

    iii) Show that this root also satisfies the equation
    $$x = \frac{1}{3} \ln (7 - 2x^2)$$ **(1)**

    iv) Use an iteration process based on the equation in part **(iii)**, with a suitable starting value, to find the root correct to 2 decimal places. Give the result of each step of the process to 4 decimal places. **(3)**

8. The equation of a curve is

    $3x^2 - 2xy + y^2 - 24 = 0$.

    i) Show that the tangent to the curve at the point $(2, 6)$ is parallel to the x-axis. **(4)**

    ii) Find the equation of the tangent to the curve at the other point on the curve for which $x = 2$, giving your answer in the form $y = mx + c$. **(5)**

1. Solve the equation

   $7^{x+1} = 7^x + 5$

   giving your answer correct to 3 significant figures. **(4)**

2. **i)** Expand $\dfrac{1}{\sqrt{1-2x}}$ in ascending powers of x, up to and including the term
   in x^3, simplifying the coefficients. **(3)**

   ii) Hence find the coefficient of x^3 in the expansion of $\dfrac{2-3x}{\sqrt{1-2x}}$. **(2)**

3. The sequence of values given by the iterative formula

   $$x_{n+1} = \frac{4x_n}{5} + \frac{10}{x_n^2}$$

   with initial value $x = 3.7$ converges to α.

   i) Use this iterative formula to find α correct to 2 decimal places, giving
   the result of each iteration to 4 decimal places. **(3)**

   ii) State an equation satisfied by α and hence find the exact value of α. **(2)**

4. $x^2 + y^2 - 4xy + 11 = 0$. Find the two exact values of $\dfrac{dy}{dx}$ when $x = -3$ **(6)**

5. The polynomial $x^3 - 4x^2 + ax + b$, where a and b are constants, is denoted by p(x).

   It is given that $(x - 2)$ is a factor of p(x) and that when p(x) is divided by
   $(x - 3)$ the remainder is 4.

   i) Find the values of a and b. **(5)**

   ii) When a and b have these values, find the other two linear factors of p(x). **(3)**

6. **i)** Express $8\sin x - 6\cos x$ in the form $R\sin(x - \alpha)$, where $R > 0$ and

   $0° \leq \alpha \leq 90°$, giving the value of α correct to 2 decimal places. **(3)**

   ii) Hence, solve the equation

   $8\sin 2\theta - 6\cos 2\theta = 5$,

   giving all solutions in the interval $0° \leq \theta \leq 180°$ **(4)**

   iii) State the greatest value of $24\sin 2\theta - 18\cos 2\theta$ as θ varies. **(1)**

7. The complex number is defined by $w = 4 + 3i$

   a) Find the modulus and argument of w. **(2)**

   b) On a sketch of an Argand diagram, shade the region whose points
   represent the complex numbers which satisfy the inequalities $|z - w| \leq 3$
   and $arg\,(z - 1) \leq \dfrac{\pi}{4}$ **(5)**

   c) Calculate the greatest possible value of $|z|$ in the shaded region. **(2)**

8. The two lines l and m have equations $\mathbf{r} = \begin{pmatrix} 2 \\ 3 \\ 0 \end{pmatrix} + \lambda \begin{pmatrix} 1 \\ 4 \\ -1 \end{pmatrix}$ and $\mathbf{r} = \begin{pmatrix} 1 \\ -1 \\ 1 \end{pmatrix} + \mu \begin{pmatrix} 2 \\ -5 \\ -3 \end{pmatrix}$ respectively.

    i)   Show that the lines l and m intersect and state the coordinates of the point
        of intersection. **(2)**

    ii) Calculate the acute angle between the directions of the lines l and m. **(3)**

    iii) Find the equation of the plane which contains both of the lines l and m,
        giving your answer in the form $ax + by + cz = d$. **(4)**

9. **i)**   Express $\dfrac{25}{y^2(5-y)}$ in partial fractions **(4)**

    ii) Given that $y = 1$ when $x = 0$, solve the differential equation

$$\frac{dy}{dx} = \frac{1}{25}y^2(5-y)$$

       Obtaining an expression for x in terms of y. **(6)**

10.

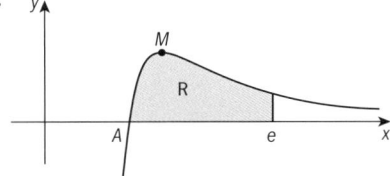

The diagram shows the curve $y = \dfrac{\ln x}{x^3}$ and its maximum point M.
The curve cuts the x-axiis at A.

    i)   Write down the x-coordinate of A **(1)**

    ii) Find the exact coordinates of M. **(5)**

    iii) Use integration by parts to find the exact area of the shaded region enclosed
        by the curve, the x-axis and the line $x = e$. **(5)**

1. Solve the inequality $|x + 2| \geq |3x - 2|$ **(4)**

2. **i)** Express $\cos^2 3x$ in terms of $\cos 6x$ **(2)**

   ii) Hence find the exact value of $\int_0^{\frac{1}{18}\pi} \cos^2 3x \, dx$ **(4)**

3. Use the substitution $u = 1 + e^x$ to show that $\int_0^1 \dfrac{e^{3x}}{1 + e^x} \, dx = 2.10$ correct to 2 decimal places. **(6)**

4. The parametric equations of a curve are
   $$x = \ln(4 - 2t), \quad y = \frac{3t + 5}{t + 1}$$

   i) Express $\dfrac{dy}{dx}$ in terms of t, simplifying your answer. **(4)**

   ii) Find the gradient of the curve at the point where the curve cuts the y-axis. **(3)**

5. Given that $2 + i$ is a root of the equation $x^3 + px^2 + qx + 15 = 0$, where p, q and r are real constants,

   i) Write down the other complex root of this equation. **(1)**

   ii) Find the value of p and the value of q **(4)**

   iii) Find the third root of the equation **(2)**

6. **i)** Prove that $\cot x - \cot y = \dfrac{\sin(y - x)}{\sin x \sin y}$ **(2)**

   ii) Solve the equation $2\sin 2\theta + \cos 2\theta = 1$ for $0° \leq \theta \leq 360°$ **(6)**

7.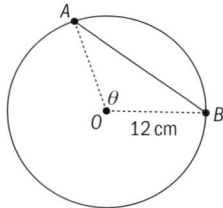

   The diagram shows a circle with centre O and radius 12 cm.

   The chord AB divides the circle into two regions whose areas are in the ratios $1 : 5$.

   It is required to find the length of AB. The angle AOB is in radians.

   i) Show that $\theta = \dfrac{\pi}{3} + \sin\theta$ **(3)**

   ii) showing all your working, use an iterative formula based on the equation in part **(i)**, with an initial value of 2, to find θ correct to 2 decimal places. Hence find the length of AB correct to 1 decimal place. **(5)**

8. **i)** Show that the vector $2\mathbf{i} - \mathbf{j} + \mathbf{k}$ is perpendicular to the plane which passes through the points $A(1, 5, 0)$, $B(2, 1, -6)$ and $C(-1, 2, 1)$. (3)

   ii) Hence find the equation of the plane which passes through the points $A(1, 5, 0)$, $B(2, 1, -6)$ and $C(-1, 2, 1)$. Give your answer in the form $ax + by + cz = d$, where a, b, c and d are integers. (2)

   iii) The perpendicular line from the point with position vector $2\mathbf{i} + \mathbf{k}$ to the plane $2x - y + z = 2$ meets the plane at point P. Find the position vector of P. (4)

9. **i)** Express in partial fractions $\dfrac{4 - 4x + 6x^2}{(1 - x)(2 + x^2)}$ (5)

   ii) Hence obtain the expansion of $\dfrac{4 - 4x + 6x^2}{(1 - x)(2 + x^2)}$ in ascending powers of x, up to and including the term in x^3. (5)

10. A certain substance is formed during a chemical reaction. The mass of substance formed t seconds after the start of the reaction is x grams.
    At any time, the rate of formation of the substance is proportional to $(30 - x)$.
    When $t = 0$, $x = 0$ and $\dfrac{dx}{dt} = 3$.

    i) Show that x and t satisfy the differential equation $\dfrac{dx}{dt} = 0.1\,(30 - x)$. (2)

    ii) Find, in any form, the solution of this differential equation. (5)

    iii) Find x when $t = 10$, giving your answer correct to 1 decimal place. (2)

    iv) State what happens to the value of x as t becomes very large. (1)

Answers

The answers given here are concise. However, when answering exam-style questions, you should show as many steps in your working as possible.

Cambridge International Examinations bears no responsibility for the example answers to questions taken from its past paper questions which are contained in this publication.

1 Algebra

Skills check page 2

1. **a)** 32 **b)** 73 **c)** 47 **d)** 58
2. **a)** 20 **b)** 6 **c)** 22 **d)** 20

Exercise 1.1 page 6

1. **a)** $x = -1$ or $x = 2$ **b)** $x = 1$
   c) $x = \frac{2}{7}$ or $x = \frac{2}{3}$ **d)** $x = \frac{1}{4}$ or $x = \frac{9}{4}$
   e) $x = 3$ or $x = -\frac{1}{3}$ **f)** $x = \frac{4}{3}$ or $x = 4$
   g) $x = -5$ or $x = \frac{3}{5}$ **h)** $x = 1$ or $x = -7$
   i) $x = \frac{3}{2}$ or $x = -\frac{5}{4}$ **j)** $x = 1$ or $x = \frac{1}{2}$
   k) $x = \frac{5}{2}$ or $x = 7$ **l)** $x = -\frac{5}{6}$ or $x = \frac{5}{2}$

2. **a)** $1 < x < 3$ **b)** $x \le -3$ or $x \ge 5$
   c) $-\frac{5}{3} \le x \le 1$ **d)** $x < -3$ or $x > 3$
   e) $x \le 0$ or $x \ge 6$ **f)** $-\frac{1}{2} < x < 3$
   g) $-1 < x < \frac{1}{3}$ **h)** $x \le -2$ or $x \ge 3$
   i) $-\frac{1}{3} < x < 5$ **j)** $0 \le x \le 1$
   k) $x < -\frac{4}{3}$ or $x > 0$ **l)** $-\frac{1}{2} \le x \le 2$

Exercise 1.2 page 7

1. **a)** $x^2 + x + 1$ **b)** $x^2 - 3x + 9$
   c) $x^2 - x + 4$ **d)** $x^2 - x - 1$
   e) $2x^2 + 7$ **f)** $3x^2 - 2x - 2$
2. **a)** -5 **b)** 4 **c)** -8
   d) $-8x - 1$ **e)** 2 **f)** -27
4. **b)** $(x - 1)(x - 2)(x - 3)$
6. $x^2 - x + 1$
7. quotient $= x^2 + x + 1$ remainder $= -2x + 6$

8. quotient $= 2x^2 - x - 2$ remainder $= 9$
9. **b)** $(2x - 1)(2x + 3)(3x + 1)$
10. $k = 10$

Exercise 1.3 page 9

1. **a)** 127 **b)** 4 **c)** $-\frac{49}{8}$
   d) -1 **e)** 4 **f)** 48
2. $a = 3$ 3. $a = 12$ 4. $a = 5$
5. $a = 3$ and $b = -6$
6. **b)** $a = -20$ 8. $a = -7$ and $b = 20$
9. $a = 2$ 10. $a = \frac{1}{6}$

Exercise 1.4 page 13

1. **a)** $(x - 3)(x + 1)(2x - 1)$
   b) $(x - 1)(x - 2)(x - 3)$
   c) $(5x - 1)(x + 1)(x + 2)$
   d) $(x + 4)(x - 2)(2x - 1)$
   e) $(x + 2)(x + 1)(x - 2)$
   f) $(3x + 2)(x + 2)(2x - 1)$
2. **a)** $x = \frac{3}{2}$ or $x = -1$ or $x = -4$
   b) $x = \frac{1}{2}$ or $x = -2$ or $x = 4$
   c) $x = -1$ or $x = 2$ or $x = 5$
   d) $x = 2$ or $x = -1$ or $x = -4$
   e) $x = 4$ or $x = 5$ or $x = -\frac{3}{2}$
   f) $x = -1$ or $x = 2$ or $x = -\frac{1}{3}$
4. $(x + 1)(x - 1)(x - 2)(x + 3)$
5. $a = 7$
6. **b)** $(2x - 5)(x - 3)(2x + 1)$
7. $a = 3$ or $a = -3$ 8. $a = 3$ and $b = 8$
9. 225 $(a = -6, b = 5)$
10. $(x - 1)(x + 1)(x + 2)(2x + 1)$

Summary exercise 1 page 14

1. $x = 6$ or $x = -1$ 2. $x = -1$ or $x = \frac{9}{5}$
3. $x \le -2$ or $x \ge 1$ 4. $\frac{4}{3} < x < 4$
5. $x < 1$ or $x > 3$ 6. $1 \le x \le 9$
7. $x \le -\frac{5}{2}$, $x \ge -\frac{1}{4}$ 8. $x < \frac{a}{4}$
9. $2x^3 - 3x^2 + 4x - 3$
10. quotient $= x^2 - x + 4$ remainder $= 9$

11. b) $(x-4)(x+3)(x-2)$

12. $k = 1$

13. $a = 3, \quad b = 2$

14. a) $a = 4, \quad b = -1$

 b) $(x+1)(x-2)(x+2)(x+3)$

15. a) $x^2 - 16$

 b) $x = -4, x = 4, x = -3, x = 2$

16. a) $a = 12, \quad b = 20$

 b) $(3x+2)(x-2)(2x-5)$

17. a) $A = -14$

 b) i) $x = 1 \quad$ or $\quad x = 4 \quad$ or $\quad x = 9$

  ii) $x = \pm 1 \quad$ or $\quad x = \pm 2 \quad$ or $\quad x = \pm 3$

18. a) $A = -18 \quad B = 32$

 b) i) $x = -2 \quad$ or $\quad x = 1 \quad$ or $\quad x = 16$

  ii) $x = \pm 1 \quad$ or $\quad x = \pm 2$

19. $x < \dfrac{1}{3}, x > 1$

20. b) $x = -1$ and $x = -\dfrac{1}{2}$

2 Logarithms and exponential functions

Skills check page 16

1. a) x^5 **b)** x^5 **c)** $x^{\frac{7}{3}}$

2. 2^{ab}

Exercise 2.1 page 19

1. a) 614.4   **b)** 972   **c)** 80.1 (3 s.f)

 d) 2.09×10^{-4}

2. a) 252 ml   **b)** 7.1 ml

3. \$2.82 million

4. After 12 hours there is $100 \times (0.8)^{12} = 6.9$ mg so need further injections to be $100 - 6.9 = 93.1$ mg to bring total to 100 mg.

5. 9 am the next morning is 13 hours later. At 9 am the volume is $3000 \times (0.98)^{13} = 2307$ ml $= 2.3$ l so there is enough (>2 litres) for him to drive the car to the garage.

Exercise 2.2 page 22

1. a) -4 **b)** $\dfrac{3}{2}$ **c)** $\dfrac{7}{2}$ **d)** 8

 e) 4.89 **f)** $-\dfrac{5}{3}$ **g)** $\dfrac{5}{4}$ **h)** -2

 i) $-\dfrac{1}{2}$ **j)** 5 **k)** -3 **l)** $-\dfrac{3}{2}$

2. a) $256 = 2^8$ **b)** $3^{-3} = \dfrac{1}{27}$ **c)** $4^{2.5} = 32$

 d) $3^{2.5} = 9\sqrt{3}$ **e)** $a^q = x$ **f)** $s^u = t$

3. a) $-3 = \log_7\left(\dfrac{1}{343}\right)$

 b) $9 = \log_{10}(1\,000\,000\,000)$

 c) $-4 = \log_2\left(\dfrac{1}{16}\right)$ **d)** $2.5 = \log_5\left(25\sqrt{5}\right)$

 e) $-3 = \log_t v$ **f)** $x = \log_p m$

4. a) 2 **b)** $\dfrac{1}{2}$ **c)** $\dfrac{2}{3}$ **d)** $-\dfrac{1}{2}$

 e) $5\dfrac{1}{2}$ **f)** $-\dfrac{7}{3}$ **g)** 0

5.

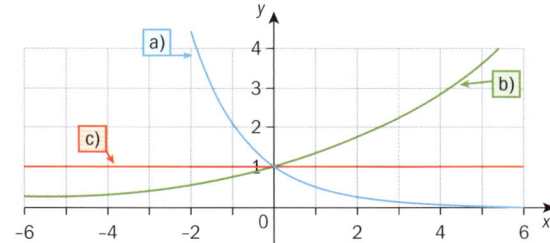

6. a) $\log_5 24$ **b)** $\log_6 70$ **c)** 2

 d) $\log_5 36$ **e)** $\log_5 54$ **f)** 0

 g) 0 **h)** 2

7. a) $\log_5 75$ **b)** $\log_6 2160$

 c) $\log_{10}(1000x)$ **d)** $\log_5(125x^2)$

 e) $\log_{10}\left(\dfrac{x^2(x+3)}{8}\right)$ **f)** $\log_{10}\left(\dfrac{800}{x}\right)$

8. a) $\log x + \log y - 2\log z$

 b) $5\log x + 6\log y - 3\log z$

 c) $2\log x + \log y - 2\log z$

9. a) $x + 2y$ **b)** $z + 2y - x$

 c) $z - 3x$ **d)** $-3x - 3y$

Exercise 2.3 page 26

1. a) 1.365 **b)** -1.635 **c)** 0.434

 d) 1.872 **e)** -5.036

2. a) 1.152 **b)** 0.878 **c)** 0.461

3. a) i) 15.1 kg of TNT

  ii) 15300000 kg of TNT

 b) 2890000 kg of TNT

4. 17.3 grams

Exercise 2.4 page 29

1. Solve
   a) 2.86   **b)** 0.613   **c)** 2.51   **d)** −1
   e) 0.638   **f)** 0.712   **g)** 1.20   **h)** 2.37
   i) 2   **j)** −2   **k)** 6   **l)** −3

2. **a)** 7.64   **b)** 4.96   **c)** 12.1
   d) −1   **e)** 5.45   **f)** 1.31

3. **a)** $x \geq 3.36$ **b)** $x \leq -0.125$ **c)** $x < 1.76$
   d) $x < -0.756$ **e)** $x \leq 0.414$ **f)** $x < 4.96$

4. **a)** 0.827   **b)** 1.277   **c)** 2.104

5. Solve
   a) 2.113   **b)** 1.272

6. 13.2; 14 terms   **7.** 19.5; 20 terms

8. 9.45; 10 terms   **9.** 11.7; 12 terms

10. 14.3; 15 terms   **11.** 17.23; 18 terms

12. 8.38; 9 terms

13. **a)** $\dfrac{1}{3}$ **b)** $\dfrac{2}{3}$ **c)** $\dfrac{1}{2}$
    d) $\dfrac{4}{99}$ **e)** $\dfrac{1}{1+e}$

14. **a)** $-3, \dfrac{2}{5}$ **b)** $\dfrac{2}{3}; \dfrac{3}{2}$ **c)** $\dfrac{6}{5}$ **d)** $\dfrac{1}{2}; \dfrac{4}{5}$

15. **a)** 0.248 grams   **b)** 76 years

16. 27 years.   **17.** 3 hours and 10 minutes.

Exercise 2.5 page 33

1. **a)** $\log p = \log 3 + t \log b$; $\log p$ and t.
   b) $\log y = \log K - 2 \log x$; $\log y$ and $\log x$.
   c) $\log y = \log \alpha + \dfrac{1}{2} \log x$; $\log y$ and $\log x$.

2. **a)** $y = 10x^2$ **b)** $V = \dfrac{4\pi}{3} r^3$ **c)** $y = 10^{(0.1+1.3x)}$

3. $y = Ax^n \Rightarrow \log y = \log A + n\log x$
   n (the gradient) is approximately 0.6, and the intercept
   ($\log A$) is approximately 0.75, so A is approximately
   $10^{0.75} = 5.6$

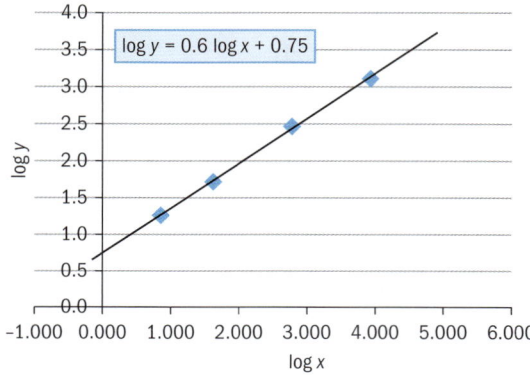

4. $-\ln k$ (the gradient) is approximately −0.675, so k is
   approximately $e^{0.675} = 2.0$ and the intercept ($\log A$) is
   approximately 3.36, so A is approximately $e^{3.36} = 28.8$

5. n (the gradient) is approximately −0.5, and the
   intercept ($\log A$) is approximately 3.5, so A is
   approximately $10^{3.5} = 3200$.

   > Note that if you use natural logarithms, the
   > graph will look different but the estimates will
   > be the same – to within the accuracy possible in
   > estimating from a graph.

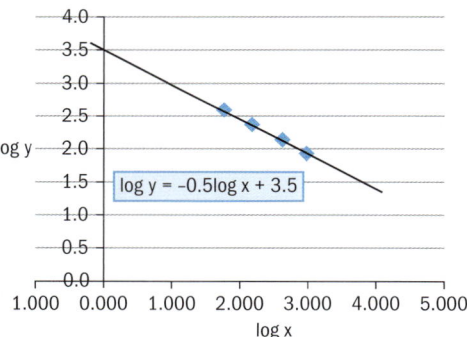

6. n (the gradient) is approximately 0.5, and the intercept
   ($\log k$) is approximately 2.8, so k is approximately
   $10^{2.8} = 630$.

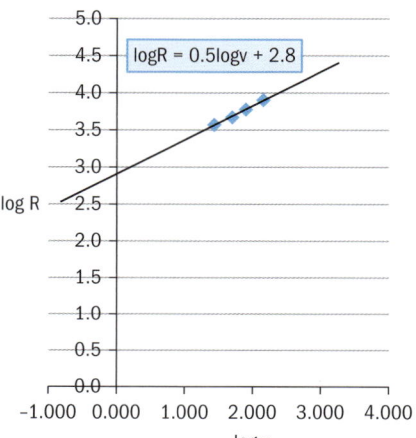

7. n (the gradient) is approximately 1.5, and the intercept ($\log k$) is approximately -0.7, so k is approximately $10^{-0.7} = 0.2$.

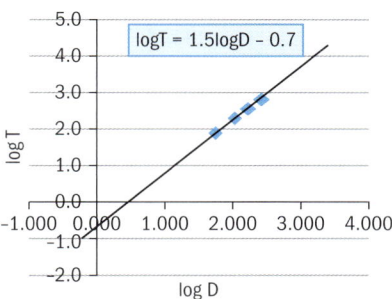

8. t (the gradient) is approximately 0.25, and the intercept ($\log k$) is approximately 6.3, so A is approximately $e^{6.3} = 545$.

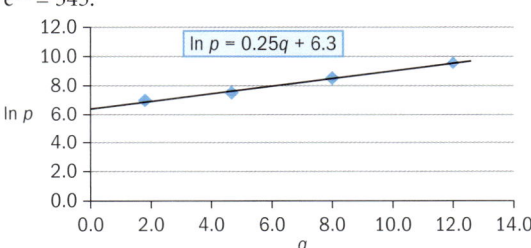

9. gradient $= 3\dfrac{\log 4}{\log 7}$ and intercept $= 2\dfrac{\log 4}{\log 7}$

Summary exercise 2 page 35

1. **a)** -5 **b)** $\dfrac{3}{2}$ **c)** $\dfrac{1}{4}$

2. **a)** $4^4 = 256$ **b)** $2^{-5} = \dfrac{1}{32}$ **c)** $2^{5.5} = 32\sqrt{2}$

3. **a)** $-3 = \log_6 \dfrac{1}{216}$ **b)** $7 = \log_{10} 10\,000\,000$

4. **a)** 4 **b)** $\dfrac{3}{2}$ **c)** $\dfrac{7}{4}$

5. **a)** $\log_4 14$ **b)** $\log_{10} 204$ **c)** $\ln 13.5$
   d) 2 **e)** 2 **f)** $\log_5 (25x^3)$
   g) $\ln(2x^4 + 3x^3)$ **h)** $\ln\left(\dfrac{4e^2}{x^2}\right)$

6. $3 \log x + 10 \log y + 5 \log z$ **7.** $z - 3x - y$

8. **a)** 2.27 **b)** -0.748 **c)** 0.139

9. **a)** 6 **b)** 1.03 **c)** 1.09

10. **a)** $x \geq 2.34$ **b)** $x \leq -0.22$ **c)** $x < 3.38$

11. 9.01; 10 terms   **12.** 19.932; 20 terms

13. **a)** $\dfrac{1}{3}$ **b)** $\dfrac{1}{33}$ **c)** -2 **d)** $x = \dfrac{3}{2 + e^2}$

14. **a)** $-2, \dfrac{1}{3}$ **b)** $-\dfrac{1}{2}, \dfrac{1}{5}$

15. **a)** 23.6 grams   **b)** 108 years

16. 4 hours 59 minutes.

17. $\ln V = \ln A + T \ln r$. $-\ln r$ (the gradient) is approximately -0.06, so r is approximately $e^{0.06} = 1.06$ (interest rate of 6% pa) and the intercept ($\ln A$) is approximately 8.1, so A is approximately $e^{8.1} = 3300$.

18. $\log P = \log k + t \log r$. $\log r$ (the gradient) is approximately 0.11, so r is approximately $10^{0.11} = 1.29$, and the intercept ($\log k$) is approximately 0.95, so k is approximately $10^{0.95} = 8.9$.

19. gradient $= 2\dfrac{\log 6}{\log 5}$; intercept $= -5\left(\dfrac{\log 6}{\log 5}\right)$

20. $\ln y = \ln A + x \ln k$. $-\ln k$ (the gradient) is approximately 0.4, so r is approximately $e^{0.4} = 1.49$ and the intercept ($\ln A$) is approximately 0.37, so A is approximately $e^{0.37} = 1.45$.

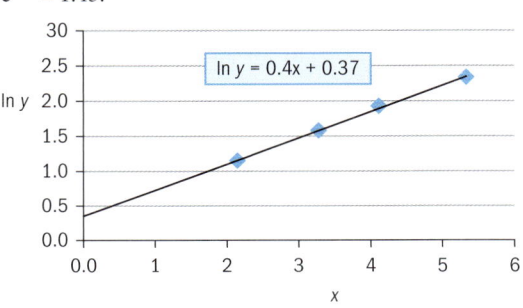

21. $1.97 < A < 2.03$, $n = 0.25$ **22.** 2.32

3 Trigonometry

Skills check page 38

1. a) 0 b) $\dfrac{\sqrt{3}}{2}$ c) $-\dfrac{1}{\sqrt{3}}\left(=-\dfrac{\sqrt{3}}{3}\right)$

   d) $\dfrac{\sqrt{3}}{2}$ e) $\dfrac{\sqrt{3}}{2}$ f) -1

2. a) i) $\dfrac{\pi}{6}$ ii) $-\dfrac{\pi}{4}$

   b) i) $31.8°$ ii) $-11.5°$

3. a) $26.6°, 206.6°$ b) $\dfrac{\pi}{3}, \dfrac{2\pi}{3}, \dfrac{4\pi}{3}, \dfrac{5\pi}{3}$

Exercise 3.1 page 43

1. a) $\dfrac{2}{\sqrt{3}}\left(=\dfrac{2\sqrt{3}}{3}\right)$ b) 2 c) $\dfrac{1}{\sqrt{3}}\left(=\dfrac{\sqrt{3}}{3}\right)$

   d) $\sqrt{2}$ e) $-\dfrac{2}{\sqrt{3}}\left(=-\dfrac{2\sqrt{3}}{3}\right)$

   f) $\sqrt{2}$ g) $-\dfrac{1}{\sqrt{3}}\left(=-\dfrac{\sqrt{3}}{3}\right)$

   h) $-\dfrac{2}{\sqrt{3}}\left(=-\dfrac{2\sqrt{3}}{3}\right)$

2. a) $-\sqrt{2}$ d), e) and g)   b) 2 c) $\sqrt{3}$

   d) 1 e) $-\dfrac{2}{\sqrt{3}}\left(=-\dfrac{2\sqrt{3}}{3}\right)$ f) -1

   g) $-\dfrac{2}{\sqrt{3}}\left(=-\dfrac{2\sqrt{3}}{3}\right)$ h) -1

3. a), b), c), f) and h) are defined
   d), e) and g) are not defined

4. a) 1.60 b) 1.06 c) -1.22 d) 1.08
   e) 1.70 f) -0.727 g) -3.24 h) -3.24

5. a) $45°, 225°$ b) $45°, 315°$ c) $60°, 120°$
   d) $30°, 150°$ e) $150°, 330°$ f) $180°$

6. a) $\dfrac{\pi}{2}$ b) $0, 2\pi$ c) $\dfrac{3\pi}{4}, \dfrac{7\pi}{4}$

   d) $\dfrac{\pi}{6}, \dfrac{11\pi}{6}$ e) $\dfrac{\pi}{2}, \dfrac{3\pi}{2}$ f) $\dfrac{4\pi}{3}, \dfrac{5\pi}{3}$

7. a) $70.5°, 289.5°$ b) $14.5°, 165.5°$
   c) $48.0°, 228.0°$ d) $11.3°, 191.3°$
   e) $231.1°, 308.9°$ f) $114.1°, 245.9°$

8. a) $81.8°, 278.2°$ b) $81.8°, 98.2°, 261.8°, 278.2°$

9. $33.7°, 213.7°$

10. $3.87, 5.55$

11. a) $0°, 180°, 360°$
    b) $7.2°, 82.8°, 187.2°, 262.8°$
    c) $12.3°, 72.3°, 132.3°, 192.3°, 252.3°, 312.3°$

12. a) $180°$ b) $143.1°$
    c) $23.9°, 96.1°, 143.9°, 216.1°, 263.9°, 336.1°,$

13. a)

   b)

   c)

   d)

e)

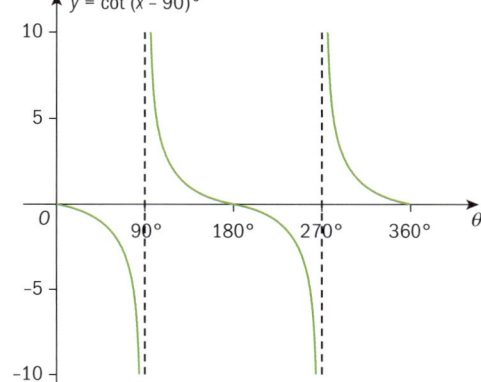

$y = \cot(x - 90)°$

f)

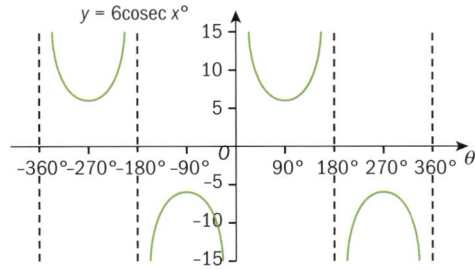

$y = 3\sec 2x°$

14. a)

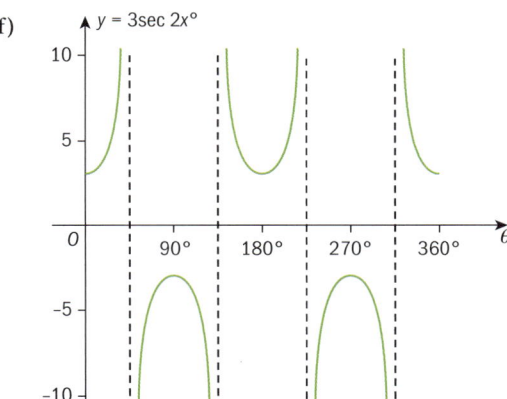

$y = 6\cosec x°$

b)

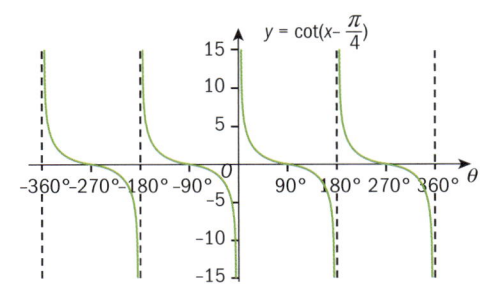

$y = \cot\left(x - \dfrac{\pi}{4}\right)$

c)

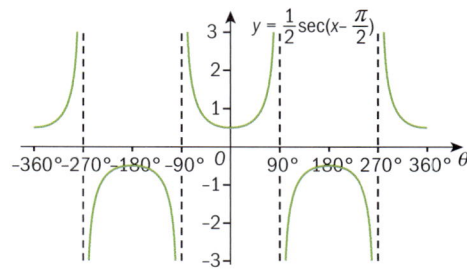

$y = \dfrac{1}{2}\sec\left(x - \dfrac{\pi}{2}\right)$

d)

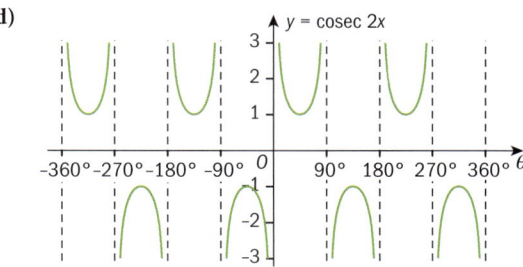

$y = \cosec 2x$

Exercise 3.2 page 46

1. **a)** 1   **b)** 1   **c)** $\sin\theta$

   d) 1   **e)** $\sin^3\theta$ **f)** $\tan^2\theta$

2. 26.6°, 116.6°, 206.6°, 296.6°

3. 0.666, 2.48

4. 70.5°, 120°, 240°, 289.5°

10. 18.4°, 135°, 198.4°, 315°

11. 64.3°, 140.1°, 219.9°, 295.7°

12. $\dfrac{3\pi}{4}, \dfrac{7\pi}{4}$

14. 45°, 56.3°, 225°, 236.3°

15. 51.8°, 308.2°

Exercise 3.3 page 51

1. **a)** $\dfrac{\sqrt{2}}{4}(\sqrt{3} - 1)$ (or $\dfrac{1}{2\sqrt{2}}(\sqrt{3} - 1)$) **b)** $\dfrac{\sqrt{2}}{4}(\sqrt{3} + 1)$

   c) $\dfrac{1+\sqrt{3}}{1-\sqrt{3}}$ (or $-2 - \sqrt{3}$)   **d)** $\dfrac{4 + \sqrt{6} - \sqrt{2}}{4 - \sqrt{6} + \sqrt{2}}$

3. **a)** $\cos\alpha = \dfrac{3}{5}$, $\cos\beta = \dfrac{12}{13}$

   b) $\sin(\alpha + \beta) = \dfrac{63}{65}$, $\sin(\alpha - \beta) = \dfrac{33}{65}$, $\cos(\alpha + \beta) = \dfrac{16}{65}$,

   $\cos(\alpha - \beta) = \dfrac{56}{65}$

   c) $\tan(\alpha + \beta) = \dfrac{63}{16}$, $\tan(\alpha - \beta) = \dfrac{33}{56}$

4. a) $\frac{\sqrt{2}}{4}(\sqrt{3}+1)$ (or $\frac{1}{2\sqrt{2}}(\sqrt{3}+1)$)

b) $\frac{\sqrt{2}}{4}(1-\sqrt{3})$ (or $\frac{1}{2\sqrt{2}}(1-\sqrt{3})$)

c) $\frac{\sqrt{3}-1}{\sqrt{3}+1}$ (or $2-\sqrt{3}$)

d) $\frac{\sqrt{2}}{4}(1-\sqrt{3})$ (or $\frac{1}{2\sqrt{2}}(1-\sqrt{3})$)

5. a) $\frac{\sqrt{3}}{2}$ **b)** $\frac{\sqrt{3}}{2}$ **c)** 1   **d)** $\sqrt{3}$

6. a) $2\sin\alpha\cos\beta$ **b)** $2\cos\alpha\cos\beta$ **c)** $-\sin\beta$

9. $\cot(\theta+\alpha)=\dfrac{\cot\theta\cot\varphi-1}{\cot\theta+\cot\varphi}$

11. a) $\frac{11}{13}$ **b)** $-\frac{63}{16}$ **c)** $-\frac{5}{4}$

12. 29.3°, 105.7°, 209.3°, 285.7°

13. $0, \pi, 2\pi$

14. 45°, 225°

Exercise 3.4 page 55

1. a) 30°, 90°, 150°, 270°

b) 60°, 90°, 120°, 270°

c) 60°, 300°

d) 48.6°, 131.4°, 210°, 330°

e) 0°, 60°, 120°, 180°, 240°, 300°, 360°

f) 0°, 45°, 135°, 180°, 225°, 315°, 360°

2. a) 0, 0.420, 2.72, 3.14, 3.56, 5.86, 6.28

b) 0, 1.37, 3.14, 4.91, 6.28   **c)** $\frac{\pi}{3}, \frac{5\pi}{3}$

3. a) $\frac{7}{25}$ **b)** $\frac{24}{7}$ **c)** $-\frac{336}{527}$

4. a) $\frac{240}{289}$ **b)** $\frac{240}{161}$ **c)** $\frac{289}{161}$

5. a) $-\frac{3}{\sqrt{10}}$ **b)** $-\frac{1}{\sqrt{10}}$ **c)** $-\frac{3}{4}$ **d)** $-\frac{4}{5}$

6. a) $1-\sin 2x$ **b)** $\cos^3 2x$

7. $12\cos^3\theta-8\cos^5\theta-4\cos\theta$

8. a) i) $3\sin\theta-4\sin^3\theta$
     ii) 0°, 45°, 135°, 180°, 225°, 315°, 360°

b) $\frac{3\tan\theta-\tan^3\theta}{1-3\tan^2\theta}$

9. $\frac{\cot^2\theta-1}{2\cot\theta}$

Exercise 3.5 page 58

1. a) $\sqrt{13}\sin(\theta+33.69°)$ **b)** $\sqrt{149}\sin(\theta+34.99°)$

c) $\sqrt{5}\sin(\theta+26.57°)$ **d)** $\sqrt{1681}\sin(\theta+12.68°)$

2. a) i) $\sqrt{89}\sin(\theta-57.99°)$ **ii)** $25\sin(\theta-73.74°)$

    iii) $\sqrt{5}\sin(\theta-26.57°)$ **iv)** $\sqrt{19}\sin(\theta-66.59°)$

b) i) $\sqrt{89}, -\sqrt{89}$ **ii)** $25, -25$

    iii) $\sqrt{5}, -\sqrt{5}$ **iv)** $\sqrt{19}, -\sqrt{19}$

3. a) $\sqrt{2}\cos(\theta+45°)$ **b)** $\sqrt{6}\cos(\theta+35.26°)$

c) $5\cos(\theta+36.87°)$ **d)** $\sqrt{6}\cos(\theta+24.09°)$

4. a) i) $5\cos(\theta-53.13°)$ **ii)** $13\cos(\theta-67.38°)$

    iii) $\sqrt{3}\cos(\theta-35.26°)$ **iv)** $\sqrt{2}\cos(\theta-45°)$

b) i) $5, -5$ **ii)** $13, -13$

    iii) $\sqrt{3}, -\sqrt{3}$ **iv)** $\sqrt{2}, -\sqrt{2}$

c) i) 53.1°   **ii)** 67.38°

    iii) 35.26°   **iv)** 45°

5. a) $\sqrt{5}\sin(\theta+63.43°)$ **b)** 26.6°, 153.4°

6. a) $2\sin(\theta-30°)$ **b)** 60°, 180°   **c)** 4, 0

7. a) $\sqrt{8}\cos(\theta+45°)$ **b)** 69.3°, 290.7°

8. a) $\sqrt{52}\cos(\theta-56.31°)$ **b)** 10.2°, 102.4°   **c)** 57, 5

9. a) 257.6°, 349.8°   **b)** 163.7°

c) 53.1°, 323.1°   **d)** 29.6°, 256.7°

10. a) $\sqrt{5}\sin(2x+26.57°)$

b) 0, 63.4°, 180°, 243.4°, 360°

c) $10+\sqrt{5}, 10-\sqrt{5}$

11. a) $\sqrt{17}\sin(x+1.326)$

b) 1.00, 5.77

c) $16, -1$

12. 79.7°, 153.4°

Summary exercise 3 page 61

1. 52.5°, 232.5°

2. 0.841, 2.09, 4.19, 5.44

3. 23.6°, 156.4°

6. b) $\frac{1}{1+\sqrt{2}}$ $(=\sqrt{2}-1)$

8. 1

10. a) $13\sin(\theta + 67.38°)$

b) i) 13   **ii)** $22.6°$

11. a) $\sqrt{41}\cos(\theta + 51.34°)$ **b)** $56.9°, 200.5°$ **c)** 20

12. a) $25\sin(\theta + 73.74°)$ **b)** $69.4° < \theta < 323.1°$

14. $0, 56.3°, 180°, 236.3°, 360°$

15. $120°, 240°, 300°$

17. a) $\sin\dfrac{\theta}{2} = \dfrac{t}{\sqrt{1+t^2}}, \cos\dfrac{\theta}{2} = \dfrac{1}{\sqrt{1+t^2}}$ **c)** $68.7°, 164.4°$

18. b) $x = 22.5°$

19. $30°, 150°$

20. a) $R = 25; \alpha = 73.74°$ **b)** $126.9°; 20.6°$

21. b) $27.2°, 152.8°, 207.2°, 332.8°$

Review exercise A – Pure 2 page 65

1. $x = 1.22$

2. a) $\alpha = 18.43°$ **b)** $34.6°, 163.8°, 214.6°, 343.8°$

3. b) $x = 10.9°, x = -169.1°$

4. 2.58

5. b) $x = 1.58$ or $1.585, x = 0$

6. a) $a = -4$ and $b = 1$ **b)** $(x - 3)$ and $(x + 1)$

7. a) $a = 2$ and $b = -6$ **b)** $x - 1$ and $x - 3$

8. $x = 1.87$

9. b) $\tan x = \dfrac{\sqrt{3}}{5}$

10. $\theta = 60°$

11. $x = 2.41$

12. a) $a = -17$ and $b = 12$ **b)** $x = -3$ and $x = \dfrac{4}{3}$

13. $A = 8.5, b = 1.6$

14. a) $a = 2$ **b) i)** $(x - 2)(2x + 3)(x - 1)$ **ii)** 6

15. $x < -1, x > 4$

16. a) $56.31°$ **b)** $211.7°$ **c)** $60; 8$

17. $x = \dfrac{3}{5}$

18. $x = 3.11$

19. a) $\tan\theta = 6$ **b) i)** $\dfrac{5}{7}$ **ii)** $-\dfrac{12}{35}$

20. $x < -2$ or $x > -\dfrac{3}{2}$

21. $\dfrac{1}{5} < x < \dfrac{7}{5}$

22. $-3 < x < 1$

23. $1.46, 0.631$

24. a) $\alpha = 8$ **b)** $3x^2 + 2x + 4$

25. $26.6°, 146.3°, 206.6°, 326.3°$

26. 4.11

27. $\dfrac{1}{3} < x < 1$

28. a) $\alpha = 11.31°$ **b)** $27.0°, 310.4°$

Review exercise A – Pure 3 page 68

1. $48.6°, 131.4°, 270°$

3. a) $a = -6$

b) $x^2 + x - 2 = 0$ has two real roots;
$x^2 - x + 3$ has no real roots

4. a) $\dfrac{-1}{x - 2} + \dfrac{3x - 1}{x^2 + 3}$ **b)** $\dfrac{1}{6} + \dfrac{5}{4}x + \dfrac{17}{72}x^2$

5. $x = 0.481$

6. b) $26.6°, 206.6°$

7. $x = 0.313$

8. $a = 4$

9. a) $-\dfrac{2}{3}$ and $\dfrac{4}{5}$ **b)** -0.161

10. $\theta = 105.9°$

11. $x < 2a$

12. a) $a = 1$ **b)** $x = -3, x = -1$

13. a) $(x + 2)(4x + 3)(3x - 2)$ **b)** -0.369

14. $-1 < x < -\dfrac{1}{7}$

16. 0.631 and -0.369

17. $1.83 < x < 1.95$

4 Differentiation

Skills check page 72

1. a) $12x^3 - \dfrac{1}{x^3}$ **b)** $8 + \dfrac{9}{\sqrt{x}}$

c) $x^4 - \dfrac{8}{x^5}$ **d)** $-\dfrac{5}{6}x^{\frac{3}{2}}$

2. a) $\dfrac{5}{(4 - x)^2}$ **b)** $\dfrac{-7}{(2x + 3)^{\frac{3}{2}}}$

c) $\dfrac{3 - 2x}{(x^2 - 3x + 7)^2}$ **d)** $\dfrac{-30}{\sqrt{(5x + 1)^5}}$

Exercise 4.1 page 75

1. a) $2e^{2x}$ **b)** $-5e^{-5x}$ **c)** $3e^{3x+9}$

d) $-e^{8-x}$ **e)** $-7e^x$ **f)** $10xe^{x^2}$

g) $6e^{2x-1}$ **h)** $-8xe^{6+x^2}$ **i)** $10xe^{5x^2}$

j) $\dfrac{-3e^{\frac{3}{x}}}{x^2}$ **k)** $\dfrac{3e^{\sqrt{x}}}{\sqrt{x}}$ **l)** $\dfrac{-2}{e^{2x}}$

2. **a)** $-2e^{-2x}$ **b)** e^{5+x} **c)** $27e^{3x}$

d) $-\dfrac{4e^{\sqrt{x}}}{\sqrt{x}}$ **e)** $3e^x - 8e^{4x}$ **f)** $-4e^{-8x}$

g) $4e^{4x} - e^{-x}$ **h)** $-6e^{-x} - 12e^{3x}$ **i)** $2e^{2x} + e^x$

j) $24e^{6x} - 12e^{3x}$ **k)** $\dfrac{-1}{e^x}$ **l)** $-4e^{-4x+2} - 14e^{5-2x}$

3. 10   **4.** $e - 3$

5. (0, 1)   **6.** Maximum T/P (0, −1)

Exercise 4.2 page 77

1. **a)** $\dfrac{1}{x}$ **b)** $\dfrac{2x}{x^2-1}$ **c)** $\dfrac{5}{3+5x}$

d) $\dfrac{3x^2}{x^3+2}$ **e)** $\dfrac{e^x}{e^x-7}$ **f)** $\dfrac{1}{2x}$

g) $\dfrac{24}{6x-3}$ **h)** $\dfrac{4x+1}{2x^2+x}$ **i)** $-\dfrac{2}{x}$

j) $\dfrac{3}{x}$ **k)** $-\dfrac{2}{x}$ **l)** $\dfrac{162}{9x-2}$

2. **a)** $\dfrac{8}{x}$ **b)** $\dfrac{3}{x}$ **c)** $\dfrac{2}{2x-1}$

d) $\dfrac{6}{x}$ **e)** $\dfrac{1}{2x+1}$ **f)** $\dfrac{10}{x}$

g) $\dfrac{3e^x-2}{3e^x-2x}$ **h)** $\dfrac{18(2x+3)}{x^2+3x}$ **i)** $\dfrac{10}{x(7x-2)}$

j) $\dfrac{4(x-4)}{x(x-8)}$ **k)** 5   **l)** $-\dfrac{6}{x+3}$

3. 1   **4.** $\dfrac{3}{5}$

5. (2, ln 6)   **6.** Minimum T/P (1, 1)

Exercise 4.3 page 79

1. **a)** $e^x(x+1)$ **b)** $2(6x^2 - x - 10)$

c) $\dfrac{x(3x^2+2)}{\sqrt{x^2+1}}$ **d)** $\dfrac{2+\ln x}{2\sqrt{x}}$

e) $2x(2x+5)^6 (9x+5)$ **f)** $\dfrac{(x-3)(5x-3)}{2\sqrt{x}}$

g) $y = 2e^{2x} (4x+3)$ **h)** $(x+3)^3(5x+3)$

i) $x + 2x \ln x$ **j)** $x(x+3)^2(5x+6)$

k) $e^x (2x-1)^3(2x+7)$ **l)** $\dfrac{1}{(x+1)^2}$

m) $\dfrac{3e^{\sqrt{x}}(2\sqrt{x}+x+4)}{\sqrt{x}}$ **n)** $4x^2 e^{2x} (2x+3)$

o) $2e^x \left(\dfrac{1}{x} + \ln 2x\right)$ **p)** $\dfrac{4}{(x+2)^2}$

q) $\dfrac{17}{(1-5x)^2}$ **r)** $2x[1 + \ln(x^2+1)]$

2. $\dfrac{1}{e}$

3. $y = 725x - 1200$

4. Proof

5. Minimum T/P $(-2, -e^{-2})$

Exercise 4.4 page 82

1. **a)** $\dfrac{6}{(x+3)^2}$ **b)** $\dfrac{e^{2x}(2x-1)}{x^2}$

c) $\dfrac{x(x+8)}{(x+4)^2}$ **d)** $\dfrac{1-x^2}{(1+x^2)^2}$

e) $\dfrac{6}{(2-x)^2}$ **f)** $y = \dfrac{3x(4e^x - 2xe^x - x)}{(2e^x - x)^2}$

g) $\dfrac{11}{(2x+1)^2}$ **h)** $\dfrac{-3x-7}{(x+1)^3}$

i) $\dfrac{6x(4-x)}{(2-x)^2}$ **j)** $\dfrac{4(2x+1)}{(1-x)^4}$

k) $\dfrac{x^2-4x-1}{(x-2)^2}$ **l)** $\dfrac{5x(2\ln x - 1)}{(\ln x)^2}$

m) $\dfrac{x^2(3-4x^2)}{(1-2x^2)^{\frac{3}{2}}}$ **n)** $\dfrac{4x^5 + 4x^3 - 6x^2 + 3}{(2x^2+1)^2}$

o) $\dfrac{2+\sqrt{x}}{2(1+\sqrt{x})^2}$ **p)** $\dfrac{2(x^2-7x-6)}{(2x-7)^2}$

q) $\dfrac{(3x-2)\sqrt{(x+1)^3}}{2x^2}$ **r)** $\dfrac{-2x(2x^2+3)}{(1+2x^2)^{\frac{3}{2}}}$

2. $x = 0$ and $x = 3$

3. $3x + 4y = 22$

4. $\dfrac{2}{(x-1)^3}$

Exercise 4.5 page 84

1. **a)** $5 \cos 5x$ **b)** $3 \sec^2 \dfrac{x}{2}$

c) $-2(x+1)\sin(x^2+2x)$ **d)** $8 \tan x \sec^2 x$

e) $3x^2 \cos(x^3 - 7)$ **f)** $20 \sec^2 4x$

g) $-3\cos^2 x \sin x$

h) $2\cos 2x \cos x - \sin 2x \sin x$

i) $x(3x \sec^2 3x + 2 \tan 3x)$

j) $\dfrac{-2x \sin 2x - 2\cos 2x}{x^3}$

k) $\sin^2 3x\,(9 \cos x \cos 3x - \sin x \sin 3x)$

l) $\dfrac{2 \sec^2 \sqrt{x}}{\sqrt{x}}$

3. −3

4. $x + y = \pi - 1$

6. $\left(\dfrac{\pi}{6}, \dfrac{3}{2}\right)$ max, $\left(\dfrac{\pi}{2}, 1\right)$ min, $\left(\dfrac{5\pi}{6}, \dfrac{3}{2}\right)$ max

7. Proof.   Minimum value is $\dfrac{\pi}{3} - \sqrt{3}$

Exercise 4.6 page 87

1. a) $\dfrac{x}{y-1}$ b) $\dfrac{\cos y - 1}{2y + x\sin y}$

    c) $-\dfrac{x}{4y}$ d) $\dfrac{5 - 3x^2 - y^2}{2xy}$

    e) $\dfrac{2x}{1 - 2y}$ f) $\dfrac{2y + 7}{3 + 2y - 2x}$

    g) $\dfrac{y(1 - \ln y)}{x}$ h) $\dfrac{y - x^2}{y^2 - x}$

    i) $\dfrac{x+1}{2-y}$ j) $\dfrac{2x^3 + xy^2}{2y^3 - yx^2}$

    k) $\dfrac{y\cos x - \sin y}{x\cos y - \sin x}$ l) $\dfrac{2}{e^{x+y}} - 1$

    m) $\dfrac{y^2 - 2xy}{x^2 - 2xy}$ n) $\dfrac{1 - 2x}{1 + 3y^2}$

    o) $\dfrac{-e^y}{xe^y + 2}$ p) $\dfrac{y\ln y}{y - x}$

    q) $\dfrac{3x^2}{3 - 2y + 3y^2}$ r) $\dfrac{2xy^3 + \sin x}{2 - \cos y - 3x^2 y^2}$

2. $-\dfrac{1}{3}$ and $\dfrac{1}{3}$

3. $x + y = 2$

Exercise 4.7 page 89

1. a) $\dfrac{4}{3t}$ b) $\dfrac{1}{3t^2}$

    c) $-\dfrac{2}{3}\cot t$ d) $\dfrac{1}{t}$

    e) $\dfrac{\sin t}{1 - \cos t}$ f) $2t(1 - t)^2$

    g) $\dfrac{1}{4t^{\frac{3}{2}}}$ h) $-\tan\theta$

    i) $\dfrac{3}{2t + 3}$ j) $-\csc\theta$

    k) $-\dfrac{b}{a}\cot\theta$ l) $\dfrac{3t^2 - 4}{2t}$

6. 1

7. $9y - x + 80 = 0$

9. 0 or $\dfrac{4}{3}$

10. a) $\dfrac{-2\cos 3t}{5\sin 3t}$ b) $(0, 2)$

Summary exercise 4 page 90

1. a) $-5e^{-5x}$ b) $\dfrac{8x}{4x^2 + 5}$

    c) $(1 + 4x)(1 + x)^2$ d) $\dfrac{-1}{(2x - 3)^2}$

    e) $\dfrac{\cos x - \sin x}{e^x}$ f) $(x + 1)(x - 2)^2(5x - 1)$

    g) $\dfrac{x(x^2 - 2)}{(x^2 - 1)^{\frac{3}{2}}}$ h) $\dfrac{9}{3x - 4}$

    i) $\dfrac{3x^2 - y^2}{2xy - 1}$ j) 14

    k) $\dfrac{e^x(x\ln x + \ln x - 1)}{(\ln x)^2}$ l) $\dfrac{3x^3 + 2x}{\sqrt{(1 + x^2)}}$

    m) $-\dfrac{1}{3t^2}$ n) $y = -\dfrac{e^{\sqrt{x}}}{\sqrt{x}}$

    o) $\dfrac{2x^2 + y^2}{y^2 - 2xy}$ p) $\dfrac{4e^{2x} - 9}{e^x}$

    q) $2e^{\tan^2 x}\tan x \sec^2 x$ r) $xe^{3x-2}(3x + 2)$

    s) $3\cot\theta$ t) $\dfrac{3}{3x + 1}$

    u) $\dfrac{12t^2 - 1}{2(t + 2)^3}$ v) $1 + \ln x$

    w) $\dfrac{1 + 3x^2}{2\sqrt{x}(1 - x^2)^2}$ x) $\dfrac{3x^2 + y^2}{3y^2 - 2xy}$

2. $\left(-1, \dfrac{-3}{e^2}\right)$

4. a) $\dfrac{t - 1}{t + 1}$ b) $(4, 0)$

5. $y = -26x + 60$ 6. $-\dfrac{e}{2}$

7. $5x - y + 9 = 0$ 8. $3x - 2y - 1 = 0$

9. $\dfrac{3\pi}{4}$ 12. $x - 2y + 4 = 0$

13. $\dfrac{25}{8}$ 14. $\left(\dfrac{1}{3e}, -\dfrac{1}{3e}\right)$

19. a) $\dfrac{5 - 2x}{3 - 2y}$ b) Minimum $= -\dfrac{3}{2}$ and maximum $= \dfrac{9}{2}$

20. $(49, -9)$ 21. $1.83 < x < 1.95$ 22. b) 0.5

5 Integration

Skills check page 94

1. a) $2e^{2x+1} - e^{-x}$ b) $4\cos 4x + 6\sin 2x$ c) $\dfrac{3}{(3x + 7)}$

Exercise 5.1 page 97

1. a) $\dfrac{1}{3}e^{3x-1} + c$ b) $\dfrac{1}{4}e^{4x+1} + c$ c) $2e^{0.5x+1} + c$

    d) $\dfrac{1}{\pi}e^{\pi x+3} + c$ e) $\dfrac{1}{4}e^{4(x+1)}$ f) $2e^{0.5x-0.5} + c$

    g) $-e^{2-x} + c$ h) $-\dfrac{1}{2}e^{3-2x} + c$ i) $-2e^{0.5-0.5x} + c$

2. a) $\frac{1}{3}e^3 - \frac{1}{3}$ b) $\frac{1}{4}e - \frac{1}{4}$ c) $e - 1$

   d) $e^4 - 1$ e) $4e^3 - 4$ f) $\frac{1}{2}e^8 - \frac{1}{2}$

   g) $1 - \frac{1}{e}$ h) $1 - \frac{1}{e^2}$ i) $2e^{0.5} - 2e^{-0.5}$

3. a) 1   b) 8   c) 1   d) 80

   e) $16 - e^2$ f) $\frac{4}{5}$ g) 1.82   h) $\frac{7}{4\ln 2}$

   i) $\frac{242}{81\ln 3}$

4. $e^3 - 1$ 5. $\frac{1}{2}(1 - e^{-4})$

6. 5.52   7. $1 - e^{-k}; 1$

8. a) $\frac{1}{3}$ b) $\frac{1}{4}e^{-4} = 0.0046$ c) 1

   d) 1   e) $2e^3 = 40.2$ f) $2e = 5.44$

9. $\frac{\pi}{2}(1 - e^{-6}) = 1.57$

Exercise 5.2 page 101

1. a) $\frac{1}{2}\ln|2x - 1| + c; \ x \neq \frac{1}{2}$

   b) $\ln|2x - 1| + c; \ x \neq \frac{1}{2}$

   c) $\frac{1}{7}\ln|7x + 3| + c; \ x \neq -\frac{3}{7}$

   d) $-5\ln|5 - 2x| + c; \ x \neq \frac{5}{2}$

   e) $2\ln|1 - 3x| + c; \ x \neq \frac{1}{3}$

   f) $-\frac{1}{3}\ln|6x + 5| + c; \ x \neq -\frac{5}{6}$

   g) $-3\ln|x + 1| + c; \ x \neq -1$

   h) $-\ln|7 - x| + c; \ x \neq 7$

   i) $2\ln|x + e| + c; \ x \neq -e$

   j) $-\frac{1}{4}\ln|2x + 1| + c; \ x \neq -\frac{1}{2}$

   k) $-\frac{4}{3}\ln|9x + 16| + c; \ x \neq -\frac{16}{9}$

   l) $-\frac{1}{5}\ln|10x - 3| + c; \ x \neq \frac{3}{10}$

2. a) $\frac{1}{2}\ln\left(\frac{5}{3}\right)$ b) $\ln\left(\frac{13}{9}\right)$ c) $\ln\left(\frac{10}{3}\right)$

   d) $2\ln\left(\frac{5}{3}\right)$ e) $\frac{9}{2}\ln 2$ f) $7\ln\left(\frac{3}{5}\right)$

   g) $2\ln\left(\frac{3}{5}\right)$ h) $2\ln\left(\frac{15}{19}\right)$ i) $\ln 16 = 2.77$

3. $\ln 2$

4. $2\ln\left(\frac{3}{5}\right) = -1.02$ 5. $\frac{1}{2}\ln|21| = 1.52$

Exercise 5.3 page 104

1. a) $\frac{1}{2}\sin 2x + c$ b) $-\frac{1}{2}\cos\left(2x + \frac{\pi}{4}\right) + c$

   c) $\frac{1}{3}\tan\left(3x + \frac{\pi}{12}\right) + c$ d) $-\sin(\pi - x) + c$

   e) $-4\sin\left(\frac{\pi}{2} - \frac{x}{4}\right) + c$ f) $\frac{1}{2}\cos\left(2x - \frac{\pi}{4}\right) + c$

   g) $-\frac{1}{5}\tan\left(\frac{\pi}{12} - 5x\right) + c$ h) $\cos\left(\frac{1}{2}x\right) + c$

   i) $-3\sin\left(4 - \frac{1}{3}x\right) + c$

2. a) $\frac{1}{2}$ b) $\frac{1}{\sqrt{2}}$ c) 0.488

   d) 0   e) -1.5 f) $\frac{\sqrt{2}}{3}$

   g) 0.770   h) -0.510 i) 1.04

3. $y = \tan x + 1$ 4. 2

5. 1   6. $\frac{1}{2\sqrt{3}} = 0.289$

Exercise 5.4 page 109

1. a) $\frac{1}{8}\sin 4x + \frac{1}{2}x + c$ b) $\frac{1}{2}x - \frac{1}{12}\sin 6x + c$

   c) $\frac{1}{2}\tan 2x + c$ d) $-\frac{1}{4}\cos 2x + c$

   e) $\frac{1}{8}x - \frac{1}{32}\sin 4x + c$ f) $x - \frac{1}{2}\cos 2x + c$

2. a) $\frac{\pi}{8}$ b) $\frac{\pi}{4}$ c) $\frac{1}{2\sqrt{3}}$ d) $\frac{3\pi}{4}$

   e) $3\pi + 8 = 17.4$ f) $\frac{\pi}{2} - 1 = 0.571$

3. $\Rightarrow y = \frac{1}{2}x + \frac{1}{4}\sin 2x$ 4. 0.631

5. $\frac{1}{2}$ 6. b) $-\frac{1}{32}$

7. b) $4x^2 \sin 2x$ 8. b) $\frac{\pi^2 - 4}{8}$

9. b) $\left(\frac{\pi}{12} - \frac{\sqrt{3}}{16}\right) - (0 - 0) = 0.154$

10. b) $\left(\frac{\pi}{12} + \frac{\sqrt{3}}{16}\right) - (0 - 0) = 0.370$ c) $\frac{\sqrt{3}}{2} - 0 = 0.866$

Exercise 5.5 page 115

1. a) 0.335   b) 1.42   c) 0.695

   d) 1.46   e) 1.13   f) 0.429

2. a) 0.333   b) 1.53   c) 0.663

   d) 1.50   e) 1.11   f) 0.416

3. a) 2.97

b) The approximation in **i)** is an overestimate as the gradient of the curve is increasing over the interval so the chords are always above the curve and give a (slightly) greater area in each strip than the curve itself would.

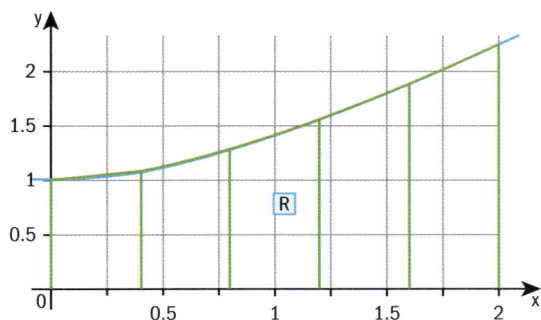

4. a) 0.590

b) The approximation in **i)** is an underestimate as the chords are always below the curve and give a (slightly) greater area in each strip than the curve itself would.

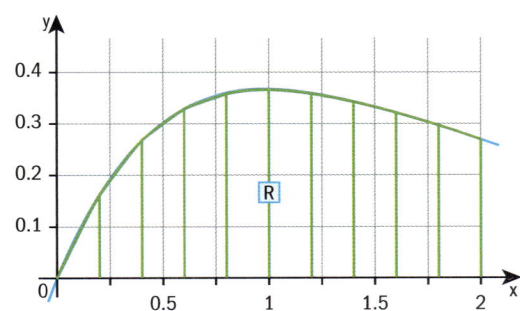

5. a) i) 1.17 (3 s.f.)   **ii)** 1.10 (3 s.f.)

b) 1.10 (3 s.f.)

Summary exercise 5 page 117

1. a) $\frac{1}{2}e^{2x-3} + c$ **b)** $\frac{1}{2}\ln|6x+2| + c$

c) $-\frac{1}{4}\cos 4x + c$ **d)** $\frac{x}{2} - \frac{\sin 12x}{24} + c$

e) $-2e^{\left(1-\frac{1}{2}x\right)} + c$ **f)** $4\ln|3x-4| + c$

g) $\frac{5}{2}x + \cos 2x + \frac{3}{4}\sin 2x + c$

h) $-2\ln|3x+5| + c$ **i)** $-\frac{1}{3}\sin\left(3x + \frac{\pi}{3}\right) + c$

2. a) $\frac{1}{5}e^{10} - \frac{1}{5}$ **b)** $\ln 3$ **c**

d) $\frac{11\pi}{8} - 3$ **e)** $2\cos\left(\frac{\pi}{8}\right) - \sqrt{2}$ **f)** $\frac{5}{2}$

g) $-\frac{\sqrt{3}}{4}$ **h)** $-\frac{1}{2}(e^4 - e^{-2})$ **i)** $\frac{5}{3}\ln\frac{1}{4}$

3. $\frac{1}{2}e - \frac{1}{2}e^{-3}$

4. $\ln 1.5$

5. $y = \frac{1}{2}x - \frac{1}{4}\sin 2x - \frac{\pi}{8}$

6. $\ln\left(\frac{5\sqrt{7}}{2}\right) = 1.89$

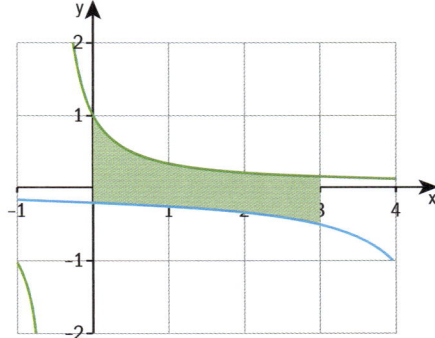

7. 5.29

8. a) $\frac{1}{2}$ **b)** $\frac{2}{\sqrt{e}}$ **c)** $\frac{2}{\sqrt{e}}$

9. $4\pi(e^{-4} - e^{-8}) = 0.226$

10. b) $\frac{3\sqrt{3}}{14} = 0.371$

11. b) 2.43

12. b) $\sqrt{3}$

13. a) 3.61

b) Underestimates the true value as the curve always lies above the trapezium.

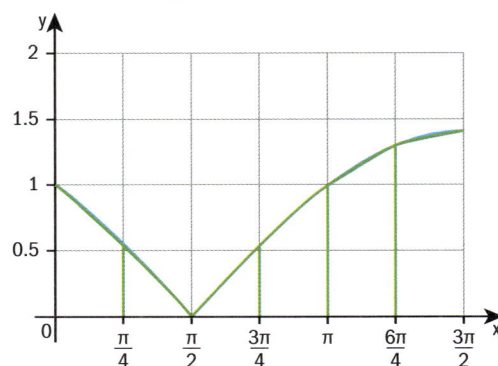

14. a) An integral gives the *signed* area – so integrating between 0 and 2 would subtract the area below the axis – between 0 and 1 – from the area above – between 1 and 2.

b) 0.824

> **Note:** The value of the function at 0 is taken as zero in using the trapezium rule to estimate R. To do the integral analytically would require it to be treated as an improper integral.

15. a) $x = \frac{1}{2}; y = 2e$ **b)** 15.4   **c)** $\frac{1}{4}\sqrt{3}$

16. a) $2\sec^2 2x$ **b)** $\frac{1}{2}\sqrt{3} - \frac{1}{6}\pi$ **c)** $\frac{1}{4}\sqrt{3}$

6 Numerical solutions of equations

Skills check page 120

1. a) 1.80   **b)** 2.24   **c)** 1

2. a) 2.31   **b)** 2.62   **c)** Not defined

3. $x = \frac{1}{2}\left(\ln(y+3) - 1\right)$

Exercise 6.1 page 125

1. Root is 1.2 to 1 d.p.   **2.** Root is 1.0 to 1 d.p.

3. Root is 1.0 to 1 d.p.   **4.** Root is 0.9 to 1 d.p

5. a) Root is 0.3 to 1 d.p.

    b) Root is −1.9 to 1 d.p. and 1.5 to 1 d.p.

6. Root is 0.77 to 2 d.p.   **7.** Root is 0.87 to 2 d.p.

8. Root is 3.96 to 2 d.p.

Exercise 6.2 page 130

1. a) $x^2 + 3x - 1 = 0$ **b)** $x^4 + 5x^3 + 9x^2 - 2 = 0$

   c) $e^{-2x} - x = 0$ **d)** $2x - e^{3-x} = 0$

   e) $\cos x + \frac{\pi}{4} - x = 0$ **f)** $x^{10} - x^6 - 3 = 0$

2. The lists below should not be taken as exhaustive of all possibilities:

a) $x = \sqrt[4]{3x - 2}$

$x = \frac{x^4 + 2}{3}$

$x = \frac{3x - 2}{x^3}$

$x = \sqrt{\frac{3x - 2}{x^2}}$

$x = \sqrt[3]{3 - \frac{2}{x}}$

b) $x = \sqrt[7]{4x^3 - 1}$

$x = \frac{x^7 + 1}{4x^2}$

$x = \sqrt[3]{\frac{x^7 + 1}{4}}$

$x = \sqrt[3]{\frac{1}{4 - x^4}}$

c) $x = \frac{x^2 + 1}{x^4}$

$x = \sqrt[5]{x^2 + 1}$

$x = \frac{x^5 - 1}{x} = x^4 - \frac{1}{x}$

$x = \sqrt{x^5 - 1}$

$x = \frac{1}{x(x^3 - 1)}$

$x = \sqrt{\frac{1}{(x^3 - 1)}}$

d) $x = \frac{x^2 - e^{-x}}{3}$

$x = \frac{3x + e^{-x}}{x}$

$x = \sqrt{3x + e^{-x}}$

$x = -\ln\left(x^2 - 3x\right)$

$x = \frac{e^{-x}}{x - 3}$

e) $x = \frac{1}{3}e^{x^2}$

$x = \sqrt{\ln 3x}$

$x = \frac{\ln 3x}{x}$

f) $x = \tan^{-1}\left(\frac{1}{x^2 - 3}\right)$

$x = \frac{\cot x + 3}{x}$

$x = \sqrt{\cot x + 3}$

3. a) $x_1 = 0.5$

$x_2 = 0.28571$

$x_3 = 0.30434$

$x_4 = 0.30263$

$x_5 = 0.30278$

b) $x_1 = 0.5$

$x_2 = 0.41649$

$x_3 = 0.42134$

$x_4 = 0.42100$

$x_5 = 0.42103$

c) $x_1 = 0.5$

$x_2 = 0.36787$

$x_3 = 0.47914$

$x_4 = 0.38355$

$x_5 = 0.46435$

d) $x_1 = 1.75$

$x_2 = 1.74723$

$x_3 = 1.74881$

$x_4 = 1.74791$

$x_5 = 1.74843$

e) $x_1 = 1.1$

$x_2 = 1.23899$

$x_3 = 1.11114$

$x_4 = 1.22903$

$x_5 = 1.12054$

f) $x_1 = 1.2$

$x_2 = 1.19595$

$x_3 = 1.19474$

$x_4 = 1.19439$

$x_5 = 1.19428$

4. c) root is 1.47 to 2 decimal places

5. c) $x_2 = 2.42037$

$x_3 = 2.38492$

$x_4 = 2.36911$

$x_5 = 2.36205$

6. b) $\theta_2 = 1.70670\ldots$

$\theta_3 = 1.74946..$

$\theta_4 = 1.70724\ldots$

$\theta_5 = 1.74893\ldots$

c) The odd numbered values are decreasing and the even numbered values are increasing – moving towards one another, eventually will converge to the root.

7. b) 1.37 to 2 d.p.

8. b) $x_{n+1} = \cos x_n$

$x_1 = 0.75$

$x_2 = 0.73168\ldots$

$x_3 = 0.74404..$

$x_4 = 0.73573\ldots$

$x_5 = 0.74133\ldots$

so this iterative relationship converges to 0.74 (2 d.p.)

$x_{n+1} = \cos^{-1} x_n$

$x_1 = 0.75$

$x_2 = 0.72273\ldots$

$x_3 = 0.76304..$

$x_4 = 0.70278\ldots$

$x_5 = 0.79149\ldots$

so this iterative relationship does not converge

9. b) i) $x_{n+1} = \sqrt{\dfrac{5}{x_n^2 - 1}}$

$x_1 = 1.7$

$x_2 = 1.6265\ldots$

$x_3 = 1.74315..$

$x_4 = 1.56610\ldots$

$x_5 = 1.85523\ldots$

so this iterative relationship does not converge

ii) $x_{n+1} = \dfrac{x_n^2 + 5}{x_n^3}$

$x_1 = 1.7$

$x_2 = 1.60594\ldots$

$x_3 = 1.82988..$

$x_4 = 1.36249\ldots$

$x_5 = 2.71075\ldots$

so this iterative relationship does not converge either

iii) $x_{n-1} = \sqrt[4]{x_n^2 + 5}$

$x_1 = 1.7$

$x_2 = 1.67598\ldots$

$x_3 = 1.67165..$

$x_4 = 1.67088\ldots$

$x_5 = 1.67074\ldots$

so the root is 1.67 correct to 2 decimal places

10. c) −0.68 (2 d.p.)   **11. d)** 0.61 (2 d.p.)

Summary exercise 6 page 135

1. 0.3 (1 d.p.)   **2.** 2.6 (1 d.p.)

3. i) 1.3 (1 d.p.)   **ii)** 1.3 (1 d.p.)

4. 1.3 (1 d.p.)

5. i) 0   **ii)** 0

6. These are not exhaustive lists of rearrangements.

i) $x = \sqrt[4]{4x^2 + 3}$

$x = \dfrac{4x^2 + 3}{x^3}$

$x = \sqrt{\dfrac{4x^2 + 3}{x^2}}$

ii) $x = \dfrac{x^5 + 2}{3x}$

$x = \sqrt{\dfrac{x^5 + 2}{3}}$

$x = \dfrac{3x^2 - 2}{x^4}$

$x = \sqrt[5]{3x^2 - 2}$

7.
i) $x_1 = 0.5$

$x_2 = 0.22222\ldots$

$x_3 = 0.20930\ldots$

$x_4 = 0.20873\ldots$

$x_5 = 0.20871\ldots$

ii) $x_1 = 0.5$

$x_2 = 0.34015\ldots$

$x_3 = 0.33662\ldots$

$x_4 = 0.33659\ldots$

$x_5 = 0.33659\ldots$

8. 2.87

10. ii) 1.36 radians

11. ii) 17.4

12. iv) 1.13

13. i) 2.78

ii) $\alpha = \sqrt[4]{60}$

14. $x = 0.80$

15. ii) 5.20

16. iii) 1.03

17. iv) 1.90

18. $\dfrac{1}{6}\pi - \dfrac{1}{4}\sqrt{3}$

Review exercise B – pure 2 page 139

1. i) 4

ii) $\dfrac{2}{25}$

3. iv) 1.26

4. $(1, 6)$

5. iii) 1.56

6. Maximum at $x = 0$, and minimum at $x = \ln 4$

7. iv) 0.76

8. ii) $(\ln 3, -2)$

9. ii) 0.678

11. iii) $(e^{-6}, 4e^{-2} + 3)$

13. ii) $x\sin x$

iii) 1

14. ii) $2x - 5y + 8 = 0$

15. iv) 0.27

16. iii) $-\dfrac{1}{2}\cot x + c$

17. i) $\left(\dfrac{1}{e}, -\dfrac{1}{e}\right)$

ii) Minimum

18. i) $p + e^{-p} - 1$

iii) 1.84

19. i) $(1, 1)$ and $(1, -3)$

ii) $2x + y + 1 = 0$

20. i) 1.21

ii) Over-estimate

21. i) $\left(-\dfrac{1}{2}, -\dfrac{1}{(2e)}\right)$

iii) 1.35

22. i) $y = 3x - 3$

ii) **a)** $A = 4$

b) $\dfrac{4}{3}\ln(3x - 2)$

23. $(1, -e)$

24. iv) 0.58

25. iii) $\sqrt{2}$

26. ii) $(2, 1), (-2, -1)$

Review exercise B – pure 3 page 144

2. i) $-\dfrac{1}{2}$

ii) $-\dfrac{5}{6}$

3. ii) 0.76

4. $k = \dfrac{1}{2}(e^x + 1)$

5. $y = x$

6. iii) 1.276

iv) -2.24 and 0.906

7. iv) 1.13

8. i) $\dfrac{2\cos 2x}{1 + \sin 2x}$

ii) $\dfrac{x\sec^2 x - \tan x}{x^2}$

9. ii) $k = 5$ and $c = 68$

10. iii) 1.38

11. $0.365, 1.206$

7 Further algebra

Skills check page 148

1. $A = 1$ $\quad B = -4$ $\quad C = -9$

2. $A = 2$ $\quad B = -4$ $\quad C = 0$

3. $A = -7$ $\quad B = \dfrac{2}{3}$ $\quad C = -2$

4. $A = 3$ $\quad B = \dfrac{1}{2}$ $\quad C = 0$

5. $A = 3$ $\quad B = 0$ $\quad C = -1$

Exercise 7.1A page 150

1. $\dfrac{3}{x - 1} - \dfrac{2}{x - 2}$

2. $\dfrac{1}{5(x + 1)} + \dfrac{4}{5(x - 4)}$

3. $\dfrac{1}{x + 3} + \dfrac{2}{x - 1}$

4. $\dfrac{-1}{x} + \dfrac{4}{2x + 1}$

5. $\dfrac{1}{x + 2} - \dfrac{2}{x + 3} + \dfrac{3}{x - 3}$

Exercise 7.1B page 152

1. $\dfrac{1}{x - 2} + \dfrac{2}{(x - 2)^2}$

2. $\dfrac{3}{x} - \dfrac{9}{3x - 1} + \dfrac{9}{(3x - 1)^2}$

3. $\dfrac{1}{x - 1} - \dfrac{1}{x - 2} + \dfrac{2}{(x - 2)^2}$

4. $\dfrac{1}{4(x + 1)} - \dfrac{1}{4(x - 1)} + \dfrac{1}{2(x - 1)^2}$

5. $\dfrac{1}{x} - \dfrac{1}{x - 1} + \dfrac{1}{(x - 1)^2}$

Exercise 7.1C page 153

1. $\dfrac{5}{x^2 + 2} - \dfrac{2}{1 - x}$

2. $\dfrac{1}{x} - \dfrac{3x}{2 + x^2}$

3. $\dfrac{1 - x}{x^2 + 4} + \dfrac{1}{x + 1}$

4. $\dfrac{2}{1 - 2x} + \dfrac{x + 1}{1 + x^2}$

5. $\dfrac{3 - 2x}{3 - x^2} + \dfrac{2}{x - 2}$

Exercise 7.1D page 155

1. $1 - \dfrac{9}{x - 2} + \dfrac{14}{x - 3}$

2. $1 + \dfrac{1}{2(x + 1)} - \dfrac{1}{2(x - 1)}$

3. $3 + \dfrac{5x}{x^2 + 4} - \dfrac{1}{x - 3}$

4. $-1 - \dfrac{2}{x + 3} + \dfrac{4}{(x + 3)^2}$

5. $A = -2, B = 3, C = -1$

Exercise 7.1E page 156

1. $\dfrac{2}{x+3} - \dfrac{1}{x+1}$

2. $1 + \dfrac{8}{x+1} - \dfrac{15}{x+2}$

3. $\dfrac{2}{x-5} + \dfrac{3}{(x-5)^2}$

4. $\dfrac{5}{1-5x} + \dfrac{x-2}{1+x^2}$

5. $\dfrac{3}{x+1} - \dfrac{2}{x-2}$

6. $3 - \dfrac{1}{x} + \dfrac{15}{x-3}$

7. $\dfrac{1}{x} + \dfrac{2}{x-1} + \dfrac{4}{(x-1)^2}$

8. $\dfrac{7x}{2(x^2+2)} - \dfrac{3}{2x}$

9. $\dfrac{-1}{x} + \dfrac{2}{x+1}$

10. $1 + \dfrac{5}{2(x-3)} - \dfrac{1}{2(x+3)}$

11. $\dfrac{5}{x-2} + \dfrac{3}{x+1} - \dfrac{2}{(x+1)^2}$

12. $-\dfrac{1}{x-1} + \dfrac{x+2}{x^2+1}$

13. $\dfrac{1}{x-1} - \dfrac{1}{x-2} + \dfrac{2}{(x-2)^2}$

14. $1 + \dfrac{1}{x-1} - \dfrac{1}{x+1}$

15. $\dfrac{1}{x+2} + \dfrac{x}{x^2+3}$

Exercise 7.2 page 158

1. **a)** $1 + 4x + 10x^2 + 20x^3$ for $-1 < x < 1$

   b) $1 - 15x + 135x^2 - 945x^3 + \dots$ for $-\frac{1}{3} < x < \frac{1}{3}$

   c) $1 + \frac{1}{3}x - \frac{1}{9}x^2 + \frac{5}{81}x^3 + \dots$ for $-1 < x < 1$

   d) $1 - \frac{3}{2}x + \frac{3}{2}x^2 - \frac{5}{4}x^3 + \dots$ for $-2 < x < 2$

   e) $x + x^2 + x^3$ for $-1 < x < 1$

   f) $1 + 10x + 75x^2 + 500x^3$ for $-\frac{1}{5} < x < \frac{1}{5}$

2. $1 - \frac{1}{2}x^2 - \frac{3}{8}x^4 + \dots$ for $-1 < x < 1$

3. **a)** $1 + 2x + 3x^2 + 4x^3$ for $-1 < x < 1$ **b)** $50x^{49}$

4. **a)** $1 - x - \frac{1}{2}x^2 - \frac{1}{2}x^3 - \frac{5}{8}x^4$ for $-\frac{1}{2} < x < \frac{1}{2}$

   b) 0.8944

5. $1 - \frac{3}{2}x + \frac{7}{8}x^2 + \dots$ for $-1 < x < 1$

6. $1 - 8x^2 + 48x^4 - 256x^6 + \dots$ for $-\frac{1}{2} < x < \frac{1}{2}$

7. $1 + 4x + 7x^2 + 10x^3$ for $-1 < x < 1$

8. **a)** $1 + \frac{1}{2}x - \frac{1}{8}x^2 + \frac{1}{16}x^3 + \dots$ for $-1 < x < 1$

   b) 1.0392

9. **a)** $a = -1, n = -4$ **b)** 20

10. $1 - 3x - x^2 + 3x^3 + x^4$ for $-1 < x < 1$

11. **a)** $k = -\frac{3}{2}, n = -8$ **b)** $405 x^3$

12. $1 - 2x - 2x^2 + 4x^3 + 4x^4 + \dots$

Exercise 7.3 page 160

1. **a)** $\frac{1}{5}(1-2x)^{-\frac{1}{2}}$ **b)** $\left(1 - \frac{x}{3}\right)^{-2}$ **c)** $3\left(1 + \frac{4x}{27}\right)^{\frac{1}{3}}$

2. $\frac{1}{32} - \frac{5}{64}x + \frac{15}{128}x^2 - \frac{35}{256}x^3 \dots$ for $-2 < x < 2$

3. $2 - \frac{1}{4}x + \frac{1}{64}x^2 + \dots$ for $-4 < x < 4$

4. $\frac{1}{27}$

5. $\frac{1}{2} - \frac{x^2}{4} + \frac{x^4}{8} - \frac{x^6}{16} + \dots$ for $-\sqrt{2} < x < \sqrt{2}$

6. **a)** $\frac{1}{16} + \frac{1}{32}x + \frac{3}{256}x^2 + \frac{1}{256}x^3 \dots$ for $-4 < x < 4$

   b) $-\frac{1}{256}$

7. $3 + \frac{53}{6}x - \frac{109}{216}x^2 - \frac{55}{3888}x^3 + \dots$

8. $\frac{1}{2} - \frac{1}{8}x^2 + \frac{3}{64}x^4 - \frac{5x^2}{256} + \dots$ for $-\sqrt{2} < x < \sqrt{2}$

9. **a)** $p = -24, q = 108$ **b)** $1728x^3$

10. $k = 2$

Exercise 7.4 page 162

1. **a)** $\dfrac{2}{(1-x)} + \dfrac{3}{(1+3x)}$

   b) $5 - 7x + 29x^2$ for $-\frac{1}{3} < x < \frac{1}{3}$

2. **a)** $\dfrac{1}{(1+2x)} - \dfrac{1}{(1+x)^2}$ **b)** coeff: -4

3. **a)** $\dfrac{3}{2(x-1)} - \dfrac{1}{2(x+1)}$

   b) $-2 - x - 2x^2 - x^3$ for $-1 < x < 1$

4. **a)** $\dfrac{2}{(1-x)} + \dfrac{1}{(1-x)^2} + \dfrac{2}{(2-x)}$

   b) $4 + \frac{9}{2}x + \frac{21}{4}x^2$ for $-1 < x < 1$

5. **a)** $-\dfrac{1}{(1-x)} - \dfrac{1}{(x-2)}$ **b)** $-\frac{7}{8}$

6. **a)** $\dfrac{1}{2(1-x)} + \dfrac{x+1}{2(1+x^2)}$

   b) $1 + x + x^4$ for $-1 < x < 1$

7. **a)** $\dfrac{2}{(1-2x)} + \dfrac{1+x}{(1+x^2)}$

   b) $3 + 5x + 7x^2 + 15x^3 + \dots$

Summary exercise 7 page 163

1. $\dfrac{4}{x-3} - \dfrac{1}{x+2}$ 2. $\dfrac{3x+1}{2(x^2+1)} - \dfrac{3}{2(x+1)}$

3. $1 + \dfrac{3}{x-2} + \dfrac{1}{x+1}$ 4. $\dfrac{5}{9(x-1)} + \dfrac{4}{3(x-1)^2} - \dfrac{5}{9(x+2)}$

5. $\dfrac{3}{5(x-2)} - \dfrac{1}{2(x+3)} - \dfrac{1}{2x-1}$

6. $\dfrac{1}{x-4} - \dfrac{6}{x+4} + \dfrac{4}{(x+4)^2}$ 7. $\dfrac{3x-1}{x^2-2} + \dfrac{4}{x+3}$

8. $A = 5, B = -3, C = 2, D = -2$

9. $\frac{1}{4} + \frac{5}{4}x + \frac{75}{16}x^2 + \frac{125}{8}x^3 + \dots$ for $-\frac{2}{5} < x < \frac{2}{5}$

10. a) $1 - \frac{1}{2}x - \frac{1}{8}x^2 - \frac{1}{16}x^3 + \dots$ for $-1 < x < 1$

    b) 0.9487

11. $\frac{1}{64} - \frac{9}{256}x + \frac{27}{512}x^2$ for $-\frac{4}{3} < x < \frac{4}{3}$

12. a) $\frac{1}{2} - \frac{1}{8}x - \frac{3}{64}x^2 - \frac{5}{256}x^3 + \dots$ for $-2 < x < 2$

    b) 0.49875467

13. $2 - 3x - \frac{1}{2}x^3 + \dots;$ for $-\frac{1}{2} < x < \frac{1}{2}$

14. a) $\frac{3}{(1 + x)} - \frac{1}{1 + x}$

    b) $\frac{5}{2} - \frac{11}{4}x - \frac{23}{8}x^2 + \frac{47}{16}x^3$ for $-1 < x < 1$

15. a) $\frac{2}{(1 + x)} - \frac{1}{(1 - x)} + \frac{3}{(1 - x)^2}$

    b) $4 + 3x + 10x^2$ for $-1 < x < 1$

16. i) $\frac{1}{1 - x} + \frac{2}{1 + 2x} + \frac{-4}{2 + x}$ **ii)** $1 - 2x + \frac{17}{2}x^2$

17. i) $(1 + x) - (1 - x)$

18. i) $\frac{2}{x + 2} + \frac{x - 1}{x^2 + 1}$ **ii)** $\frac{1}{2}x + \frac{5}{4}x^2 - \frac{9}{8}x^3$

8 Further integration

Skills check page 166

1. a) $1 + \frac{1}{x^2 + 1} + \frac{3}{x - 5}$

    b) $\frac{-5}{9(x + 2)} + \frac{5}{9(x - 1)} + \frac{4}{3(x - 1)^2}$

    c) $\frac{3x - 1}{x^2 - 2} + \frac{4}{x + 3}$

2. $\frac{1}{2x + 1}$

Exercise 8.1 page 171

1. a) $\ln\left|\frac{2x - 1}{x + 3}\right|$ **b)** $\ln\left|\frac{x + 5}{x + 7}\right|$ **c)** $\ln\left|\frac{x - 2}{x + 5}\right|$

    d) $\ln\left|\frac{\sqrt{2x + 3}}{x + 5}\right|$ **e)** $\frac{1}{3}\ln|2 + 3x| - \frac{1}{4}\ln|1 - 4x|$

    f) $\ln\left|\left(x - 2\sqrt{3 - 2x}\right)\right|$ **g)** $2\ln\left(\frac{15}{14}\right)$

    h) $\ln 6$ **i)** $\ln\left(\frac{98}{9}\right)$

2. a) $\ln\left|\frac{x}{x - 1}\right| - \frac{1}{(x - 1)} + c$ **b)** $\ln\left|\frac{x}{x + 1}\right| - \frac{1}{(x + 1)} + c$

    c) $\ln\left|\frac{x + 1}{x - 3}\right| - \frac{12}{(x - 3)} + c$ **d)** -0.493

    e) $\ln\left(\frac{6}{7}\right) + \frac{2}{7} = 0.132$ **f)** $= \ln\left(\frac{2}{3}\right) + 1 = 0.595$

3. a) $x + \ln\left|\frac{(x - 2)^2}{(x + 1)^3}\right| + c$ **b)** $2x + \ln\left|\frac{(x - 3)^4}{(x + 1)}\right| + c$

    c) $1 + \ln\left(\frac{2^2 \times 6^4}{7^4}\right) = 1.77$ **d)** $5 + \ln\left(\frac{4^2 \times \sqrt{3}}{3^2}\right) = 6.13$

4. $\ln\left(\frac{3}{2}\right)$ **5.** $2 + \frac{1}{2}\ln\left(\frac{5}{12}\right)$

6. $\ln\left(\frac{4}{3}\right)$ **7.** 4.61

Exercise 8.2 page 174

1. a) $2\ln|x^3 + 3| + c$ **b)** $\frac{1}{2}\ln|1 + \sin 2x| + c$

    c) $\frac{1}{2}\ln|x^2 - 4x + 3| + c$ **d)** $\frac{1}{6}\ln|4 + 2e^{3x}| + c$

    e) $\frac{1}{2}\ln|x^2 + 4| + c$ **f)** $\ln|\sin x| + c$

    g) $\frac{1}{2}\ln 1.5(-0)$ **h)** $\frac{1}{2}\ln\left(\frac{1 + e^2}{2}\right)$

    i) $\ln 2$

2. 0

3. b) $\ln|x + \sin^2 x| + c$

4. b) 2.82

5. b) 1.38

6. b) $\frac{1}{2}\ln|e^{2x} + e^{-2x}| + c$

7. b) $\frac{1}{2}\ln 2$

8. b) 0.186

Exercise 8.3 page 180

1. a) $2xe^{2x} - e^{2x} + c$ **b)** $\frac{3}{2}x\cos 2x + \frac{3}{4}\sin 2x + c$

    c) $(2x - 1)e^x + c$ **d)** $\frac{1}{3}(x + 3)\sin 3x + \frac{1}{9}\cos 3x + c$

    e) $\frac{1}{3}xe^{3x - 1} - \frac{1}{9}e^{3x - 1} + c$ **f)** $\frac{1}{2}x^2\ln 5x - \frac{1}{4}x^2 + c$

2. a) $1 + e^2$ **b)** $-\frac{3\pi}{4}$ **c)** $3e^1 + 1$

    d) 1.05 **e)** 23.8 **f)** 1

3. a) $6x^2e^{3x} - 4xe^{3x} + \frac{4}{3}e^{3x} + c$

3. a) $6x^2e^{3x} - 4xe^{3x} + \frac{4}{3}e^{3x} + c$

b) $\frac{3}{2}x^2\sin 2x + \frac{3}{2}x\cos 2x - \frac{3}{4}\sin 2x + c$

c) 31.9   **d)** 0.718   **e)** 1.14

f) 0.319   **g)** 3.65   **h)** 0.0741

i) 1.39

4. 1

5. a) 0.227

6. a) 0.102   **b)** 0.0745

Exercise 8.4 page 186

1. a) $\left(\sqrt{2x+1}\right)^3 - 3\sqrt{2x+1} + c$

b) $\frac{2}{9}(1 + x^3)^{\frac{3}{2}} + c$ **c)** $\sin^{-1}\left(\frac{x}{3}\right) + c$

d) $\frac{2}{3}(x+2)^{\frac{3}{2}} - 4(x+2)^{\frac{1}{2}} + c$

e) $\frac{1}{10}\tan^{-1}\left(\frac{2x}{5}\right) + c$ **f)** $\ln|e^{2x} + x| + c$

2. a) 2   **b)** $5\frac{7}{9}$ **c)** $\frac{\pi}{6}$

d) $\sqrt{2} - 1$ **e)** $\sqrt{3}$ **f)** 2.13

g) 4.5   **h)** 1

3. 1.14   **4.** $\frac{1}{5}$

5. 1.08   **6.** 216

Summary exercise 8 page 188

1. a) $\ln\left|A\left(\frac{x-2}{x+3}\right)\right|$ **b)** $\ln\left|A\sqrt{2x-1}(x+4)^2\right|$

c) $\ln\left|A\left(\frac{x+1}{x}\right)^2\right| - \frac{3}{x+1}$ **d)** $\ln\left|A\frac{x+1}{\sqrt{2x-1}}\right| + \frac{1}{(x+1)}$

e) $3x - \ln A(x-1)(x+4)^4$

2. $\ln\left|\frac{Ax}{\sqrt{x^2+4}}\right|$

3. a) $\ln|x^4 + 6| + c$ **b)** $-\frac{1}{6}\ln|3 - 2\sin 3x| + c$

c) $\frac{1}{2}\ln|3x^2 - 4x + 7| + c$

d) $-\ln|6 - 5e^{5x}| + c$ **e)** $\ln|x^2 + 9| + c$

f) $\frac{1}{3}\ln|x^3 + 10| + c$ **g)** $\frac{1}{3}\ln\frac{3}{2}$

h) $\frac{1}{3}\ln\left(\frac{2 + e^3}{3}\right)$ **i)** $\ln 3$

---

4. a) $\frac{1}{3}xe^{3x} - \frac{1}{9}e^{3x} + c$ **b)** $2x\sin 2x + \cos 2x + c$

c) $(2x - 5)e^x + c$

d) $\frac{1}{3}(x^2 - x + 3)\sin 3x + \frac{1}{9}(2x - 1)\cos 3x - \frac{2}{27}\sin 3x + c$

e) 0.264   **f)** 13.9

5 a) $\frac{4}{15}\left(\sqrt{3x-1}\right)^3 + \frac{4}{5}\left(\sqrt{3x-1}\right) + c$

b) $\frac{2}{3}(4 + x^4)^{\frac{3}{2}} + c$ **c)** $\frac{1}{4}\sin^{-1}\left(\frac{4}{5}x\right) + c$

d) $\frac{1}{6}(2x + 7)^{\frac{3}{2}} - \frac{7}{2}(2x + 7)^{\frac{1}{2}} + c$

e) $\frac{16}{3}$ **f)** 1.06   **g)** $\frac{1}{7}$ **h)** $\sqrt{3} - 1$

7. i) $R = 2, \alpha = \frac{1}{3}\pi$

Review exercise C – pure 3 page 190

1. $1 + 2x - \frac{3}{2}x^2$

2. $\frac{2}{x} + \frac{5x + C}{x^2 + 1}$

3. $a = \frac{1}{\sqrt{2}}$

4. i) $1 + 2x + 6x^2$ **ii)** 5

5. $x = 1.10$

6. i) $\frac{-1}{1+x} + \frac{2x+1}{1+2x^2}$ **ii)** $3x - 3x^2 - 3x^3$

7. i) $a = -3$ **ii)** $-\frac{10}{3}x^3$

8. b) $\frac{11}{96}$ **11. i)** $\frac{3}{2-x} + \frac{4x}{4+x^2}$

13. i) $\left(\exp\left(-\frac{1}{3}\right), \frac{-1}{3e}\right)$ **ii)** $4\ln 2 - \frac{15}{16}$

14. i) $\frac{1}{x+1} + \frac{-1}{x+3}$ **15. ii)** $\frac{1}{32}\left(2\pi - \sqrt{3}\right)$

16. i) $(1, 0)$ **ii)** $x = e^2$

17. i) $x = \sqrt{\frac{2}{3}}$ **iii)** $\frac{1}{16}\pi$

18. i) $x = \frac{1}{2}$ **ii)** $\pi\left(2\sqrt{e} - 3\right)$

22. i) $\frac{2}{2-x} + \frac{2x+4}{1+x^2}$ **ii)** $5 + \frac{5}{2}x = \frac{15}{4}x^2 - \frac{15}{8}x^3$

9 Vectors

Skills check page 195

1. a) $\begin{pmatrix} -5 \\ 3 \end{pmatrix}$ **b)** $\begin{pmatrix} 6 \\ -5 \end{pmatrix}$

2. a) $-5\mathbf{i} - 4\mathbf{j} + \mathbf{k}$ **b)** $5\mathbf{i} + 4\mathbf{j} - \mathbf{k}$

3. a) $16.1°$ **b)** $126.6°$

Exercise 9.1 page 198

1. **a)** **i)** $(2, 3, -1)$ **ii)** $(6, -2, 9)$
      iii) $(-10, 18, -31)$

   b) **i)** $(-4, 6, -3)$ **ii)** $\left(\frac{1}{2}, \frac{3}{2}, 0\right)$
      iii) $(14, -11, 9)$

2. **a)** $\mathbf{r} = \begin{pmatrix} 4 \\ -1 \end{pmatrix} + \lambda \begin{pmatrix} -3 \\ 5 \end{pmatrix}$, or another form

   b) $\mathbf{r} = \begin{pmatrix} 2 \\ 5 \end{pmatrix} + \lambda \begin{pmatrix} 4 \\ -6 \end{pmatrix}$ **c)** $\mathbf{r} = \lambda \begin{pmatrix} 1 \\ 0 \end{pmatrix}$

   d) $\mathbf{r} = \begin{pmatrix} 2 \\ -5 \end{pmatrix} + \lambda \begin{pmatrix} 2 \\ 8 \end{pmatrix}$ **e)** $\mathbf{r} = \lambda \begin{pmatrix} 1 \\ 3 \end{pmatrix}$

   f) $\mathbf{r} = \begin{pmatrix} 0 \\ 1 \end{pmatrix} + \lambda \begin{pmatrix} 1 \\ -1 \end{pmatrix}$

3. **a)** $\mathbf{r} = \begin{pmatrix} 2 \\ 1 \\ 5 \end{pmatrix} + \lambda \begin{pmatrix} -1 \\ -1 \\ 1 \end{pmatrix}$ **b)** $\mathbf{r} = \begin{pmatrix} 5 \\ 2 \\ -1 \end{pmatrix} + \lambda \begin{pmatrix} -1 \\ 4 \\ -4 \end{pmatrix}$

   c) $\mathbf{r} = \lambda \begin{pmatrix} 0 \\ 1 \\ 0 \end{pmatrix}$ **d)** $\mathbf{r} = \begin{pmatrix} -3 \\ 7 \\ 0 \end{pmatrix} + \lambda \begin{pmatrix} -5 \\ 10 \\ 0 \end{pmatrix}$

   e) $\mathbf{r} = \lambda \begin{pmatrix} 1 \\ 1 \\ 1 \end{pmatrix}$ **f)** $\mathbf{r} = \begin{pmatrix} -3 \\ 2 \\ 6 \end{pmatrix} + \lambda \begin{pmatrix} 0 \\ 0 \\ 1 \end{pmatrix}$

4. **a)** $\mathbf{r} = \begin{pmatrix} 1 \\ 2 \\ 3 \end{pmatrix} + \lambda \begin{pmatrix} -2 \\ 0 \\ -4 \end{pmatrix}$ **b)** $\mathbf{r} = \begin{pmatrix} 1 \\ 2 \\ 3 \end{pmatrix} + \lambda \begin{pmatrix} 5 \\ 0 \\ 6 \end{pmatrix}$

   c) $\mathbf{r} = \begin{pmatrix} 0 \\ 2 \\ 1 \end{pmatrix} + \lambda \begin{pmatrix} 2.5 \\ 0 \\ 3 \end{pmatrix}$

5. P and Q

6. **a)** $\mathbf{r} = \begin{pmatrix} -2 \\ 1 \\ 0 \end{pmatrix} + \lambda \begin{pmatrix} 6 \\ -1 \\ 4 \end{pmatrix}$

7. **a)** $\mathbf{r} = \begin{pmatrix} 2 \\ -1 \\ 3 \end{pmatrix} + \lambda \begin{pmatrix} -1 \\ 2 \\ -4 \end{pmatrix}$ **b)** $\alpha = -3, \beta = 7$

8. **a)** $\mathbf{r} = \begin{pmatrix} 2 \\ 3 \\ 4 \end{pmatrix} + \lambda \begin{pmatrix} -1 \\ 1 \\ -1 \end{pmatrix}$ **b)** $x = -2, y = 7$

Exercise 9.2 page 203

1. **a)** Parallel   **b)** Parallel   **c)** Not parallel
   d) Not parallel   **e)** Not parallel   **f)** Not parallel
   g) Parallel   **h)** Parallel

2. **a)** Intersect at $\begin{pmatrix} 3 \\ 2 \\ -2 \end{pmatrix}$ **b)** No intersection

   c) Intersect at $\begin{pmatrix} 10 \\ 1 \\ 11 \end{pmatrix}$ **d)** No intersection

3. **a)** $\begin{pmatrix} 1 \\ 1 \\ 0 \end{pmatrix}$ **b)** $\begin{pmatrix} \frac{19}{2} \\ \frac{3}{2} \\ 1 \end{pmatrix}$

4. $\begin{pmatrix} -1 \\ 6 \\ 0 \end{pmatrix}$

7. $c = -2$; intersect $= \begin{pmatrix} 1 \\ 5 \\ -3 \end{pmatrix}$

Exercise 9.3 page 206

1. **a)** $83.8°$ **b)** $84.1°$ **c)** $45.6°$
   d) $72.0°$ **e)** $22.2°$

3. **a)** $35.1°$ **b)** $50.5°$ **c)** $66.3°$

Exercise 9.4 page 210

1. **a)** $\mathbf{r}.\begin{pmatrix} 1 \\ 4 \\ -3 \end{pmatrix} = 0, x + 4y - 3z = 0$

   b) $\mathbf{r}.\begin{pmatrix} 1 \\ -3 \\ 5 \end{pmatrix} = -1, x - 3y + 5z = -1$

   c) $\mathbf{r}.\begin{pmatrix} -1 \\ -3 \\ 4 \end{pmatrix} = 0, -x - 3y + 4z = 0$

   d) $\mathbf{r}.\begin{pmatrix} 2 \\ -3 \\ 1 \end{pmatrix} = -1, 2x - 3y + z = -1$

2. a) $\begin{pmatrix} 2 \\ -1 \\ 3 \end{pmatrix}$ **b)** $2x - y + 3z = 0$

c) $2x - y + 3z = 17$

3. a) $\begin{pmatrix} 2 \\ -1 \\ 3 \end{pmatrix}$ **b)** $2x - y + 2z = 9$

4. r. $\begin{pmatrix} 4 \\ -3 \\ 1 \end{pmatrix} = -2$ **5.** $2x - 5y + 3z = -2$

6. a) $AB = \begin{pmatrix} -3 \\ 2 \\ 0 \end{pmatrix}$, $BC = \begin{pmatrix} 0 \\ -2 \\ 4 \end{pmatrix}$ **b)** $\begin{pmatrix} 4 \\ 6 \\ 3 \end{pmatrix}$

c) $4x + 6y + 3z = 12$

7. a) $\begin{pmatrix} 0 \\ 0 \\ 1 \end{pmatrix}$, $z = 1$ **b)** $\begin{pmatrix} 1 \\ 1 \\ 0 \end{pmatrix}$, $x + y = 1$

c) $\begin{pmatrix} 1 \\ -1 \\ 0 \end{pmatrix}$, $x - y = 0$ **d)** $\begin{pmatrix} 1 \\ 1 \\ 1 \end{pmatrix}$, $x + y + z = 3$

e) $\begin{pmatrix} 1 \\ 1 \\ 1 \end{pmatrix}$, $x + y + z = 2$

Exercise 9.5 page 213

1. $\left(\dfrac{7}{2}, -\dfrac{1}{2}, 0 \right)$ **2.** $19.5°$

4. a) $11.7°$ **b)** $17.6°$

   c) $32.5°$ **d)** $34.1°$

5. a) intersects   **b)** parallel

   c) intersects   **d)** parallel

7. a) $\left(1, -\dfrac{13}{5}, \dfrac{12}{5} \right)$ **b)** $59.0°$ **8.** $\dfrac{5}{3}$

Exercise 9.6 page 217

1. a) $6x - 2y + 4z = 12$ **b)** $3x - y + 2z = 0$

   c) $x + 3y = 7$

3. $75.8°$ **4.** $\mathbf{i} - 3\mathbf{j} + 2\mathbf{k}$

5. a) $79.0°$ **b)** $\mathbf{r} = \dfrac{1}{15}\begin{pmatrix} 37 \\ 1 \\ 0 \end{pmatrix} + \lambda \begin{pmatrix} 14 \\ 2 \\ 15 \end{pmatrix}$

6. a) $\mathbf{r} = \mathbf{i} + \lambda(\mathbf{i} + \mathbf{j} + \mathbf{k})$ **b)** $79.1°$

8. r. $\begin{pmatrix} 4 \\ -1 \\ 3 \end{pmatrix} = -3$ **9.** $x - 17y - 7z = -42$

Exercise 9.7 page 221

1. 3 **2.** $\dfrac{25}{\sqrt{38}}$

3. $\dfrac{22}{\sqrt{94}}$ **4.** 3

5. $\dfrac{\sqrt{78}}{3}$ **6.** $\dfrac{2\sqrt{150}}{3}$

7. $3x - y + 2z = 14$

8. $1 - \dfrac{1}{2\sqrt{6}}$ $(= 0.796)$

Summary exercise 9 page 222

1. $\mathbf{r} = \begin{pmatrix} 1 \\ 2 \\ -1 \end{pmatrix} + \lambda \begin{pmatrix} 2 \\ 2 \\ 0 \end{pmatrix}$

2. $\dfrac{1}{10}$

3. a) $(1, 1, 1)$ **b)** $54.7°$

4. $22.4°$

5. a) $(1, 1, 2)$ **b)** $70.5°$

6. $3x - y + 2z = 14$

8. $\mathbf{r} = 9\mathbf{i} - 5\mathbf{j} + 12\mathbf{k}$

9. $\begin{pmatrix} \frac{11}{9} \\ \frac{26}{9} \\ \frac{-7}{9} \end{pmatrix}$ **10. r.** $\begin{pmatrix} -7 \\ 5 \\ 4 \end{pmatrix} = -6$

11. b) $2x - y + z = 5$

12. $\mathbf{i} - \mathbf{j} + 2\mathbf{k}$

13. $x - y + z = 0$

14. a) $\mathbf{r}.\dfrac{1}{3}\begin{pmatrix} 2 \\ 2 \\ 1 \end{pmatrix} = 3$ **b)** 3

15. a) $x + 3y + 5z = 8$ **b)** $\dfrac{15}{\sqrt{35}} \left(= \dfrac{3\sqrt{35}}{7} \right)$

16. a) $\mathbf{r}.\begin{pmatrix} 3 \\ 4 \\ 5 \end{pmatrix} = \dfrac{50}{\sqrt{2}}$ or $\mathbf{r}.\begin{pmatrix} 3 \\ 4 \\ 5 \end{pmatrix} = 25\sqrt{2}$

   b) $\mathbf{r}.(2\mathbf{i} + 3\mathbf{j} + 4\mathbf{k}) = 29$

17. $2x + 7y + 2z = -10$ or $\mathbf{r}.\begin{pmatrix} 2 \\ 7 \\ 2 \end{pmatrix} = -10$

18. $2x + 2y + z = 5$ or $\mathbf{r}.(2\mathbf{i} + 2\mathbf{j} + \mathbf{k}) = 5$

19. a) $\dfrac{38}{7}, \dfrac{20}{7}, \dfrac{12}{7}$ **b)** $\dfrac{48}{7}, \dfrac{5}{7}, \dfrac{17}{7}$

20. $\dfrac{2}{\sqrt{35}} (5\mathbf{i} - \mathbf{j} - 3\mathbf{k})$ or $\dfrac{2}{\sqrt{35}} \begin{pmatrix} 5 \\ -1 \\ -3 \end{pmatrix}$

21. $36.7°$

22. $(1, -2, -1)$

23. a) $22.2°$ **b)** $\mathbf{r} = \begin{pmatrix} 4 \\ 1 \\ 0 \end{pmatrix} + \lambda \begin{pmatrix} 2 \\ 1 \\ -1 \end{pmatrix}$

    c) $\dfrac{2\sqrt{3}}{3}$ or $\dfrac{2}{\sqrt{3}}$ **d)** $x - y + z = 3$

24. $\mathbf{r} = \dfrac{1}{11} (27\mathbf{i} + 19\mathbf{j} + 14\mathbf{k}) + \lambda(\mathbf{i} - 3\mathbf{j} - \mathbf{k})$

25. i) $r = \begin{pmatrix} -1 \\ 3 \\ 5 \end{pmatrix} + \begin{pmatrix} 3 \\ -1 \\ -4 \end{pmatrix}$ **ii)** $\begin{pmatrix} 3 \\ -1 \\ -4 \end{pmatrix}$

    iii) $7x - 11y + 8z = 0$

26. i) $4\mathbf{i} - 2\mathbf{j} - \mathbf{k}$ **ii)** $24.1°$

27. ii) $6x + y - 8z = 6$

10 Differential equations

Skills check page 226

1. a) $c = 4$ **b)** $c = -103$
    c) $c = -3$ **d)** $c = 2$

2. $1 - 2 \sin^2\theta$

3. $\dfrac{1}{x-1} - \dfrac{1}{x+1}$

Exercise 10.1 page 231

1. $\dfrac{\mathrm{d}T}{\mathrm{d}t} = -k(T - T_0)$

2. $\dfrac{\mathrm{d}x}{\mathrm{d}t} = k(30 - x)$

3. $\dfrac{\mathrm{d}N}{\mathrm{d}t} = kN(P - N)$

4. $\dfrac{\mathrm{d}N}{\mathrm{d}t} = kN; \quad k = 0.07$

5.

Exercise 10.2 page 235

1. $y^2 = x^2 + d$ **2.** $2y^2 = x^2 + d$

3. $2y^3 = 2x^3 + 3x^2 + 6x + d$ **4.** $y = Ae^{\frac{1}{2}x^2 + 3x}$

5. $\sin y = 2x + c$ **6.** $\dfrac{1}{2} y^2 = \ln x + c$

7. $y^2 = x^4 + c$ **8.** $\dfrac{1}{4y^2} = \dfrac{1}{x} + d$

9. $\ln|y| = \sin x + c$

10. $e^{-y} = -x^2 + c$

11. $\sin y = x^2 + c$

12. $-\dfrac{1}{y} = \dfrac{1}{2}\ln\left|\dfrac{x}{x+2}\right| + c$

13. $\sin y = \sin x + c$

14. $\tan y = \ln|x| + c$

15. $\ln\left|\dfrac{y-1}{y+1}\right| = x^4 + c$

16. $\tan y = \tan x + c$

17. $\sin y = xe^x - e^x + c$

18. $\left|\dfrac{y-1}{y+1}\right| = A\left|\dfrac{x}{x+1}\right|$

19. $-\dfrac{1}{y} = \tan x + c$

20. $\ln(1 + y^4) = \dfrac{1}{2}x^2 + c$

21. $\dfrac{1}{y} = \ln\left|A\dfrac{(x+1)^2}{x^3}\right|$

22. $\dfrac{1}{y^2} = \tan^{-1}(e^x) + c$

23. $\dfrac{1}{2}y^2 = -\dfrac{1}{3}\cos^3 x + \dfrac{1}{5}\cos^5 x + c$

24. $\dfrac{1}{2}y^2 = 9\theta + c = 9\tan^{-1}\left(\dfrac{x}{3}\right) + c$

Exercise 10.3 page 238

1. $y^2 = x^2 + 3$

2. $y^2 = 2x^4 + 2$

3. $\dfrac{1}{3}y^3 = \dfrac{1}{3}x^3 + \dfrac{1}{2}x^2 + x + 9$

4. $y = 5e^{\left(\frac{1}{2}x^2 + 3x\right)}$

5. $\sin y = 2x + 1$

6. $\dfrac{1}{2}y^2 = \ln x + 2$

7. $y^2 = x^4 - 7$

8. $y^2 = \dfrac{x}{3x+1}$

9. $y = e^{(\sin x - 1)}$

10. $y^2 = x^2 - 4$

11. $y^2 = \ln\left(\dfrac{x^2}{e}\right)$

12. $y^2 = x^2 + x + 2$

13. $\sqrt{y} = \sqrt{x+1} + 1$

14. $\sin y = \left(-\dfrac{1}{2x^2}\right) + 1$

15. $e^y = 2\tan x - 1$

16. $\dfrac{2}{3}y^{\frac{3}{2}} = \left(2x^{\frac{1}{2}}\right) + 14$

17. $e^y = \ln|x^2 + 1| + 1$

18. $y = \left|\dfrac{6x}{x+1}\right|$

19. $e^{-y} = \dfrac{1}{3}e^{3x} + \dfrac{2}{3}$

20. $\tan y = \left(-\dfrac{1}{2}x^{-2}\right) + c$

21. $y^2 = e^{x^2} + 8$

Exercise 10.4 page 245

1. **a)** $\dfrac{dT}{dx} = -0.16x$ **b)** $T = 400 - 0.08x^2$

  c) $200°\,C$ **d)** $68.9\,\text{cm}$

2. **a)** $\dfrac{1}{y}\dfrac{dy}{dt} = -3e^{-3t}$ **b)** $\ln|y| = 3e^{-3t} + \ln(20e^{-3})$

  c) $t \to \infty,\ \ln|y| \to \ln(20e^{-3}),\ y \to 20e^{-3}$

  Initially there was 20 grams of Y, so the proportion which will remain is 5.0%.

3. **a)** $\dfrac{1}{m}\dfrac{dm}{dt} = -k$ **c)** $k = 1.21 \times 10^{-4}$

  d) $40\,000$ years

4. **a)** $\left(\dfrac{1}{d-150}\right)\dfrac{dd}{dt} = -10^{-5}$ **b)** $|d-150| = 142e^{-0.00001t}$

  c) $d = 8.21$ metres

5. **a)** $\dfrac{1}{N}\dfrac{dN}{dt} = k$ **b)** $N = 2e^{0.0372t}$

  c) 43.3 hours

6. **a)** $\dfrac{1}{v}\dfrac{dv}{dt} = -k$ **b)** $v = 25e^{-0.0223t}$

  c) 31 hours

7. **a)** $\dfrac{dn}{dt} = kn(P - n)$ **b)** $\dfrac{n}{n-P} = \dfrac{1}{19}e^{\frac{t}{3}\ln\frac{19}{9}}$

  c) 11.8 days

  d) As $t \to \infty$, $n \to P$ so the whole population will be infected.

Summary exercise 10 page 248

1. a) $\frac{1}{4}y^4 = \frac{1}{3}x^3 + c$
   b) $\tan y = \frac{1}{2}x^2 + c$

   c) $\sin y = \frac{1}{3}e^{3x} + c$
   d) $A = x\cos y$

   e) $e^y = \frac{1}{3}e^{3x} + c$
   f) $y = A\sqrt{x^2+1}$

   g) $y = A(x^2+6)^4$
   h) $A(x+2) = \sqrt[3]{\sin\theta}$

2. a) $y^4 = x^3 + \frac{1}{2}x^2 - 9$
   b) $y = \frac{1}{3+2\cos x}$

   c) $|y-1| = 6|x+2|$
   d) $|y| = \left|\frac{3e^x(x-1)}{x+1}\right|$

   e) $e^y = \frac{1}{3}e^{3x} + \frac{2}{3}$
   f) $2v^{\frac{1}{2}} = -e^{-t} + 7$

   g) $e^y = xe^x + e^x$

   h) $y\ln y - y = 2\ln|x+1| - \ln x + 1 - \ln 4$

3. $\ln y = 2(x+1)^{\frac{1}{2}} - 2\sqrt{2}$

4. $\sin y = xe^x - e^x + 1.5$

5. $ye^y - e^y = \tan x - 2$

6. $\frac{2}{3}y^{\frac{3}{2}} = \ln\left|\frac{3(x-1)}{x+1}\right|$

7. $2y^2 = 2x + \sin 2x + 1$

8. $\ln(y^2 + 1) + \tan^{-1} y = \ln x$

9. a) $T - 20 = Ae^{-kt}$
   b) $T = 20 + 70e^{-0.14t}$

   c) $37.3°\,C$
   d) 18.9 minutes

10. a) $v = \frac{1}{kt + 0.02}$
    b) $v = \frac{300}{t+6}$

    c) $18.75\,\text{ms}^{-1}$
    d) 24 seconds

11. a) $\frac{1}{N(2000-N)}\frac{dn}{dt} = k$
    b) $N = 2000\left(\frac{f(t)}{1+f(t)}\right)$

    c) the time for a majority to have heard the rumour will be when $N = 1001$
    5.7 hours

12. d) $v = 8\,\text{ms}^{-1}$

13. a) $\frac{1}{m}\frac{dm}{dt} = -0.05$
    b) $m = 12e^{-0.05t}$

    c) 8.1 minutes

15. i) $\ln x = -\frac{1}{2}kt^2 + \ln 100$ ii) $t = 51.3$

16. i) $\frac{1}{4}\ln y - \frac{1}{4}\ln(4-y)$ ii) $y = \frac{4}{3e^{-4x} + 1}$

    iii) y approaches 4 as x becomes very large.

11 Complex numbers

Skills check page 253

1. a) $2 + 2\sin\theta$ b) $\sin^4\theta$ c) 1
3. a) 17   b) 81   c) 73

Exercise 11.1 page 255

1. a) real roots   b) real roots   c) no real roots
   d) no real roots   e) real roots

2. a) $\pm 2i$ b) $\pm 10i$ c) $\pm\sqrt{13}i$
   d) $\pm\sqrt{75}i$ or $\pm 5\sqrt{3}i$

3. a) i b) $-i$ c) $-i$
   d) i e) 1

4. a) -1 b) $-128i$ c) $-27i$
   d) -1024

5. a) ± 8 b) $\pm 8i$ c) $\pm\sqrt{19}i$
   d) $\pm\sqrt{32}i$ and $\pm 4\sqrt{2}i$

6. a) $\pm 7i$ b) $\pm\sqrt{18}i$ or $\pm 3\sqrt{2}i$
   c) $\pm\sqrt{2}i$ d) $\pm\sqrt{5}i$

7. a) -2 b) 49
   c) $-\frac{1}{27}i$ d) $\frac{1}{256}$

8. a) $x + y = 10, xy = 40$

Exercise 11.2 page 257

1. a) $5 + i$ b) $4 - \sqrt{6}i$
   c) $-2 + 2\sqrt{10}i$ d) $-1 - 9i$

2. a) $4 + 2i$ b) $-5 - i$
   c) $-9 + 7i$ d) $-6 - 3i$

3. a) $z^* = 10 - 12i$ b) $z = 3 + 3i$
   c) $z^* = -5i$ d) $z = 8$

4. a) 2 real roots   b) 2 complex roots
   c) 2 real roots   d) 2 real roots
   e) 2 complex roots   f) 2 complex roots

5. a) $2 \pm i$ **b)** $-\frac{1}{4} \pm \frac{\sqrt{3}}{4}i$

c) $\frac{9 \pm \sqrt{105}}{4}$ **d)** $\frac{2}{3}, -1$

e) $\frac{1}{6} \pm \frac{\sqrt{11}}{6}i$ **f)** $-\frac{1}{4} \pm \frac{\sqrt{7}}{4}i$

g) $\pm\frac{\sqrt{3}}{2} - \frac{1}{2}i$

6. a) $1 \pm \sqrt{2}i$ **b)** $-1 \pm \sqrt{\frac{7}{3}}i$

c) $2 \pm \frac{1}{\sqrt{2}}$ **d)** $\frac{3}{2} \pm \sqrt{\frac{23}{12}}$

e) $-1 \pm i$ **f)** $-3 \pm i$

g) $\pm 2 + i$

7. $\frac{1}{i}$ or $\frac{-2}{i}$

Exercise 11.3 page 260

1. a) $5 + 6i$ **b)** $11 + 4i$
c) -3 **d)** $3 + 3i$
e) $1 + 5i$ **f)** $-1 - 8i$

2. a) $2 + 2i$ **b)** $2 + 6i$
c) $9 - 6i$ **d)** $11 - 8i$
e) $-6 + 2i$ **f)** $2 + 6i$

3. a) $21 + 11i$ **b)** $-2 - 10i$
c) $-5 - 15i$ **d)** $24 + 41i$
e) $3 - 33i$ **f)** $-2 - 5i$

4. a) $9 + 7i$ **b)** $23 - 2i$
c) $1 + 7i$ **d)** $5 + 10i$
e) $-14 + 10i$ **f)** $15 + 45i$

5. a) $\frac{14}{5} + \frac{3}{5}i$ **b)** $\frac{16}{5} - \frac{3}{5}i$

c) $-\frac{11}{61} + \frac{60}{61}i$ **d)** $\frac{1}{4}$

e) $\frac{4}{5} - \frac{3}{5}i$ **f)** $\frac{8}{25} + \frac{6}{25}i$

g) $\frac{1}{2} - \frac{3}{2}i$ **d)** $\frac{1}{7} - \frac{4\sqrt{3}}{7}i$

6. a) $1 - 2\sqrt{2}i$

b) i) $2\sqrt{2}$ **ii)** $-4i$

iii) 15 **iv)** $-\frac{2\sqrt{2}}{3} - \frac{1}{3}i$

7. a) $8 - 16i$ **b)** $-2 + 8i$

c) $-33 - 56i$ **d)** $\frac{63}{169} + \frac{16}{169}i$

8. a) $3 - 4i$ **b)** $2 - 11i$
c) $-7 - 24i$ **d)** $-278 + 29i$

9. a) $\frac{3}{10} + \frac{1}{10}i$ **b)** $\frac{23}{65} - \frac{11}{65}i$

10. a) i **b)** $\frac{3}{17} - \frac{5}{17}i$ **c)** $\frac{3}{5} + \frac{4}{5}i$

d) $\frac{a}{a^2 + b^2} - \frac{b}{a^2 + b^2}i$ **e)** $\frac{52}{25} + \frac{33}{25}i$

11. a) $\text{Re}(u^2) = a^2\text{Re}(u^2) = a^2 - b^2, \text{Im}(u^2) = 2abb^2,$
$\text{Im}(u^2) = 2ab$
b) $\text{Re}(u - u^*) = 0, \text{Im}(u - u^*) = 2b$
c) $\text{Re}(u^3 - (u^*)^3) = 0, \text{Im}(u^3 - (u^*)^3) = 6a^2b - 2b^3$

12. a) $(\cos^2\theta - \sin^2\theta) + (2\sin\theta\cos\theta)i$
b) $\cos\theta - i\sin\theta$ **c)** $(\cos^2\theta - \sin\theta) + (2\sin\theta\cos\theta)i$

Exercise 11.4 page 264

1. a) $\pm 20i$ **b)** $\pm 2\sqrt{15}i$
c) $1 + 2i, -1 - 2i$ **d)** $2 + i, -2 - i$
e) $2 + 3i, -2 - 3i$ **f)** $5 + 2i, -5 - 2i$
g) $2 - 5i, -2 + 5i$ **h)** $6 - i, -6 + i$
i) $-3\sqrt{2} - 3\sqrt{2}i, 3\sqrt{2} + 3\sqrt{2}i$

2. a) $2 - 1, -2 + i$ **b)** $-i, -2$

3. $2 + 3i$

4. square roots of i are $\frac{\sqrt{2}}{2} + \frac{\sqrt{2}}{2}i$ and $-\frac{\sqrt{2}}{2} - \frac{\sqrt{2}}{2}i$

square roots of $-i$ are $\frac{\sqrt{2}}{2} - \frac{\sqrt{2}}{2}i$ and $-\frac{\sqrt{2}}{2} + \frac{\sqrt{2}}{2}i$

5. a) $\frac{5\sqrt{2}}{2} + \frac{5\sqrt{2}}{2}i, \frac{-5\sqrt{2}}{2} - \frac{5\sqrt{2}}{2}i$

b) $-2\sqrt{2} + 2\sqrt{2}i, 2\sqrt{2} - 2\sqrt{2}i$ **c)** $\pm 2i$

6. $a = 16, b = -2$

7. $1 + 4i$

8. a) $\frac{1}{2} + \frac{1}{2}i$ **b)** $\pm\frac{\sqrt{3}}{2} + \frac{1}{2}i$

9. $3 - i$

10. $a = 5, b = 30$ or $a = -5, b = -30$

Exercise 11.5 page 265

1. a) $1 - \sqrt{3}i$ **b)** $z^3 - 4z^2 + 8z - 8 = 0$

2. a) $\sqrt{2} \pm \sqrt{2}i$ **b)**

3. a) $i, -5i$ **b)** $-\frac{1}{2}i$ **c)** $\pm\frac{\sqrt{3}}{6} - \frac{1}{2}i$

4. $z = 3 + i$, $w = 2 - 3i$

5. $z = 4i$, $w = -3i$

6. **a)** $z = 1$ **b)** $z - 1$ **c)** $1, -\dfrac{1}{2} \pm \dfrac{\sqrt{3}}{2}i$

7. $3, -\dfrac{3}{2} \pm \dfrac{3\sqrt{3}}{2}i$

8. **a)** -4 **b)** $1 \pm i, -1 \pm i$

9. $-\dfrac{5}{2}, 5 \pm 2i$

10. **a)**   **b)** $2, -1 \pm 2i$

11. **a)**   **b)** $\pm 1, 2 \pm 3i$

12. **a)** -2 **b)**   **c)** $-2, \pm i, 1 \pm 2i$

Exercise 11.6 page 268

2. $u = 5 + 3i$, $u^* = 5 - 3i$, $2u = 10 + 6i$, $-u = -5 - 3i$

3. $z_1 = 4 + 5i$, $z_2 = 3i$, $z_3 = 2 - 6i$, $z_4 = -2i$, $z_5 = -4$,
$z_6 = -6 - 4i$, $z_7 = -3 + 4i$

4. **a)** $3 + 4i, -4 + 3i, 4 - 3i, 4 + 3i$

5. **a)** $-2 + 5i$

   b) $iz_1 z_2 = -13 - 11i$

6. **a)** $z + w = 5 + 5i$, $z - w = 3 - i$

   b) $z + w = -2 + 6i$, $z - w = 6 + 4i$

   c) $z + w = -3$, $z - w = -7 - 6i$

   d) $z + w = 4 - 3i$, $z - w = 2 - 5i$

7. $\dfrac{1}{z} = \dfrac{1}{4} + \dfrac{1}{4}i$, $z + \dfrac{1}{z} = \dfrac{9}{4} - \dfrac{7}{4}i$

Exercise 11.7 page 274

1. **a)** $1, 0$ **b)** $1, \pi$ **c)** $\sqrt{2}, \dfrac{\pi}{4}$

   d) $\sqrt{2}, -\dfrac{\pi}{4}$ **e)** $3, \dfrac{\pi}{2}$ **f)** $13, 1.18$ rad

   g) $7, -1.03$ rad   **h)** $1, \dfrac{\pi}{4}$

2. **a)** $1, \dfrac{\pi}{3}$ **b)** $1, -\dfrac{\pi}{3}$ **c)** $1, \dfrac{2\pi}{3}$

   d) $1, -\dfrac{2\pi}{3}$ **e)** $1, -\dfrac{2\pi}{3}$ **f)** $1, \pi$

   g) $1, 0$ **h)** 0, not defined   **i)** $1, -\dfrac{\pi}{3}$

   j) $2, -\dfrac{2\pi}{3}$

3. **a)** **i)** 5 **ii)** 25
       iii) 125
   b) **i)** 0.634 rad   **ii)** 1.29 rad
       iii) 1.93 rad

4. **a)** $2i$ **b)** $-5 + 0i$ **c)** $2 + 2\sqrt{3}i$

   d) $\dfrac{3\sqrt{3}}{2} + \dfrac{3}{2}i$ **e)** $5\sqrt{2} + 5\sqrt{2}i$ **f)** $-\dfrac{1}{2}i$

   g) $-\dfrac{\sqrt{2}}{2} - \dfrac{\sqrt{2}}{2}i$ **h)** $-3 + 3\sqrt{3}i$ **i)** $7 + 0i$

6. **a)** $|wz| = 2\sqrt{2}$, $\arg(wz) = -\dfrac{\pi}{12}$

   b) $\left|\dfrac{w}{z}\right| = \sqrt{2}$, $\arg\left(\dfrac{w}{z}\right) = -\dfrac{7\pi}{12}$

   c) $|iz| = \sqrt{2}$, $\arg(iz) = \dfrac{3\pi}{4}$

   d) $\left|\dfrac{1}{w^*}\right| = \dfrac{1}{2}$, $\arg\left(\dfrac{1}{w^*}\right) = -\dfrac{\pi}{3}$

7. **a)** -1 **b)** $-i$ **c)** $2\sqrt{2} + 2\sqrt{2}i$

8. **a)** $\sqrt{2}e^{\frac{i\pi}{4}}$ **b)** $3e^{i\pi}$ **c)** $\sqrt{5}e^{2.03i}$

   d) $7e^{\frac{i\pi}{2}}$ **e)** $10e^{0.927i}$ **f)** $10e^{-0.927i}$

   g) $2e^{\frac{3\pi i}{4}}$ **h)** $2\sqrt{13}e^{-0.197i}$ **i)** $e^{\frac{\pi i}{3}}$

9. **a)** $\left(6, \dfrac{\pi}{2}\right)$ **b)** $\left(12, \dfrac{3\pi}{4}\right)$

   c) $\left(\dfrac{2}{3}, -\dfrac{\pi}{6}\right)$ **d)** $\left(4, \dfrac{\pi}{2}\right)$

10. **a)** $9, \dfrac{\pi}{2}$ **b)** $27, \dfrac{3\pi}{4}$ **c)** $81, \pi$

   d) $6561, 0$ **e)** $6, \dfrac{5\pi}{12}$ **f)** $12, \dfrac{7\pi}{12}$

   g) $216, -\dfrac{3\pi}{4}$ **h)** $\dfrac{3}{2}, \dfrac{\pi}{12}$ **i)** $\dfrac{1}{9}, -\dfrac{\pi}{2}$

   j) $2\sqrt{3}, 0$ **k)** $2, \dfrac{\pi}{2}$ **l)** $\dfrac{1}{6}, \dfrac{5\pi}{12}$

Exercise 11.8 page 279

2. **a)** **i)** $|z| = 3$ **ii)** $|z - (2 + 2i)| = 2$ or $|z - 2 - 2i| = 2$
       iii) $|z - (-4 - 3i)| = 5$ or $|z + 4 + 3i| = 5$

9. **b)** Greatest value of $|z|$ is 4   Least value of $|z|$ is 0

10. **b)** Greatest value of $|z|$ is 5   Least value of $|z|$ is 1

   c) Greatest value of arg $z = 1.46$ rad

Summary exercise 11 page 281

1. $x = \dfrac{9}{41}$, $y = -\dfrac{42}{41}$

2. **a)** Both roots have modulus $\sqrt{10}$ **b)** $-1.89, 1.89$

3. **b)** $-\dfrac{1}{2} + \dfrac{\sqrt{3}}{2}i, -\dfrac{1}{2} - \dfrac{\sqrt{3}}{2}i$

4. **a)** $-4 - 2\sqrt{5}i$ **b)** $p = 2$

5. **b)** Least value is $4\sqrt{2} - 3$ Greatest value is $4\sqrt{2} + 3$

7. **a)** $-4, -1$ **b)** $\pm i, \pm 2i$

8. **b)** Greatest value of $|z|$ is $6\sqrt{17}$ Least value of $|z|$ is 0

9. **b)** Greatest value of $|z|$ is 8   Least value of $|z|$ is 2
   Greatest value of arg z is 2.68 rad
   Least value of arg z is 0.464 rad

10. Greatest value of $|z - w|$ is $\sqrt{65} + 4$

Least value of $|z - w|$ is $\sqrt{65} - 4$

11. b) $1 - i$ **c)** $1 \pm i, 2 \pm 3i$

12. $2 + \sqrt{5}i, -4 \pm \sqrt{13}$

13. b) $1 \pm 2i, 1$

14. $-\frac{3\sqrt{3}}{2} + \frac{3}{2}i, \frac{3\sqrt{3}}{2} + \frac{3}{2}i, -3i$

15. a) $-\frac{1}{2} - \frac{\sqrt{3}}{2}i$ **c) i)** $\left(1, \frac{2\pi}{3}\right), \left(1, -\frac{2\pi}{3}\right)$

17. ii) $\frac{3}{5} + \frac{4}{5}i$

18. ii) $1 - 2i$

19. i) $2 + i, -2 + i$ **ii)** $\sqrt{5}; 26.6°, 153.4°$

Review exercise D – Pure 3 page 286

1. a) $u = -2 = 2i; v = 1 + 2i$ **b)** $\frac{3}{\sqrt{2}}$

2. i) $3x + y - 2z = 1$ **ii)** $\mathbf{i} + 2\mathbf{j} + 2\mathbf{k}$

3. ii) $\frac{2}{5}h^{\frac{5}{2}} + \frac{1}{150}H^{\frac{5}{2}}t + \frac{2}{5}H^{\frac{5}{2}}$ **ii)** $t = 49.4$

4. i) $12x + 6y - 6z = 48$ **ii)** 5.20

5. i) $x = \sqrt{\frac{2}{3}}$ **iii)** $\frac{1}{16}\pi$

6. i) $13\sin(x + 67.38°)$ **ii)** $27.4° + 175.2°$

7. i) $57.7°$

   ii) e.g. $\mathbf{r} = 2\mathbf{i} - \mathbf{k} + \lambda(4\mathbf{i} - 7\mathbf{j} + 5\mathbf{k})$

8. a) $7 - 2i$ **b)** $6.69e^{\frac{1}{4}\pi i}$

9. i) $1 - \sqrt{2}i$ **ii)** $-1 + i$ and $-1 - i$

10. i) $3x + z = 13$ **ii)** $3\sqrt{2}$

11. i) $-\frac{2}{5} + \frac{11}{5}i$

12. 2.44

13. i) $\sqrt{104}$ **ii)** $-3x + 9y - z = 28$

14. $x = \sqrt{(2\sin 2\theta)} - 1$

15. ii) $\lambda = \frac{3}{8}$

16. a) $\pm(\sqrt{3} - i\sqrt{2})$ **b)** $131.8°$

17. i) $3x - 4y + 6z = 80$ **ii)** $30.8°$

18. i) $-\frac{1}{2}\pi$ **ii)** $\frac{3}{2}\sqrt{2}$ **iii)** $3\sqrt{2} + 1$

19. i) $N = \dfrac{1800e^{\frac{1}{2}t}}{5 + e^{\frac{1}{2}t}}$ **ii)** N approaches 1800

20. i) $2, \frac{1}{6}\pi$

   ii) a) $3\sqrt{3} + i$ **b)** $\frac{1}{2}\sqrt{3} + \frac{1}{2}i$

21. i) $\mathbf{r} = \mathbf{i} + 2\mathbf{j} + 2\mathbf{k} + \lambda(2\mathbf{i} + 2\mathbf{j} - 2\mathbf{k})$

   ii) $\frac{2}{3}\mathbf{i} + \frac{5}{3}\mathbf{j} + \frac{7}{3}\mathbf{k}$ **iii)** $2x + 5y + 7z = 26$

22. $k = 0.05$

   ii) $\ln 20 - \ln(20 - x) = \frac{1}{20}t$

   iii) $x = 7.9$ **iv)** x approaches 20

23. $y^2 = 4(x^2 - 1)$

24. i) $\sqrt{8}; \frac{1}{4}\pi$ **iii)** $\sqrt{7}$

25. ii) $74.2°$ **iii)** $-5x + 3y + 4z = 2$

26. i) $\mathbf{r} = 2\mathbf{i} + \mathbf{j} - 2\mathbf{k} + \lambda(\mathbf{i} - 3\mathbf{j} + 3\mathbf{k})$

   ii) $5\mathbf{i} - 8\mathbf{j} + 7\mathbf{k}$

27. i) $k = 20$ **ii)** $-2 - i$

   iii) $\sqrt{5}; 153.4°$ **iv)** $k = 20$

28. ii) $r = \dfrac{5}{(1 - 0.4t)}$ **iii)** $0 \le t < 2.5$

29. i) $\dfrac{1}{x} + \dfrac{10}{x^2} + \dfrac{1}{10 - x}$

   ii) $t = \ln\left(\dfrac{9x}{10 - x}\right) - \dfrac{10}{x} + 10$

30. i) $4 - b - 2x = 0$

32. i) $x = \frac{1}{2}$ **ii)** $\pi(2\sqrt{e} - 3)$

33. ii) $5\mathbf{i} + 3\mathbf{j} + 4\mathbf{k}$

34. i) $x = \frac{1}{4}\pi$ **ii)** Maximum

35. i) $\ln N = 50k\sin(0.02t) + \ln 125$

   ii) $k = 0.0100479$ **iii)** 75.6

36. i) $4x + 2y + z = 8$ **ii)** $77.4°$

37. ii) $h = 9 - \left(4 - \frac{1}{15}t\right)^{\frac{3}{2}}$ **iii)** $h - 9; 60$ years

   iv) $t = 19.1$

38. $y^2 = 5e^{2x} - 1$

39. ii) $4x + y - 2z - 1$

40. i) $1 + i$ **ii)** $\sqrt{2}; 45°$ **iv)** $\sqrt{2} - 1$

41 i) $\alpha = 73.74°$ **ii)** $126.9°$

Index

Page numbers in *italics* indicate question sections.